Introduction to Digital Signal Processing Using
MATLAB with Application to Digital
Communications

K.S. Thyagarajan

Introduction to Digital Signal Processing Using MATLAB with Application to Digital Communications

 Springer

K.S. Thyagarajan
Extension Program
University of California, San Diego
San Diego, CA, USA

ISBN 978-3-030-09384-6   ISBN 978-3-319-76029-2   (eBook)
https://doi.org/10.1007/978-3-319-76029-2

Printed on acid-free paper

This Springer imprint is published by the registered company Springer International Publishing AG part of Springer Nature.
The registered company address is: Gewerbestrasse 11, 6330 Cham, Switzerland

Preface

The field of digital signal processing is well matured and has found applications in most commercial as well as household items. It started in the 1960s when computers were used only in the academic institutions. Moreover, these computers were built around vacuum tubes with limited memory and slow processing power. This situation was not conducive to rapid advancements in digital signal processing theory. As the computer technology advanced due to the invention of microprocessors and semiconductor memories, the field of digital signal processing also simultaneously progressed. Today, digital signal processing is used in a myriad of fields such as communications, medicine, forensics, imaging, and music, to name a few. It is, therefore, necessary for an aspirant to learn the basics of digital signal processing so as to be able to apply his or her knowledge in this field to career advancement.

There are many excellent textbooks on digital signal processing in the market. This book, though, is meant to serve working professionals who are looking for online courses to complete certificate programs in areas such as electrical engineering, systems engineering, communications, and embedded systems. Since these professional engineers are time-constrained, it is important that the textbook they are supposed to follow should be easy to understand, brief, and up to the point, and should contain the necessary supplements as aids to understanding the materials. With these factors in mind, this book is based on my online course in digital signal processing at the University of California Extension Program, San Diego. This book uses MATLAB tools to make understanding of the materials easier. In my experience in teaching this online course, I found that students come from different fields, but mostly from digital communications – hardware and software. Therefore, I find it appropriate to include applications of digital signal processing in digital communications. The students are required to have a college-level math background to fully understand the topics discussed in this book.

After a brief introduction to areas such as audio/speech processing, digital communications, and digital image processing, Chap. 2 starts with the discussion on discrete-time signals and systems. It characterizes the various discrete-time signals and systems in mathematical terms followed by examples to clarify the subject matter. Chapter 2 also describes the process of converting continuous-time signals to discrete-time sequences. The Z-transform is introduced in Chap. 3. Since Z-transform is very useful in both analysis and design of discrete-time systems, its properties are elaborated with several examples. Next the representation of discrete-time signals and systems in the frequency domain is discussed in Chap. 4. Here, the connection between the Z-transform and discrete-time Fourier transform is explained. Several examples are worked out to make the subject matter clearer. Since digital signal processing implies computational methods, Chapt. 5 introduces the concept of discrete Fourier transform. It also deals with the relationship between discrete-time Fourier transform and discrete Fourier transform. Again, MATLAB-based examples are included.

Once the signals and systems are described in the time and frequency domains, Chap. 6 then deals with the design of infinite impulse response (IIR) digital filters. It treats the design of IIR digital filters based on analytical methods as well as on computer-based techniques. In addition, real-life systems are simulated using MATLAB/Simulink tool. Continuing further, Chap. 7 discusses the design of finite impulse response (FIR) digital filters using both the analytical and computer-based methods. Many examples are included to aid the students in understanding the material better. It is not enough just to learn the design of IIR and FIR digital filters. A professional engineer must know how to implement these filters in various real-time applications. Therefore, Chap. 8 is included, which deals with the signal flow graphs of digital filters. It describes both canonical and noncanonical structures to implement IIR and FIR digital filters. Knowing how to draw the signal flow graphs of digital filters makes one to implement them either in software or hardware. Even though discrete Fourier transform (DFT) is introduced in Chap. 5, it does not deal with the efficient implementation of the DFTs. Chapter 9 describes efficient computational methods to calculate the DFT of a sequence. It further deals with short-time Fourier transform, zoom FFT, etc.

So far these chapters describe discrete-time signals and systems and various design techniques. In Chap. 10, the application of digital signal processing methods in wireless communications in general and digital communications in particular is discussed. The chapter deals with reducing the intersymbol interference, pulse shaping, detection of binary data using matched filters, channel equalization, phase-locked loop, orthogonal frequency division multiplexing, and software-defined radio, all using digital signal processing. Examples based on MATLAB are presented along with SIMULINK-based digital communications system. Codes for all MATLAB and SIMULINK.

I thank Tony Babaian for giving me the opportunity to teach the online courses titled *DSP I* and *DSP for wireless communications*. My sincere thanks to Sveteslav Maric for editing the book draft. I also thank the students for their feedback on the contents of the application of DSP in wireless communications. I am indebted to

Mathworks for their continued support in providing MATLAB license, which enabled me to develop this and my other books. My thanks go to Springer Publishing Company and their staff for publishing my book. I am extremely grateful to my wife Vasú, for suggesting to write this book. Without her kind and gentle encouragement, I would not have been able to even think of writing this book, let alone completing it.

San Diego, CA, USA   K.S. Thyagarajan

Contents

Chapter 1
Introduction

The field of digital signal processing is well matured and has found applications in most commercial as well as household items. It started in the 1960s when computers were used only in the academic institutions. Moreover, these computers were built around vacuum tubes with limited memory and slow processing power. This situation was not conducive to rapid advancements in digital signal processing theory. As the computer technology advanced due to the invention of microprocessors and semiconductor memories, the field of digital signal processing also simultaneously progressed. Today, digital signal processing is used in a myriad of fields such as communications, medicine, forensic, imaging, and music to name a few. It is, therefore, necessary for an aspirant to learn the basics of digital signal processing so as to be able to apply his or her knowledge in this field to career advancement.

1.1 What Is Digital Signal Processing

A signal can be considered, for example, as a voltage or current that varies as a function of time. A digital signal, on the other hand, can be any sequence of numbers that can be stored in a computer memory or a piece of hardware. Or, it may be acquired in real time from a signal source. If this sequence of numbers is related or meaningful, then, it is a useful signal or just signal. Figure 1.1 shows a signal sequence. If the sequence of numbers is random, it can be considered as noise. In Fig. 1.2 a random sequence is shown. Therefore, digital signal processing refers to any operation performed on the digital signal. This processing may be carried out in real time or non-real time depending on the application. The type of digital signal

Electronic supplementary material: The online version of this article (https://doi.org/10.1007/978-3-319-76029-2_1) contains supplementary material, which is available to authorized users.

K. S. Thyagarajan, *Introduction to Digital Signal Processing Using MATLAB with Application to Digital Communications*, https://doi.org/10.1007/978-3-319-76029-2_1

Fig. 1.1 An example of a
signal sequence

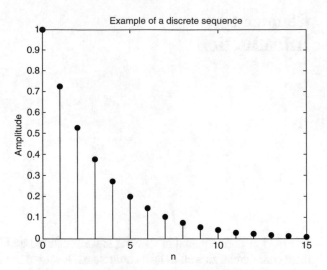

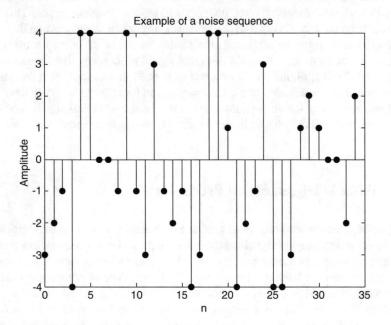

Fig. 1.2 An example of a random or noise sequence

processing depends on the particular application in hand. Filtering is a typical signal processing operation in which unwanted components or features can be removed or filtered out from an input digital signal. Consider an example of a signal, which consists of components of two sinusoidal frequencies at 1500 Hz and 4000 Hz, respectively, as shown in Fig. 1.3a. We want to remove the unwanted component at

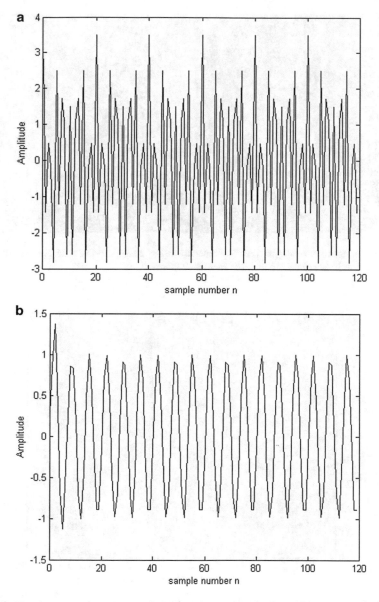

Fig. 1.3 Filtering as an example of digital signal processing. (**a**) Input signal consisting of 1500 and 4000 Hz sinusoidal components. (**b**) Filtered signal consisting of only the 1500 Hz sinusoid

4000 Hz. We then have to filter it out using a digital filter. The filtered signal will then have only the 1500 Hz component present. Figure 1.3b shows this exactly.

In digital image processing, for example, one can increase or decrease the sharpness by a suitable filtering operation. Figure 1.4a is an example of a black and white image lacking in sharpness. The same image after sharpening using a

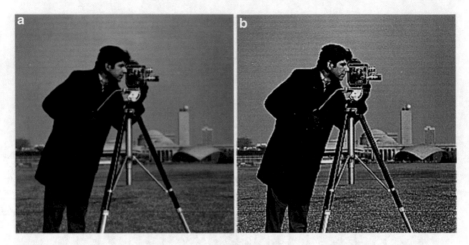

Fig. 1.4 An example of digital image sharpening using an appropriate filtering operation. (**a**) Original image lacking in sharpness, (**b**) sharpened image

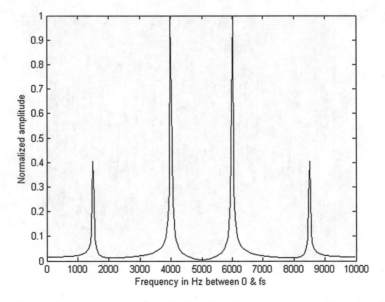

Fig. 1.5 Frequency spectrum of the signal in Fig. 1.3a. There are two frequencies present, one at 1500 Hz and the other at 4000 Hz

filtering operation is shown in Fig. 1.4b. One can notice the image details more clearly in Fig. 1.4b.

Another digital signal processing operation may be to estimate the frequency spectrum of a signal. This is useful in speech compression. An example of frequency spectrum is shown in Fig. 1.5. The signal shown in Fig. 1.3a is used as input in this example. It consists of two sinusoids at frequencies of 1500 Hz and 4000 Hz, respectively. The corresponding spectrum is shown in Fig. 1.5. As can be seen, there are two frequencies present in the signal, one at 1500 Hz and the other at 4000 Hz. Note that the spectrum is symmetric about 5000 Hz, which is half the sampling frequency. We will discuss the sampling process in detail later in the book. The amplitude of the 1500 Hz component relative to that of the 4000 Hz component is 0.4 as shown in the figure. As mentioned, there are numerous operations to choose from in processing a digital signal. It all depends on a particular application at hand.

1.2   A Few Applications of Digital Signal Processing

As mentioned above, there are numerous areas where digital signal processing is used. We will describe here a few applications of digital signal processing that will motivate the readers to go deeper into it.

1.2.1   Audio/Speech Processing

One of the most widely used aspects of speech in communications is speech compression. As landline and wireless telephony are ubiquitous, voice bandwidth is constrained to a minimum of 4 kHz to conserve bandwidth or, equivalently, to accommodate more subscribers. A voice with this bandwidth is quite intelligible and discernible. One of the voice compression methods uses what is known as *linear prediction*. Since there is a high correlation from sample to sample in speech signals, it is more efficient to store or transmit the difference between a sample and its predicted value rather than storing or transmitting the actual sample values. The predicted value of a current sample is obtained as a linear combination of the previously predicted samples. This, therefore, forms a feedback loop and can be considered a digital filtering operation.

In another form of speech compression, a bank of filters is used to separate the speech signal into different frequency bands or *subbands* and then coding each subband with a different number of bits of *quantization* based on the importance of each subband. This is known as *sub-band coding*. Of course, sub-band coding uses digital signal processing. One form of an efficient speech compression system extracts features of a speech signal first and then encodes the extracted features for storage or transmission. This process uses digital signal processing. At the receiving side, these features are used to synthesize speech, which again uses heavily digital signal processing. In music too various digital signal processing methods are used to produce sound effects. These are just a few cases where digital signal processing is involved in audio/speech signals.

1.2.2   Digital Communications

In digital communications, a signal such as voice or music to be transmitted is first converted to PCM (pulse code modulation) signals, which can then be transmitted as such or can be used to modulate a carrier signal. If no carrier modulation is used, the binary signal transmission is called baseband transmission. In baseband transmission, a sequence of binary data consisting of zeros and ones is transmitted as a sequence of two waveforms. The binary 0 corresponds to the waveform $s_0(t)$, $0 \leq t \leq T_b$, and the binary 1 corresponds to the waveform $s_1(t)$, $0 \leq t \leq T_b$, where T_b is the duration of each bit of data. During the transmission of the baseband signals through a channel, the data is corrupted by electrical noise. An optimal baseband receiver *correlates* the received signal with a replica of each waveform and decides which binary symbol was transmitted, by comparing the outputs of the two correlators. Instead of the correlator, one can also use what is known as *matched filter* to recover the transmitted binary data. Both correlators and matched filters can be realized as digital filtering operations.

Though the abovementioned receiver operation seems perfect, there is a problem due to finite bandwidth of the transmission channel. As a result, the sequence of transmitted binary waveforms can interfere with neighboring waveforms. This interference is called the *inter-symbol interference* (ISI). One way of minimizing the ISI is to design proper waveforms corresponding to the binary zero and one. These waveforms can be sampled versions and stored in memory. Another ISI mitigation method is to use linear *equalizers*, where the nonlinear *group delay* of the channel is equalized or linearized using digital FIR filters. Similar digital signal processing operations are used extensively in *spread* spectrum communications systems.

1.2.3   Digital Image Processing

Digital images and video are used worldwide through the Internet. Modern cameras are mostly digital cameras, while film-based cameras have all but disappeared. Similarly, digital TV transmission has replaced analog TV. Therefore, digital signal processing tools are an essential part of digital image and video. Digital signal processing as applied to images is used for image enhancement, image restoration, image compression, and image analysis/computer vision, to name a few.

Contrast Enhancement Captured images may suffer from poor lighting conditions, low contrast, and noise. In order to correct these defects and render a cleaner and more pleasing image, digital signal processing is a must. The effect of contrast enhancement of an image is demonstrated in Fig. 1.6. The low-contrast image is shown in Fig. 1.6a, b shows the same image after contrast enhancement. Clearly, one can see the image details much better in the enhanced image.

Fig. 1.6   Image contrast enhancement. (**a**) Original image with low contrast. (**b**) Contrast-enhanced image

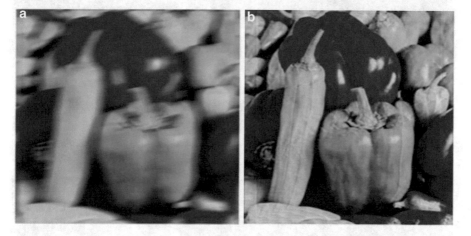

Fig. 1.7   An example of image deblurring. (**a**) Original blurred image. (**b**) Restored image

Image Restoration   Image restoration refers to that process by which a blurred image is restored to its original focused condition. Image blurring may be due to the relative motion of camera and object or due to improper lens setting. Deblurring or image restoration involves *inverse filtering* and is a digital signal processing operation. An example of image restoration is depicted in Fig. 1.7. Figure 1.7a is the blurred image due to camera movement, and the restored image is shown in Fig. 1.7b.

Image Compression   As the proverb "a picture is worth a thousand words" goes, digital image and video carry a lot of binary data. It is almost impossible to store or transmit these image and video data in raw format, especially when these are bounced around the Internet nonstop 24/7. Similarly, high-definition video carries an enormous amount of data and requires high compression for broadcasting. Image and video compression involves digital signal processing methods. In Fig. 1.8a, the

Fig. 1.8 Image compression. (**a**) Original uncompressed black and white image with 8 bits/pixel. (**b**) Compressed/decompressed image

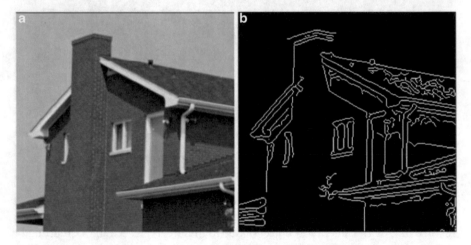

Fig. 1.9 Image boundary detection. (**a**) An original color image. (**b**) Its detected boundaries

original uncompressed image is shown, and the corresponding compressed/ decompressed image is shown in Fig. 1.8b. The quality of the reconstructed image or video depends on the amount and method of compression.

Image Analysis In order for a computer to recognize an object in a scene, it has to first segment the image into object and background. Image segmentation involves detecting object boundaries. An example of object boundary detection is illustrated in Fig. 1.9. In Figure 1.9a, the original image is seen, and its detected boundaries are shown in Fig. 1.9b. Once the object boundaries are detected, extracting certain features and comparing them to those stored in a database identify the object inside the detected boundary. All these operations involve some form of digital signal processing.

1.2.4   Digital Signal/Image Processing in Medicine

Digital signal processing is an inseparable aid in medicine. A patient's heart health is determined from his/her electrocardiogram (ECG). It may be difficult to completely assess the heart condition in terms of artery blockage or ventricular volume and blood dynamics purely from a raw ECG. It has to be processed in such a way so as to reveal hidden details that may otherwise not be visible to the naked eye. Digital signal processing is the solution. Another example of the application of digital signal processing in medicine is the ultrasound image processing. Ultrasound images are noisy and preclude correct detection of fetal defects. These images need to be properly preprocessed for further consideration. Echocardiogram technique is another tool used in monitoring heart condition. Similar to the ultrasound images, echocardiograms are also noisy and of low resolution. One can enhance an echocardiogram for proper visualization purpose. Such an example is shown in Fig. 1.10. The original echocardiogram is shown in Fig. 1.10a and its enhanced version in Fig. 1.10b.

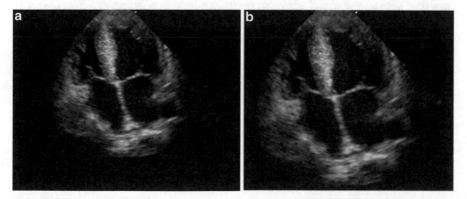

Fig. 1.10   An example of echocardiogram image processing: (**a**) Original image. (**b**) Enhanced image

1.3 A Typical Digital Signal Processing System

A typical digital signal processing system is shown in Fig. 1.11. The first component is a signal source, which is a source of analog signal. Since the signal to be processed is a digital signal, we have to first convert the input analog signal into a digital signal. This is achieved by an analog-to-digital converter or ADC, for short. An ADC is also called a DAQ for digital acquisition. The ADC samples the analog signal, holds each sample at its value until it is converted to a digital value, and then moves to the next sample and so on. The analog sample is represented as a B-bit binary number. The number of bits of binary representation is unique to each ADC. The accuracy of representing the analog sample value depends on B. For instance, an analog value of 0.723 when represented by a 4-bit binary number is equivalent to 0.6875. When the same number is represented with 8 bits, it is equivalent to 0.72265625. We will discuss in detail A/D conversion in a later chapter.

Once the analog signal is converted to a digital signal, it is then processed by the digital signal processor, which outputs to a digital-to-analog converter, which, in turn, feeds it to the analog signal sink. Similar to ADC, digital-to-analog converters are available commercially as chips. The whole process may be carried out in real time or non-real time, depending on the application in hand. The signal processing may be performed either by a special hardware or by a DSP chip. There are several DSP chips available in the market. For a given application, the DSP algorithm can be loaded into the chip. Therefore, the DSP chip is flexible. On the other hand, a special purpose DSP hardware is fixed for a particular application. In addition to DSP chips, one can also use *field-programmable gate arrays* (FPGAs) to implement a given signal processing algorithm.

After having introduced digital signal processing with a few applications, we will now deal with a brief introduction to continuous-time signals and systems. One may be wondering why the topic of continuous-time signals and systems is necessary here. There are several legitimate reasons to discuss continuous-time signals and systems. Many discrete-time signals are generated from continuous-time signals. These include speech, music, video, seismic signals, electrocardiogram, etc. Many of the properties of continuous-time systems are directly applicable to the discrete-time domain. More importantly, many of the tools used to analyze continuous-time signals and systems are applicable to the discrete-time domain. This aspect is particularly useful in digital filter design, as we will see later in the book.

However, since our interest is in the discrete-time domain, we will describe only briefly the main results of continuous-time signals and systems here.

Fig. 1.11 A block diagram of a typical digital signal processing system

1.4   Continuous-Time Signals and Systems

What follows is a brief description of mathematical representation of continuous-time signals and systems.

1.4.1   Continuous-Time Signals in the Time Domain

A continuous-time signal is denoted by $x(t)$ or $y(t)$, etc. and is defined over the interval $-\infty \leq t \leq \infty$. The signal may be purely real, imaginary, or complex. Further, the signal amplitude is a continuum. From here on, we will refer to a continuous-time signal as simply analog signal for convenience. An analog signal that is a linear combination of signals can be described as

$$x(t) = \sum_{i=1}^{N} \alpha_i x_i(t), \quad -\infty < t < \infty \tag{1.1}$$

The constants α_i in Eq. (1.1) may be real or complex. Let us look at some useful signals encountered in the continuous-time domain.

Sinusoidal Signals   A real sinusoidal waveform is described by

$$x(t) = A \cos(2\pi f t), \quad -\infty < t < \infty \tag{1.2}$$

In the above equation, A is the amplitude, and f is the frequency of the sinusoid in Hz or cycles per second. Note that $x(t)$ above is a periodic signal with a period $T = \frac{1}{f}$ because $\cos(2\pi f(t+T)) = \cos\left(2\pi f\left(t+\frac{1}{f}\right)\right) = \cos(2\pi f t + 2\pi) = \cos(2\pi f t)$. As opposed to a real sinusoidal signal, a complex exponential signal is described by

$$x(t) = B e^{i(\omega t + \phi)}, \quad -\infty < t < \infty \tag{1.3}$$

In Eq. (1.3), the amplitude B may be real or complex, $\omega = 2\pi f$ is the angular frequency in radians/sec, and ϕ is the phase offset in radians.

Unit Impulse or Dirac Delta Function   A unit impulse or simply Dirac delta function is denoted by $\delta(t)$ and defined as

$$\delta(t) = \begin{cases} 0, t \neq 0 \\ \infty, t = 0 \end{cases}, \text{ such that } \int_{-\infty}^{\infty} \delta(\tau)d\tau = 1 \tag{1.4}$$

The unit impulse is a hypothetical signal but is useful in characterizing continuous-time signals and systems. One way of visualizing the unit impulse is to consider a rectangular pulse in the time domain of width τ seconds and amplitude $\frac{1}{\tau}$ so that the area under the pulse is unity. If the pulse width is decreased, its amplitude will increase inversely keeping the area under the pulse still unity. In the limit as the pulse width approaches zero, its amplitude will approach infinity with the area under the pulse still unity. Any given analog signal can be represented in terms of the Dirac delta function as

$$x(t) = \int_{-\infty}^{\infty} x(\tau)\delta(\tau - t)d\tau \tag{1.5}$$

Unit Step Function   A unit step function $u(t)$ is defined as

$$u(t) = \begin{cases} 1, t \geq 0 \\ 0, t < 0 \end{cases} \tag{1.6}$$

The unit step function is useful in defining the time interval over which a function is valid as well as characterizing the rise and fall times of linear time-invariant systems.

1.4.2   Continuous-Time Systems

A continuous-time system is one that accepts an input signal $x(t)$ to produce an output signal $y(t)$. Mathematically, we can describe a continuous-time system as

$$y(t) = \mathcal{L}\{x(t)\} \tag{1.7}$$

Linear System   A system is said to be linear if it satisfies the superposition rule. In other words, if the response of the system to a linear combination of input signals is the same linear combination of individual responses, then the system is said to be linear. Otherwise, it is nonlinear. That is, if $y_i(t) = \mathcal{L}\{x_i(t)\}$, then

$$\mathcal{L}\left\{\sum_{i=1}^{N} \alpha_i x_i(t)\right\} = \sum_{i=1}^{N} \alpha_i \mathcal{L}\{x_i(t)\} = \sum_{i=1}^{N} \alpha_i y_i(t) \tag{1.8}$$

The beauty of a linear system is that it can be characterized in closed-form expressions, whereas a nonlinear system typically involves iterative solution. However, it is important to note that nonlinear systems are useful, and many real-world systems are nonlinear. They are linearized so that they can be analyzed easily.

Time-Invariant System   A system is said to be time-invariant if a delayed input results in a delayed response. That is, if $y(t) = \mathcal{L}\{x(t)\}$, then the system is time-invariant if

$$\mathcal{L}\{x(t - \tau)\} = y(t - \tau), for\ all\ \tau \tag{1.9}$$

Otherwise, the system is time-variant.

Before we go any further, let us work out a couple of examples to nail the concept.

Example 1.1 Consider the continuous-time system $y(t) = tx(t)$. Is the system (a) linear? (b) time-invariant?

Solution If $x(t) = \alpha x_1(t) + \beta x_2(t)$ with α and β constants, then the response of the systems is

$$y(t) = t(\alpha x_1(t) + \beta x_2(t)) = \alpha t x_1(t) + \beta t x_2(t) = \alpha y_1(t) + \beta y_2(t) \qquad (1.10)$$

Since the superposition rule is satisfied, the given system is linear. To test if the system is time-invariant, we first delay the system response by τ so that

$$y(t - \tau) = (t - \tau)x(t - \tau) \qquad (1.11)$$

If we simply delay the input to the system, the system will respond with

$$\mathcal{L}\{x(t - \tau)\} = tx(t - \tau) \neq y(t - \tau) \qquad (1.12)$$

Therefore, the system is *not* time-invariant. That is, it is time-variant. From this example we observe that a system may be linear but not necessarily time-invariant.

Example 1.2 Consider the system $y(t) = |x(t)|$. Is the system (a) linear, (b) time-invariant?

Solution To test linearity, apply the superposition rule:

$$\mathcal{L}\{\alpha x_1(t) + \beta x_2(t)\} = |\alpha x_1(t) + \beta x_2(t)| \neq \alpha |x_1(t)| + \beta |x_2(t)| \qquad (1.13)$$

Since the superposition rule failed, the given system is nonlinear. To test for time-invariance, check

$$\mathcal{L}\{x(t - \tau)\} = |x(t - \tau)| = y(t - \tau) \qquad (1.14)$$

Since the delayed input produces delayed response, the system is time-invariant. Again, we observe that a system may be nonlinear but can be time-invariant.

Linear Convolution The response to any input signal of a linear, time-invariant (LTI) system can be obtained in terms of what is known as the *impulse response* of the LTI system. When the input to an LTI system is a unit impulse, its response is denoted by $h(t)$, that is, $\mathcal{L}\{\delta(t)\} = h(t)$. A signal $x(t)$ can be represented in terms of the unit impulse as shown in Eq. (1.5). Another way of looking at Eq. (1.5) is that the signal $x(t)$ is expressed as a linear combination of unit impulses with strength corresponding to the signal amplitude at the respective time instants. Therefore, if we apply this signal as input to an LTI system, its response can be expressed as

$$\mathcal{L}\{x(t)\} = \mathcal{L}\left\{\int_{-\infty}^{\infty} x(\tau)\delta(t - \tau)d\tau\right\} = \int_{-\infty}^{\infty} x(\tau)\mathcal{L}\{\delta(t - \tau)\}d\tau \qquad (1.15)$$

In the above equation, the system operator is taken inside the integral because the integral is a linear operator. However, since the system is time-invariant, $\mathcal{L}\{\delta(t - \tau)\} = h(t - \tau)$. Therefore,

$$y(t) = \int_{-\infty}^{\infty} x(\tau)h(t - \tau)d\tau \tag{1.16}$$

Equation (1.16) is known as the *convolution integral*. The impulse response of an LTI system is unique to each system, and therefore, the system response to a given input signal is unique to that system. It is customary to denote the convolution of two signals by the symbol $\otimes$. Thus, Eq. (1.16) can be represented as

$$y(t) = x(t) \otimes h(t) = \int_{-\infty}^{\infty} x(\tau)h(t - \tau)d\tau = \int_{-\infty}^{\infty} x(t - \tau)h(\tau)d\tau \tag{1.17}$$

Example 1.3 Compute the unit step response of the LTI system with its impulse response $h(t) = e^{-0.5t}u(t)$.

Solution Since we are looking for the response to a unit step function, the input to the system is $u(t)$. Then, from the convolution integral, the unit step response is found to be

$$y(t) = \int_{-\infty}^{\infty} x(t - \tau)h(\tau)d\tau = \int_{0}^{t} e^{-0.5\tau}d\tau = 2(1 - e^{-0.5t}), t \ge 0 \tag{1.18}$$

Since the input is zero for $t < 0$, the output is also zero for $t < 0$. From (1.18) we observe that the unit step response starts from zero at $t = 0$ and reaches a final value of 2. The *rise time* of an LTI system is defined as the time it takes for its step response to change from 10% to 90% of its final value. For this system using Eq. (1.18), we find the rise time to be about 1.093 s.

Causality An LTI system is said to be causal if its impulse response $h(t) = 0$ *for* $t < 0$. In other words, an LTI system cannot anticipate future input, or equivalently, it cannot produce a response before an input is applied. Using this criterion, we find the system in Example 1.3 to be causal.

Stability Another important property of an LTI system is its stability. An LTI system is said to be stable in the bounded input, bounded output (BIBO) sense if its impulse response is absolutely integrable. That is, the LTI system is stable in the BIBO sense if

$$\int_{-\infty}^{\infty} |h(\tau)|d\tau < \infty \tag{1.19}$$

If an LTI system is unstable, then a bounded input signal will result in an output that will eventually grow out of bound. Such a system will be useless.

Example 1.4 Is the system described in Example 1.3 stable?

Solution We need to evaluate the integral of the absolute of the impulse response and verify that it is finite in magnitude.

$$\int_{-\infty}^{\infty} |h(\tau)|d\tau = \int_{0}^{\infty} e^{-0.5\tau}d\tau = \frac{-e^{-0.5\tau}}{0.5}\Big|_{0}^{\infty} = 2 < \infty \qquad (1.20)$$

Since the impulse response is absolutely integrable, the system is stable. The implication is that as long as the input is of finite amplitude, the system response will also be of finite amplitude.

LTI System and Differential Equation An LTI system in the time domain can be characterized by a linear differential equation with constant coefficients, as described by

$$y(t) + \sum_{n=1}^{p} b_n \frac{d^n y(t)}{dt^n} = \sum_{m=0}^{q} a_m \frac{d^m x(t)}{dt^m}, q \le p \qquad (1.21)$$

If the system is time-variant, then the coefficients $\{b_n\}$ and $\{a_m\}$ will be dependent on t. Otherwise, they are constants. When the input is zero, the corresponding solution is called the *complementary solution* and refers to the impulse response of the LTI system. When an input is applied, we can assume the solution or response to be of the same form as the input and determine the exact response. This is called the *particular solution*. We will not discuss further the differential equation. Later we will deal with linear *difference equation*, which governs a discrete-time system.

1.4.3 Frequency Domain Representation of Signals and Systems

So far we have described an LTI system in the time domain. We can also describe the same system equivalently in the frequency domain. Frequency domain representation of signals and systems is more helpful in the analysis and design of such systems. We are more used to visualizing a signal in terms of its frequency than in terms of its time-domain behavior. We will first look at the frequency domain representation of signals and then the systems.

Fourier Series A periodic signal can be expressed as a sum of sinusoidal signals whose frequencies are integer multiples of a fundamental frequency. Let $x(t)$ be a periodic signal with period T. Note that $x(t)$ is periodic with period T if $x(t + nT) = x(t)$, $n \in Z$. The Fourier series expansion of $x(t)$ is given by

$$x(t) = \sum_{n=-\infty}^{\infty} c_n e^{-jn\omega_0 t} \qquad (1.22)$$

where $\omega_0 = 2\pi f_0, f_0 = \frac{1}{T}$. The Fourier series coefficients $\{c_n\}$ are obtained from

$$c_n = \frac{1}{T} \int_{-\frac{T}{2}}^{\frac{T}{2}} x(t) e^{jn\omega_0 t} dt \qquad (1.23)$$

Note that if $x(t)$ is a real function, the c_n and c_{-n} must be complex conjugate of each other.

Fourier Transform If a signal is not periodic, then it can be represented in the frequency domain by an integral involving continuous sinusoidal frequencies known as the Fourier transform, as given below:

$$X(f) = \int_{-\infty}^{\infty} x(t) e^{-j2\pi ft} dt, \quad -\infty < f < \infty \qquad (1.24)$$

The function $X(f)$ is in general complex and represents the amplitude and phase of the sinusoidal component at the frequency f. The time-domain signal $x(t)$ can be recovered from its Fourier transform by its inverse Fourier transform:

$$x(t) = \int_{-\infty}^{\infty} X(f) e^{j2\pi ft} df \qquad (1.25)$$

Thus, $x(t)$ and $X(f)$ form a Fourier transform pair.

Some Properties of the Fourier Transform
When dealing with LTI systems, one can use some features or properties of the Fourier transform to solve the system more easily and elegantly. Therefore, we will list some useful properties of the Fourier transform without proof.

Scaling If $x(t) \leftrightarrow X(f)$, then $\Im\{ax(t)\} = aX(f)$, where a is a constant.

Linearity If $x_1(t) \leftrightarrow X_1(f)$ and $x_2(t) \leftrightarrow X_2(f)$, then the Fourier transform of a linear combination of $x_1(t)$ and $x_2(t)$ is a linear combination of the respective Fourier transforms:
$\Im\{ax_1(t) + bx_2(t)\} = aX_1(f) + bX_2(f)$.

Convolution in the Time Domain The Fourier transform of the convolution of $x(t)$ and $h(t)$ is the product of the respective Fourier transforms: $\Im\{x(t) \otimes h(t)\} = X(f)H(f)$.

Convolution in the Frequency Domain The Fourier transform of the product of two time-domain functions is the convolution of the respective Fourier transforms
$\Im\{x(t)h(t)\} = X(f) \otimes H(f)$.

Modulation If $x(t) \leftrightarrow X(f)$, then $x(t) e^{2\pi f_0 t} \leftrightarrow X(f - f_0)$. That is, multiplying a signal in the time domain by a complex sinusoid of a fixed frequency is equivalent to shifting its Fourier transform by that frequency.

Fourier Transform of a Derivative of a Signal The Fourier transform of the derivative of a signal in the time domain is the Fourier transform of the signal multiplied by $j2\pi f$. That is, $\Im\left\{\frac{dx(t)}{dt}\right\} = j2\pi f X(f)$.

Transfer Function From the Fourier transform property of convolution in the time domain, we obtain the Fourier transform of the output of an LTI system in terms of its input and impulse response as

$$Y(f) = \Im\{y(t)\} = \Im\{x(t) \otimes h(t)\} = X(f)H(f) \tag{1.26}$$

The transfer function of the LTI system is defined as the ratio of the Fourier transforms of its output to input:

$$H(f) \equiv \frac{Y(f)}{X(f)} \tag{1.27}$$

If we take the Fourier transform of Eq. (1.21) using the differentiation property of the Fourier transform and obtain the ratio of the output to input, we obtain

$$H(f) = \frac{a_0 + (j2\pi f)a_1 + \cdots + (j2\pi f)^q a_q}{1 + (j2\pi f)b_1 + (j2\pi f)^2 b_2 + \cdots + (j2\pi f)^p b_p} \tag{1.28}$$

We can also express the transfer function above in terms of ω as

$$H(\omega) = \frac{a_0 + (j\omega)a_1 + (j\omega)^2 a_2 + \cdots + (j\omega)^q a_q}{1 + (j\omega)b_1 + (j\omega)^2 b_2 + \cdots + (j\omega)^p b_p} \tag{1.29}$$

We observe from Eqs. (1.28) or (1.29) that the transfer function of an LTI system is a rational polynomial in ω.

Laplace Transform The Laplace transform of a continuous-time signal $x(t)$, $t \geq 0$ is defined as

$$X(s) = \int_0^\infty x(t)e^{-st}dt \tag{1.30}$$

The Laplace variable s in Eq. (1.30) is complex and is expressed as $s = \sigma + j\omega$. Laplace transform is very useful in the analysis of continuous-time systems. The signal $x(t)$ can be recovered from the inverse Laplace transform. Since the inverse Laplace transform involves complex integral, it is seldom used in practice. Instead, one uses partial fraction expansion to determine the inverse Laplace transform.

Example 1.5 Find the Laplace transform of $x(t) = e^{-at} \cos{(\omega_0 t)}u(t)$.

Solution Using the definition, the Laplace transform of the above signal is

$$X(s) = \int_0^\infty e^{-at} \cos{(\omega_0 t)}e^{-st}dt \tag{1.31}$$

By expressing the cosine function in terms of two complex exponential functions, we have

$$X(s) = \frac{1}{2}\int_0^\infty e^{-(s+a-j\omega_0)t}dt + \frac{1}{2}\int_0^\infty e^{-(s+a+j\omega_0)t}dt \tag{1.32}$$

After evaluating the two integrals and some algebraic manipulation, we arrive at

$$X(s) = \frac{s+a}{s^2 + 2as + (a^2 + \omega_0^2)} \tag{1.33}$$

From the Laplace transform of the signal $x(t)$, we observe that it is a rational function of the Laplace variable s. Similarly, if we replace $j\omega$ by the Laplace variable s in Eq. (1.29), we find that the transfer function of an LTI system to be a rational polynomial in s.

Example 1.6 Find the response to the input $x(t) = e^{-0.5t}u(t)$ of the LTI system whose impulse response is given by $h(t) = e^{-0.25t}u(t)$ using Laplace transform.

Solution We have to first find the Laplace transforms of the input and impulse response. Using the definition of Laplace transform, we have $X(s) = \frac{1}{s+0.5}$, and $H(s) = \frac{1}{s+0.25}$. Since the given system is LTI, the Laplace transform of the response of the system equals the product of the Laplace transforms of the input signal and the impulse response. Therefore,

$$Y(s) = \mathcal{L}\{y(t)\} = X(s)H(s) = \frac{1}{(s+0.5)(s+0.25)} = \frac{1}{s^2 + 0.75s + 0.125} \tag{1.34}$$

Next, we have to express the system response in partial fractions and determine the residues. So the system response in partial fractions is expressed as

$$Y(s) = \frac{A}{s+0.5} + \frac{B}{s+0.25} \tag{1.35}$$

In the above equation, A and B are called the residues. The residues are determined by first multiplying Y(s) by the respective pole factors, cancelling the corresponding pole factor, and then evaluating the remaining function at the corresponding pole. Thus,

$$A = (s+0.5)Y(s)|_{s=-0.5} = \frac{1}{s+0.25}\bigg|_{s=-0.5} = -4 \tag{1.36a}$$

and

$$B = (s+0.25)Y(s)|_{s=-0.25} = \frac{1}{s+0.5}\bigg|_{s=-0.25} = 4 \tag{1.36b}$$

Finally, the system response in the time domain is the inverse Laplace transform of Y(s). Since Y(s) is the sum of two functions, the response y(t) is the sum of the corresponding time-domain functions, as given by

$$y(t) = \mathcal{L}^{-1}\left\{\frac{A}{s+0.5}\right\} + \mathcal{L}^{-1}\left\{\frac{B}{s+0.25}\right\} = -4e^{-0.5t}u(t) + 4e^{-0.25t}u(t) \tag{1.37}$$

1.5 Summary

We have introduced the idea of digital signal processing in this chapter and briefly reviewed the areas where DSP is widely used. Some useful applications are discussed with examples to motivate the reader to go deeper into the field of DSP. Since practical digital signals are obtained from the analog counterparts, the description of continuous-time signals and systems is of necessity. Therefore, a brief introduction to signals and linear time-invariant systems in the continuous-time domain is introduced here with examples.

With the introduction of the basic tools for the analysis of continuous-time signals and LTI systems, we will move on to discuss discrete-time signals and systems in the next chapter.

1.6 Problems

1. Express the time-domain periodic pulse train $p(t) = 1, -0.5 \leq t \leq 0.5$ with period 1 s in Fourier series.
2. Determine the Fourier transform of the function $x(t) = e^{-\alpha t}u(t)$.
3. Find the Fourier transform of the time-domain function $x(t) = e^{-0.25t}\cos(500t)u(t)$.
4. Find the poles and zeros of the function $H(s) = \frac{s+1}{s^3+2.5s^2+6s+2.5}$. Is the system stable?
5. Find the frequency response of the system in Problem 4.
6. Find the step response of the system $H(s) = \frac{1}{s^3+4s^2+5s+2}$ using Laplace transform method.

References

1. Churchill RV, Brown JW (1990) Introduction to complex variables and applications, 5th edn. McGraw-Hill, New York
2. Clifford GD, Azuaje F, McSharry PE (eds) (2006) Advanced methods and tools for ECG data analysis. Artech House, Boston/London
3. Cohen A (1986) Biomedical signal processing, vol 2. CRC Press, Boca Raton
4. Oppenheim AV, Schafer RW (1975) Digital signal processing. Prentice-Hall, Englewood Cliffs
5. Proakis JG, Manolakis DG (1996) Digital signal processing: principles, Algorithms and applications, 3rd edn. Prentice-Hall, Upper Saddle River
6. Rabiner LR, Gold B (1975) Theory and application of digital signal processing. Prentice-Hall, Englewood Cliffs
7. Rabiner LR, Shafer RW (1978) Digital processing of speech signals. Prentice-Hall, Upper Saddle River
8. Sklar B (1988) Digital communications fundamentals and applications. Prentice-Hall, Englewood Cliffs
9. Thyagarajan KS (2006) Digital image processing with application to digital cinema. Focal Press/Taylor & Francis, New York/London
10. Thyagarajan KS (2011) Still image and video compression with MATLAB. IEEE Press/Wiley, Hoboken

Chapter 2
Discrete-Time Signals and Systems

2.1 Introduction

Continuous-time or analog signals are processed using analog devices such as amplifiers, filters, etc. It is impossible to process signals multiplexed from various sources using a single hardware system in the analog domain. On the other hand, digital signals can be processed using both special-purpose hardware and software systems. Worldwide use of Internet, mobile communications, etc. demands all kinds of data such as video, audio, graphics, etc. In order to receive this information on a single device, computer, for instance, it is impossible to use analog signals and techniques. In order to be able to design and implement digitally based systems, it is absolutely necessary to have an understanding of digital signals and systems. Digital signals are discrete in time and amplitude. However, we will assume discrete-time signals to have a continuum of amplitude in order to be able to analyze such signals and systems mathematically. In this chapter we will describe typical discrete-time signals mathematically and then use them to describe and analyze linear time-invariant discrete-time systems. To help the readers understand the mathematical details, we will work out examples followed by MATLAB-based examples. Since digital signals are obtained from analog sources, we will also discuss the conversion of continuous-time signals into digital signals using analog-to-digital converters (ADC).

Electronic supplementary material: The online version of this article (https://doi.org/10.1007/978-3-319-76029-2_2) contains supplementary material, which is available to authorized users.

2.2 Typical Discrete-Time Signals

A discrete-time signal is denoted by $x[n]$, $y[n]$, etc. and is defined over the interval $-\infty < n < \infty$, $n \in Z$. The amplitude of a discrete-time signal is a continuum, while its argument n is an integer. If a discrete-time signal is obtained from a continuous-time signal, then the argument of the discrete-time signal is an integer multiple of the sampling interval. We will discuss the sampling process later in the chapter. A discrete-time signal is also referred to as a sequence. When a discrete-time signal is processed by a computer in software or hardware, the signal amplitudes are represented by numbers, and so the signal is a digital signal. Even though discrete-time signals are processed by a computer, we will still assume their amplitudes to be a continuum in our discussion.

There are several discrete-time signals that are useful in characterizing other discrete-time signals as well as systems similar to those used in the continuous-time domain. We will describe them here briefly.

Unit Impulse Sequence A unit impulse sequence is denoted by $\delta[n]$ and is defined as

$$\delta[n] = \begin{cases} 1, n = 0 \\ 0, n \neq 0 \end{cases} \tag{2.1}$$

A unit impulse in the discrete-time is similar to the Dirac delta function in the continuous-time except that the unit impulse sequence is physically realizable.

Unit Step Sequence A unit step sequence is denoted by $u[n]$ and is defined as

$$u[n] = \begin{cases} 1, n \geq 0 \\ 0, n < 0 \end{cases} \tag{2.2}$$

The unit step sequence plays a similar part in the analysis of discrete-time systems as its continuous-time counterpart.

Exponential Sequence A real exponential sequence is defined as

$$x[n] = \alpha^n u[n], |\alpha| < 1 \tag{2.3}$$

In (2.3), α is a real constant.

Real Sinusoidal Sequence A real sinusoidal sequence is defined as

$$x[n] = A \cos (n\Omega_0 + \phi), \ -\infty < n < \infty, n \in Z \tag{2.4}$$

In Eq. (2.4), A is the amplitude, $\Omega_0 = 2\pi \frac{f_0}{F_s}$ is the normalized frequency in rad, f_0 is the frequency in Hz, F_s is the sampling frequency in Hz, and ϕ is the phase offset in rad.

Complex Exponential Sequence A complex exponential sequence is described by

$$x[n] = A\alpha^n exp(-jn\Omega_0)u[n], |\alpha| < 1 \tag{2.5}$$

In Eq. (2.5), the amplitude A may be complex, and α is a real constant.

Periodic Sequence A sequence $x[n]$ is said to be periodic with period N if $x[n + kN] = x[n]$, where k is an integer. From the definition we can easily verify that the sinusoidal sequence in (2.4) is periodic with period $N = \frac{kF_s}{f_0}$.

2.3 Discrete-Time Systems

A discrete-time system, $\mathcal{L}\{.\}$, is one that accepts an input sequence $x[n]$ to produce an output sequence $y[n]$. It can be formally written as

$$y[n] = \mathcal{L}\{x[n]\} \tag{2.6}$$

Linearity A discrete-time system is said to be linear if it satisfies the superposition rule. In other words, a discrete-time system is linear if the following condition holds:

$$y[n] = \mathcal{L}\{\alpha x_1[n] + \beta x_2[n]\} = \alpha y_1[n] + \beta y_2[n], \tag{2.7}$$

where $y_1[n] = \mathcal{L}\{x_1[n]\}$ and $y_2[n] = \mathcal{L}\{x_2[n]\}$. So, a linear discrete-time system responds to a linear combination of input sequences with the same linear combination of individual responses. Linear discrete-time systems are most useful because they can be solved analytically. If the above condition stated in (2.7) is not valid, then the discrete-time system is nonlinear. Nonlinear systems in general don't have closed-form solution and must be solved iteratively. Hence linear systems are preferred in practice though many practical systems may be nonlinear.

Time- or Shift-Invariant Discrete-Time Systems A discrete-time system is said to be time- or shift-invariant if a delayed input results in a delayed response:

$$\mathcal{L}\{x[n - m]\} = y[n - m], m \in Z \tag{2.8}$$

Impulse Response As in the continuous-time system, the response of a discrete-time system to a unit impulse sequence is called the *impulse response* and is denoted by $h[n]$. The impulse response is formally defined as

$$h[n] = \mathcal{L}\{\delta[n]\} \tag{2.9}$$

The impulse response is unique to a given discrete-time linear system and is very useful in calculating the system response to any given input. It is also very useful in the design of digital filters. We will deal with the design of digital filters later in the book.

Causality A discrete-time system is causal if it is non-anticipatory. That is to say that if the response of a discrete-time system at the current time index, n, does not depend on the input at a future time instant, then the system is causal.

Let us understand what we have discussed so far clearly by going through the following examples.

Example 2.1 A discrete-time system is defined by the following difference equation:

$$y[n] = x[n+1] - 2x[n] + x[n-1] \tag{2.10}$$

Is it (a) linear? (b) Is it time-invariant? (c) Is it causal?

Solution Let me make it clear. A discrete-time system may be characterized by a linear difference equation just as we described a continuous-time system by a differential equation.

(a) Let the input be a linear sum of two input sequences: $x[n] = ax_1[n] + bx_2[n]$, where a and b are constants. Then the response of the system can be written using (2.10) as

$$\begin{aligned} y[n] &= a(x_1[n+1] - 2x_1[n] + x_1[n-1]) \\ &\quad + b(x_2[n+1] - 2x_2[n] + x_2[n-1]) \\ &= ay_1[n] + by_2[n] \end{aligned} \tag{2.11}$$

In the above equation, $y_1[n] = \mathcal{L}\{x_1[n]\}$ and $y_2[n] = \mathcal{L}\{x_2[n]\}$. Since the superposition rule is satisfied, the given discrete-time system is linear.

(b) From the given difference equation, we notice that delaying the input sequence by an integer M produces a response, which is exactly the delayed version of the response by the same integer M. Hence the system is time- or shift-invariant. Note that if the coefficients a and b are dependent on the time index n, then the system will no longer be shift-invariant!

(c) We notice from the given system's input-output relationship that the response of the system at the current time index n depends on the input at the next future input. Therefore, the system is anticipatory and hence is non-causal.

Example 2.2 Determine if the discrete-time system $y[n] = K + x[n] + 0.75x[n-1]$, where K is a constant, is (a) linear, (b) time-invariant, and (c) causal?

Solution
(a) If we apply the superposition rule, we observe that it is not satisfied due to the constant K. Hence the system is not linear. However, it is piecewise linear.
(b) Since the delayed input produces the same delayed response, the system is time-invariant. It can also be inferred from the fact that the coefficients in the given input-output relationship are constants, independent of the time index.
(c) The system response at the current time index n does not depend on the input sequence at future time instants. Hence the system is causal.

Stability A discrete-time system is said to be stable if a bounded input produces a bounded output. Equivalently, we can impose the stability condition on the impulse response. We will look at it after we define the convolution sum. This definition of stability is called bounded-input bounded-output (BIBO) stability.

Example 2.3 Is the system described in Example 2.1 stable in the BIBO sense?

Solution If we assume that the absolute value of the input sequence is finite, that is, $|x[n]| \le M$ for all n, then we see that the output sequence value is also finite:

$$|y[n]| \le |x[n+1] - 2x[n-1] + x[n]| \le 4M < \infty \qquad (2.12)$$

Hence the system is stable.

2.4 Convolution Sum

The response of an LTI discrete-time system to any given input sequence can be obtained in terms of its impulse response sequence and the input sequence by what is called the *convolution sum*. We first observe that a given sequence $x[n]$ can be represented as an infinite sum of unit impulses:

$$x[n] = \sum_{k=-\infty}^{\infty} x[k]\delta[n-k] \qquad (2.13)$$

Equation (2.13) follows from the definition of the unit impulse. So, the right-hand side of (2.13) is zero except for $k = n$, in which case, the right-hand side is simply $x[n]$. Having expressed the input sequence in terms of the unit impulse sequence, we next determine the response of the LTI system as

$$y[n] = \mathcal{L}\{x[n]\} = \mathcal{L}\left\{ \sum_{k=-\infty}^{\infty} x[k]\delta[n-k] \right\} \qquad (2.14)$$

Since the system is linear, the system operator can be taken inside the summation as

$$y[n] = \sum_{k=-\infty}^{\infty} \mathcal{L}\{x[k]\delta[n-k]\} \qquad (2.15)$$

That is, the response is the linear sum of the responses to individual impulses $\delta[n-k]$. However, $x[k]$ is a constant, and since the system is linear, the response to constant times an input is equal to the constant times the response to the input. Therefore, (2.15) can be rewritten as

$$y[n] = \sum_{k=-\infty}^{\infty} x[k]\mathcal{L}\{\delta[n-k]\} \qquad (2.16)$$

We have also assumed the system to be time-invariant. Therefore, $\mathcal{L}\{\delta[n-k]\} = h[n-k]$. Hence,

$$y[n] = \sum_{k=-\infty}^{\infty} x[k]h[n-k] \qquad (2.17)$$

Equation (2.17) is known as the convolution sum of the sequences $x[n]$ and $h[n]$ and is usually abbreviated as $x[n] \otimes h[n]$. In (2.17), if we substitute $m = n - k$, then we can also write the convolution sum as

$$y[n] = \sum_{m=-\infty}^{\infty} h[m]x[n-m] \qquad (2.18)$$

Procedure to Calculate the Convolution Sum

We can list a graphical procedure to calculate the convolution sum given in Eq. (2.17) as follows:

1. Flip the impulse response $h[n]$ about the origin, and label the abscissa with the integer variable k.
2. Multiply the input sequence and the flipped impulse response sequence point by point, and sum them over the entire interval. This sum gives the response at n = 0.
3. Slide the flipped impulse response to the right one sample at a time.
4. Multiply the input sequence and the flipped impulse response sequence point by point, and sum them over the entire interval. The sum gives the system response at subsequent time instants.
5. For negative integer values of n, repeat steps 3 and 4, except that the impulse response sequence is slid to the left instead of right.

Example 2.4 Consider the LTI discrete-time system with an impulse response $h[n] = \alpha^n u[n]$, $|\alpha| < 1$. Determine its unit step response.

Solution Using the graphical interpretation of the convolution sum, we first label the abscissa with the index k. Next we flip the impulse response about the ordinate. This is shown in black color in Fig. 2.1a with no shift. In the same plot, the input sequence in red color is shown, which is a unit step. There is only one sample that overlaps the two sequences for n = 0, and the corresponding product of the two sequences results in the response at n = 0. As can be seen from Fig. 2.1b, any shift of the impulse response sequence to the left, that is, for n < 0, will leave no overlapping of the two sequences. Hence the system response will be zero for n < 0. Figure 2.1c shows the flipped impulse response shifted to the right by 3, that is, n = 3. Now there are four overlapping samples. We multiply the overlapping samples and sum them to obtain the response at n = 3. This can be continued for each shift of the impulse response to the right. To obtain the response in closed form, we resort to the convolution sum in Eq. (2.17).

Fig. 2.1 Graphical interpretation of the convolution sum: (**a**) flipped impulse response for no shift, (**b**) flipped impulse response shifted to the left by two samples, and (**c**) flipped impulse response shifted to the right by three samples

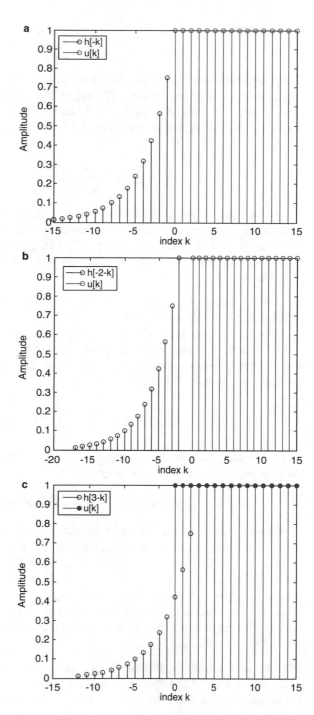

$$y[n] = \sum_{k=-\infty}^{\infty} x[k]h[n-k] = \sum_{k=0}^{n} \alpha^{n-k} = \alpha^n \sum_{k=0}^{n} \alpha^{-k} \qquad (2.19)$$

In (2.19) since the input is zero for $n < 0$, the lower limit of the summation is zero. Also, the response is zero for $n < 0$. The upper limit is n corresponding to the current time instant. The sequence on the right-hand side of (2.19) is an exponentially decreasing sequence. After simplifying (2.19) we get the unit step response of the system,

$$y[n] = \frac{\alpha^{n+1} - 1}{\alpha - 1}, n \geq 0 \qquad (2.20)$$

The impulse response and the system response are shown in Figs. 2.2a and b, respectively. The impulse response sequence is assumed to be $(0.75)^n u[n]$. The rise time of the LTI system is defined as that interval in which the step response changes from 10% to 90% of its final response. The final value of the response is found to be 4.

Causality Revisited Earlier we said that an LTI system is causal if it is non-anticipatory. We can also impose causality on the impulse response of the LTI system. To this end, recall that the convolution sum represents the response of an LTI discrete-time system to an input sequence. If the input to an LTI discrete-time system is assumed to be zero for n less than zero, then the response $y[n]$ in terms of the impulse response $h[n]$ is written as

$$y[n] = x[n] \otimes h[n] = \sum_{k=0}^{\infty} x[k]h[n-k] \qquad (2.21)$$

For instance, let us evaluate the response at $n = 1$. Expanding the summation on the right-hand side of (2.21), we have

$$y[1] = x[0]h[1] + x[1]h[0] + x[2]h[-1] + x[3]h[-2] + \cdots \qquad (2.21a)$$

Since the system is assumed to be causal, $y[n = 1]$ should not be dependent on $x[n]$ for $n > 1$. However, $x[n]$ is the input sequence and is not zero. Therefore, $h[-1] = h[-2] = \cdots = 0$. That is to say that $h[n] = 0$ for $n < 0$. Generalizing, we say that for the system to be causal, $h[n - k] = 0$, *for* $k > n$. Otherwise it will be anticipatory. Let $m = n - k$. Then, for the system to be causal, $k > n \Rightarrow m < 0$. Hence for the system to be causal, $h[n] = 0$ *for* $n < 0$, which implies that the upper limit of the summation in (2.21) is n. An LTI discrete-time system is causal if and only if its impulse response is zero for $n < 0$. Otherwise it is non-causal.

Stability in Terms of the Impulse Response An LTI discrete-time system is stable in the BIBO sense if its response is finite for a finite input. Let the input sequence $x[n]$ be bounded, that is, $|x[n]| \leq M < \infty$, for all n. Using the convolution sum in (2.17), we can bound the output as given below:

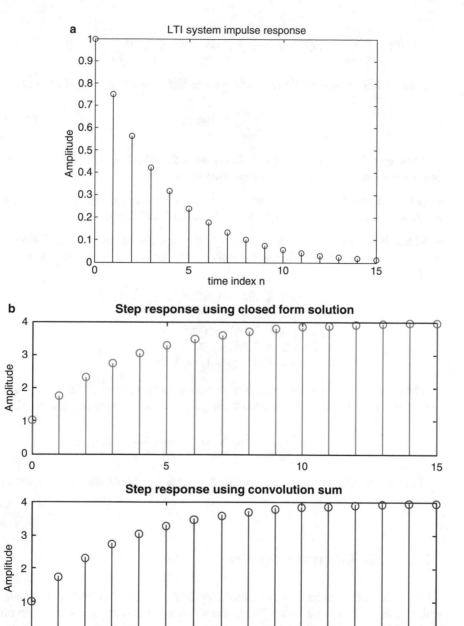

Fig. 2.2 Stem plot of the impulse response: (**a**) impulse response and (**b**) unit step response

$$|y[n]| = \left| \sum_{k=-\infty}^{\infty} x[k]h[n-k] \right| = \left| \sum_{k=-\infty}^{\infty} x[n-k]h[k] \right| \leq M \sum_{k=-\infty}^{\infty} |h[k]| \qquad (2.22)$$

From (2.22) we see that the absolute value of the response is finite if and only if

$$\sum_{n=-\infty}^{\infty} |h[n]| < \infty \qquad (2.23)$$

Thus, an LTI discrete-time system is stable in the BIBO sense if and only if its impulse response sequence is absolutely summable.

Example 2.5 An LTI discrete-time system is described by $y[n] = x[n] + 0.5y[n-1]$, $y[-1] = 0$. Determine if the system is stable in the BIBO sense.

Solution First, we need to find the impulse response of the system. Using $x[n] = \delta[n]$, in the above system definition and the fact that $y[-1] = 0$, we obtain the following:

$$\begin{aligned} y[0] &= \delta[0] + 0.5y[-1] = 1; \\ y[1] &= \delta[1] + 0.5y[0] = 0.5; \\ y[2] &= 0.5y[1] = 0.5^2; \\ y[3] &= 0.5y[2] = 0.5^3, \ldots \\ y[n] &= 0.5y[n-1] = 0.5^n \end{aligned} \qquad (2.24)$$

Thus, we find the impulse response sequence to be $h[n] = 0.5^n u[n]$. For this system to be stable, the impulse response sequence must be absolutely summable:

$$\sum_{n=-\infty}^{\infty} |h[n]| = \sum_{k=0}^{\infty} |0.5^n| = \sum_{k=0}^{\infty} 0.5^n = \frac{1}{1-0.5} = 2 < \infty$$

Since the impulse response sequence is absolutely summable, the above system is stable.

2.5   Linear Difference Equation

So far we have defined an LTI discrete-time system in terms of its impulse response, which fully defines the system. The response to any input sequence can then be obtained by convolving the input and impulse response sequences. Alternatively, one can also describe an LTI discrete-time system by a linear *difference equation* with constant coefficients. More specifically, an LTI discrete-time system can be described by

$$\sum_{k=0}^{p} b_k y[n-k] = \sum_{j=0}^{q} a_j x[n-j], b_0 = 1 \qquad (2.25)$$

The order of the difference equation (2.25) is the maximum of p and q. In Eq. (2.25), if not all b_k's are zero then the corresponding difference equation is called a *recursive* equation. It uses both feed-forward and feedback to compute the output at each time instant. On the other hand, if all but b_0 are zero, then the resulting equation is termed *non-recursive* difference equation. To make it clearer, let us rewrite (2.25) as

$$y[n] = \sum_{j=0}^{q} a_j x[n-j] - \sum_{k=1}^{p} b_k y[n-k] \qquad (2.26)$$

At each time instant n, the response is obtained by finding the weighted sum of the previous p output samples and then subtracting it from the weighted sum of the input samples, which involves the current and previous q input samples. In order to use the previous input and output samples, we need to store them. More specifically, we need to store p previous output samples and q previous input samples and retrieve them to compute the present output sample. Let us clarify this by an example.

Example 2.6 Draw a signal flow diagram to compute the response of the 2nd-order LTI discrete-time system described by the following recursive equation:

$$y[n] = a_0 x[n] + a_1 x[n-1] - b_1 y[n-1] - b_2 y[n-2] \qquad (2.27)$$

A signal flow diagram shows how the signals flow or propagate from the input to the output. It uses adders, multipliers, and delays. Each delay element corresponds to one sampling interval. Lines with arrows indicate the direction of signal flow. Figure 2.3 depicts the signal flow diagram to compute the output for a given input described by (2.27). An adder is depicted by a circle with a plus sign inscribed in it.

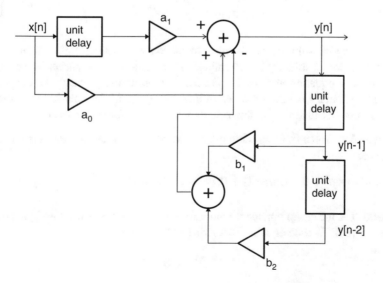

Fig. 2.3 A signal flow diagram corresponding to the difference equation (2.27)

A triangle pointing in the direction of the signal flow indicates a multiplier with the coefficient shown by the side of the triangle. A unit delay is depicted by a rectangle. The various elements are interconnected by straight lines with arrows indicating the direction of the signal flow.

The signal flow diagram shown in Fig. 2.3 is not the most efficient in terms of delay elements. It uses three delay elements. The difference equation in (2.27) uses output samples corresponding to two previous sampling intervals. Hence, it is possible to use a total of only two unit delays, which is more efficient than using three unit delays. We will discuss signal flow graphs in more detail in a later chapter.

Solving Linear Difference Equations Instead of computing the response of an LTI discrete-time system at each time instant by solving the difference equation recursively, one can also obtain the system response in closed form by means of analytical solution to the difference equation. The general solution to a constant coefficient linear difference equation consists of two parts: *complementary* solution $y_C[n]$ and *particular* solution $y_P[n]$. The complementary solution is the response to zero input, and the particular solution is the response to a specified input. Thus, the total solution to a linear difference equation with constant coefficients can be expressed as

$$y[n] = y_C[n] + y_P[n] \tag{2.28}$$

The complementary solution is obtained by (1) setting the input to zero, (2) assuming a solution of the type α^n, (3) substituting the solution in the zero input difference equation (2.25), and (4) solving for α. For a pth-order linear difference equation with constant coefficients, the complementary solution then takes the form

$$y_C[n] = \sum_{i=1}^{p} a_i \alpha_i^n \tag{2.29}$$

The particular solution is assumed to be some constant times the input. The constant of proportionality is determined by substituting the particular solution in the difference equation and solving the resulting equation. Finally, the constants in the complementary solution are determined using the initial conditions in the total solution. Let us illustrate the above statements by the following example.

Example 2.7 Solve the following difference equation when the input is a unit step sequence:

$$y[n] = x[n] + 0.25y[n-1] + 0.125y[n-2], \text{ with } y[-1] = 1, y[-2] = -1$$

Solution Let the complementary solution be $y_C[n] = \alpha^n$. Substituting $y_C[n]$ for $y[n]$ in the above difference equation with $x[n] = 0$, we get

$$\alpha^n = 0.25\alpha^{n-1} + 0.125\alpha^{n-2}$$

Or,

$$1 = 0.25\alpha^{-1} + 0.125\alpha^{-2} \Rightarrow \alpha_1 = 0.5, \alpha_2 = -0.25$$

Therefore, the complementary solution is $y_C[n] = a(0.5)^n + b(-0.25)^n$. Since the input is a unit step sequence, the particular solution is assumed to be $y_P[n] = cu[n]$. To find the value of the constant c, substitute $y_P[n]$ for $y[n]$ in the given difference equation with the input $x[n] = u[n]$. We, therefore, have

$$cu[n] = u[n] + c \times 0.25u[n-1] + c \times 0.125u[n-2] \Rightarrow c = 1.6$$

Therefore, the total solution to the given difference equation is expressed as

$$y[n] = 1.6u[n] + a(0.5)^n + b(-0.25)^n$$

Finally, use the initial conditions to solve for the two constants in the complementary solution. Thus, the two equations involving the constants a and b are

$$y[-1] = 1 = a(0.5)^{-1} + b(-0.25)^{-1} + 1.6 \Rightarrow 2a - 4b = -0.6$$

$$y[-2] = -1 = a(0.5)^{-2} + b(-0.25)^{-2} + 1.6 \Rightarrow 4a + 16b = -2.6$$

The solution to the above two equations gives $a = -\frac{5}{12}$ and $b = -\frac{7}{120}$. The overall solution to the given difference equation is, therefore,

$$y[n] = -\frac{5}{12}(0.5)^n - \frac{7}{120}(-0.25)^n + 1.6, n \geq 0$$

Note that the difference equation and the total solution give the same value of 1.125 at $n = 0$. Figure 2.4 shows stem plots of the response to a unit step sequence of the system in Example 2.7 using both the difference equation and the total solution. They appear to be identical. This shows that one can compute the response of an LTI discrete-time system either directly from the given difference equation or from the total solution obtained by analytical means.

Example 2.8 Let us consider the case where the input has the same form as one of the terms in the complementary solution. Specifically, we want to solve the difference equation of an LTI discrete-time system described by

$$y[n] - 0.8y[n-1] + 0.15y[n-2] = (0.5)^n u[n], y[-1] = 1, y[-2] = 0 \qquad (2.30)$$

Let the complementary solution be $y_C[n] = \alpha^n$. Substituting the complementary solution in (2.30) with input being zero, we have

$$\alpha^n - 0.8\alpha^{n-1} + 0.15\alpha^{n-2} = 0 \qquad (2.31a)$$

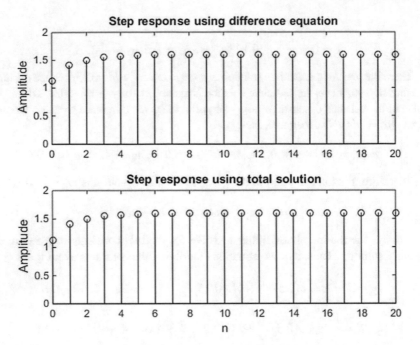

Fig. 2.4 Stem plots showing the system response to a unit step in Example 2.7. Top plot: response using the difference equation. Bottom plot: response obtained from the total solution

or,

$$\alpha^2 - 0.8\alpha + 0.15 = 0, \tag{2.31b}$$

which gives $\alpha_1 = 0.5$ and $\alpha_2 = 0.3$.

Since one of the terms in the complementary solution, namely, α_1, has the same form as the input, the particular solution must be assumed to be

$$y_P[n] = \beta n(0.5)^n u[n] \tag{2.32}$$

The complementary solution and the particular solution must be linearly independent. That is why the particular solution in (2.32) is used instead. To determine β, we solve Eq. (2.30) with $y[n]$ replaced with $\beta n(0.5)^n u[n]$:

$$\beta n(0.5)^n - 0.8\beta(n-1)(0.5)^{n-1} + 0.15\beta(n-2)(0.5)^{n-2} = 0.5^n \tag{2.33a}$$

Or,

$$\beta n - \frac{0.8\beta(n-1)}{0.5} + \frac{0.15\beta(n-2)}{0.5^2} = 1 \Rightarrow \beta = 2.5 \tag{2.33b}$$

Therefore, the total solution to the difference equation (2.30) is

$$y[n] = a(0.5)^n + b(0.3)^n + 2.5n(0.5)^n, n \geq 0 \qquad (2.34)$$

To find the values for the constants in (2.34), use the initial conditions:

$$y[-1] = 1 = \frac{a}{0.5} + \frac{b}{0.3} + \frac{2.5(-1)}{0.5} \qquad (2.35a)$$

$$y[-2] = 0 = \frac{a}{0.5^2} + \frac{b}{0.3^2} + \frac{2.5(-2)}{0.5^2} \qquad (2.35b)$$

The solution to (2.35a) and (2.35b) results in $a = 0$ and $b = 1.8$. Thus, the solution to Eq. (2.30) is

$$y[n] = 1.8(0.3)^n + 2.5n(0.5)^n, n \geq 0 \qquad (2.36)$$

Figure 2.5 shows the response calculated using the difference equation (2.30) in the top plot and the response obtained from the total solution in the bottom plot. They seem to be identical.

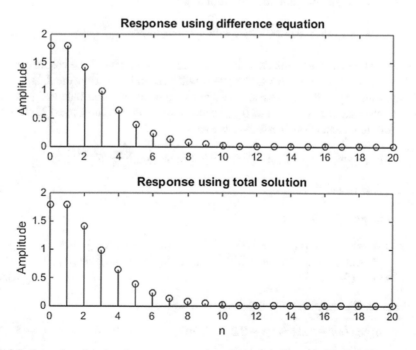

Fig. 2.5 Stem plots showing the system response to the input in Example 2.8. Top plot: response using the difference equation. Bottom plot: response obtained from the total solution

Example 2.9 An LTI discrete-time system is described by the following difference equation with initial conditions:

$$y[n] - 1.5y[n-1] + 0.5625y[n-2] = x[n], y[-1] = 1, y[-2] = 0$$

Determine the total solution to the above difference equation if $x[n] = \cos(0.2n) u[n]$.

Solution Let the complementary solution be $y_C[n] = \alpha^n$. Then, with $x[n] = 0$, the difference equation becomes

$$\alpha^n - 1.5\alpha^{n-1} + 0.5625\alpha^{n-2} = 0. \tag{2.37a}$$

Or,

$$\alpha^{n-2}(\alpha^2 - 1.5\alpha + 0.5625) = 0 \Rightarrow \alpha_1 = \alpha_2 = 0.75 \tag{2.37b}$$

Since the two roots of the characteristic equation are the same, the two terms of the complementary solution are α^n and $n\alpha^n$ and are linearly independent. Thus,

$$y_C[n] = a\alpha^n + bn\alpha^n \tag{2.38}$$

Next we assume the particular solution to be

$$y_P[n] = A\cos(0.2n + \varphi)u[n] = Re\left\{Ae^{j(0.2n+\varphi)}\right\}u[n] \tag{2.39}$$

Note that we have introduced a phase term in the argument of the cosine function of the particular solution. An LTI system will respond to a sinusoidal input of a certain frequency with the same sinusoid but with different amplitude and phase. In order to evaluate the constants of the particular solution, we substitute (2.39) in the given difference equation. Therefore, we have

$$Re\left\{Ae^{j(0.2n+\varphi)} - 1.5Ae^{j(0.2n+\varphi-0.2)} + 0.5625Ae^{j(0.2n+\varphi-0.4)}\right\}$$
$$= Re\left\{e^{j0.2n}\right\} \tag{2.40}$$

Rearranging (2.40), we get

$$A[\cos(\varphi) - 1.5\cos(\varphi - 0.2) + 0.5625\cos(\varphi - 0.4)]\cos(0.2n)$$
$$- A[\sin(\varphi) - 1.5\sin(\varphi - 0.2) + 0.5625\sin(\varphi - 0.4)]\sin(0.2n)$$
$$= \cos(0.2n) \tag{2.41}$$

From (2.41), we obtain the following two equations:

$$A[\cos(\varphi) - 1.5\cos(\varphi - 0.2) + 0.5625\cos(\varphi - 0.4)] = 1 \tag{2.42a}$$

$$A[\sin(\varphi) - 1.5\sin(\varphi - 0.2) + 0.5625\sin(\varphi - 0.4)] = 0 \tag{2.42b}$$

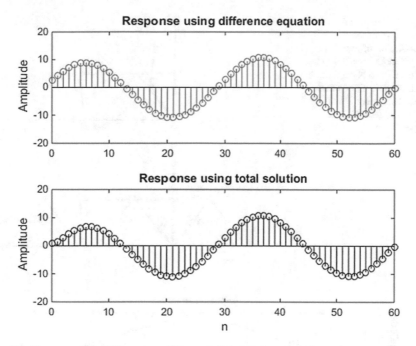

Fig. 2.6 Response of the LTI system of Example 2.9 plotted as stem plots

Solving the Eqs. (2.42a) and (2.42b) for A and φ, we get A $=$ 10.8225 and $\varphi = -1.024592$ *rad*, respectively. The total solution is, therefore,

$$y[n] = a(0.75)^n + bn(0.75)^n + 10.8225\cos(0.2n - 1.024592^r) \qquad (2.43)$$

Using the given initial conditions in the above equation and solving the resulting two equations, we find the system response to be

$$y[n] = -4.8955(0.75)^n - 2.8912 \times n(0.75)^n$$
$$+ 10.8225\cos(0.2n - 1.024592), n \geq 0 \qquad (2.44)$$

The responses obtained by recursively solving the difference equation and from the total solution to the difference equation are plotted as stem plots and are shown in Fig. 2.6. They seem to agree.

Convolution of Finite-Length Sequences Consider the two finite-length sequences $\{x[n]\}$, $0 \leq n \leq N - 1$ and $\{h[n]\}$, $0 \leq n \leq M - 1$. Since the two sequences are of finite length, the convolution of these two sequences will result in a sequence that is also of finite length. In fact, the length of the convolution of the sequences of lengths M and N is $M + N - 1$. We can demonstrate this by an example. In Fig. 2.7 top plot is shown the two sequences to be convolved. For the sake of simplicity, the sequences are shown in solid lines though they are discrete. The bottom plot of Fig. 2.7 shows $\{x[k]\}$ as well as the flipped sequences $\{h[-k]\}$,

Fig. 2.7 Graphical
illustration of the
convolution of two finite-
length sequences. Top plot:
length M and N sequences.
Bottom plot: flipped and
shifted sequence for shifts
0, M, and M + N − 1,
respectively

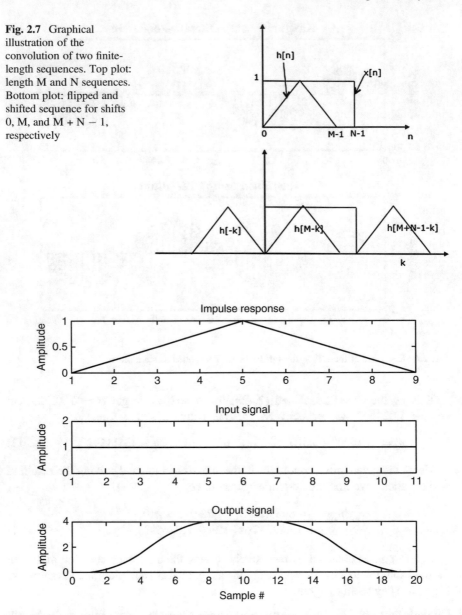

Fig. 2.8 Convolution of a length-9 triangular sequence and length-11 unit amplitude pulse

$\{h[M - k]\}$, and $\{h[M + N - 1 - k]\}$. From the figure we observe that the
convolution is zero for n < 0 and n > M + N − 1. Hence, the length of the
convolution of two sequences of length M and N is M + N − 1.

Figure 2.8 shows the convolution of a length-9 triangular sequence and a length-
11 unit amplitude pulse sequence resulting in a length- 11 + 9−1 = 19 sequence.

2.6 Sampling a Continuous-Time Signal

So far we have assumed explicitly the availability of discrete-time signals without reference to their origin. However, many discrete-time signals originate from their continuous-time counterparts. It is, therefore, necessary to understand how discrete-time signals are obtained from continuous-time signals and the implications thereof. It must be pointed out that the processed discrete-time signals must be converted back to their continuous-time versions. For example, one has to understand how many samples per second are necessary so that the discrete-time signal can be converted back to its continuous-time version without any impairment. Too many samples per second mean that the digital signal processor has to carry out a lot of arithmetic operations per second. This may impose undue constraints on the processor speed. On the other hand, fewer samples per second may cause serious distortions, which cannot be tolerated. Thus, one must determine the correct number of samples per second required for distortionless recovery of the continuous-time signal from the discrete-time signal. This can be achieved only by mathematical reasoning. In the following we will consider the process of sampling a continuous-time signal to obtain the discrete-time version and its implications. We will further ascertain the correct sampling interval for a given continuous-time signal.

Ideal Sampling A discrete-time signal is obtained from a continuous-time signal by sampling the continuous-time signal precisely at regular or uniform intervals of time. The sampled signal $x_s(t)$ can be expressed mathematically as

$$x_s(t) = x(t)|_{t=nT_s}, n \in Z \tag{2.45}$$

where T_s is the sampling interval. Figure 2.9 illustrates the ideal sampling process. The continuous-time signal is shown in cyan color and the sampled signal in red stems. At each sampling instant, the amplitude of the discrete-time signal corresponds to that of the continuous-time signal. Since the interval between any two samples is the same, the sampling is called *uniform sampling*. We also notice that the width of each sample is zero. Therefore, this type of sampling is called *ideal sampling* or *impulse sampling*. In practice, there is no such thing as ideal sampling. Each sample has a finite width, though very small. This type of sampling is called *non-ideal sampling* and has some implications, which we will consider later.

Our first task is to establish an upper limit on the sampling interval. In other words, what is the largest value of T_s and yet the continuous-time signal can be recovered from the sampled signal without any distortion? In order to answer this question, we must resort to the frequency domain representation of the signals under consideration. To this end we can rewrite Eq. (2.45) in terms of Dirac delta functions as

$$x_s(t) = \sum_{n=-\infty}^{\infty} x(t)\delta(t - nT_s) \tag{2.46}$$

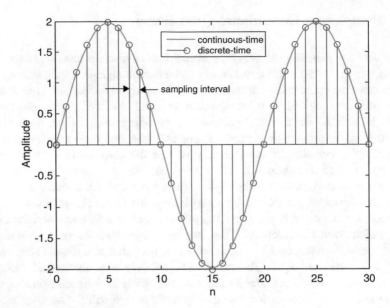

Fig. 2.9 Illustration of ideal sampling

Note that the unit impulse is zero except at time instants $t = nT_s$, at which instants its strength equals the sample value of the continuous-time signal. We can now express the Fourier transform of the sampled signal in terms of the Fourier transform of the continuous-time signal. Since the Fourier transform is the frequency domain representation of a signal, we will be able to determine the upper limit on the sampling interval. To this end, let $X(f)$ be the Fourier transform of the continuous-time signal $x(t)$. Then the Fourier transform of the sampled signal can be written as

$$X_s(f) = \mathcal{F}\{x_s(t)\} = \mathcal{F}\left\{\sum_{n=-\infty}^{\infty} x(t)\delta(t - nT_s)\right\} \tag{2.47}$$

Using the linearity and convolution in the frequency domain properties of the Fourier transform, we can express (2.47) as

$$X_s(f) = \sum_{n=-\infty}^{\infty} \mathcal{F}\{x(t)\delta(t - nT_s)\} = \sum_{-\infty}^{\infty} X(f) \otimes \delta(f - nF_s) \tag{2.48}$$

In Eq. (2.48), the sampling frequency is $F_s = \frac{1}{T_s}$. The convolution of $X(f)$ with an impulse $\delta(f - nF_s)$ results in shifting the spectrum of $X(f)$ to nF_s, that is, $X(f - nF_s)$. Therefore, we have

$$X_s(f) = \sum_{n=-\infty}^{\infty} X(f - nF_s) \tag{2.49}$$

Equation (2.49) implies that the Fourier transform of the sampled signal is an infinite sum of the Fourier transform of the continuous-time signal, replicated at integer multiples of the sampling frequency. By knowing the spectrum of the continuous-time signal, we can determine the upper limit for the sampling interval or equivalently and the lower limit on the sampling frequency. Since the continuous-time signal must be recovered from its samples, we must find a way to recover or *reconstruct* the continuous-time signal from its samples.

Sampling or Nyquist Theorem A continuous-time signal that is band limited to $|f| \leq f_c$ can be recovered or reconstructed exactly from its samples taken uniformly at a rate $F_s \geq 2f_c$. The sampling frequency $F_s = 2f_c$ is called the *Nyquist frequency*. In terms of the sampling interval, the Nyquist theorem implies $\frac{1}{T_s} \geq 2f_c$ or $T_s \leq \frac{1}{2f_c}$. That is, that the sampling interval must be less than or equal to the inverse of twice the maximum frequency of the continuous-time signal to be sampled.

From the statement of the sampling theorem, we notice that it pertains to continuous-time signals with finite bandwidth, that is, signals that are band limited. A continuous-time signal that is band limited to $|f| \leq f_c$ is the same thing as saying that its Fourier transform satisfies the condition

$$|H(f)| = \begin{cases} K, |f| \leq f_c \\ 0, otherwise \end{cases} \qquad (2.50)$$

where K is a constant. This type of magnitude response is known as the *brick wall* type of response and is an ideal case. But no physically realizable system can have such a brick wall type of frequency spectrum. So in practice, to limit the frequency spectrum to a specified frequency range, one must prefilter the continuous-time signal and then sample it.

Reconstruction of an Ideally Sampled Signal The continuous-time signal can be recovered or reconstructed *exactly* from its samples by passing the samples through an *ideal* lowpass filter having a cutoff frequency equal to half the sampling frequency. In order to prove the statement, let $h(t)$ be the impulse response of the ideal lowpass filter. Then its response $y(t)$ to the sampled signal $x_s(t)$ is the convolution of the sampled signal and the impulse response:

$$y(t) = x_s(t) \otimes h(t) \qquad (2.51)$$

Using (2.46) in (2.51), we have

$$y(t) = \left\{ \sum_{n=-\infty}^{\infty} x(nT_s)\delta(t - nT_s) \right\} \otimes h(t)$$

$$= \sum_{n=-\infty}^{\infty} x(nT_s)\{\delta(t - nT_s) \otimes h(t)\} \qquad (2.52)$$

Since the convolution of $\delta(t - nT_s)$ and $h(t)$ equals $h(t - nT_s)$, Eq. (2.52) results in

$$y(t) = \sum_{n=-\infty}^{\infty} x(nT_s)h(t - nT_s) \qquad (2.53)$$

The impulse response of the ideal lowpass filter band limited to f_c can be shown to be

$$h(t) = 2f_c \frac{\sin{(2\pi f_c t)}}{2\pi f_c t} = 2f_c sinc(2f_c t) \qquad (2.54)$$

The sinc function is defined as

$$sinc(x) = \frac{\sin{(\pi x)}}{\pi x} \qquad (2.55)$$

Using Eqs. (2.54) in (2.53), the reconstructed signal is found to be

$$y(t) = 2f_c \sum_{n=-\infty}^{\infty} x(nT_s) \frac{\sin{(2\pi f_c(t - nT_s))}}{2\pi f_c(t - nT_s)} \qquad (2.56)$$

The *sinc* function is unity at the sampling instants with an amplitude equal to the sample values of the signal. At other instants the signal amplitude is interpolated by the filter to reconstruct the continuous-time signal exactly. Thus, a continuous-time signal is recovered or reconstructed from its samples by filtering the sampled signal through an ideal lowpass filter whose cutoff frequency equals half the Nyquist frequency at most.

Aliasing Distortion What if the sampling frequency does not meet the Nyquist criterion? What happens when the sampling frequency is less than twice the maximum frequency of the continuous-time signal to be sampled? When a continuous-time signal is under-sampled, meaning the sampling frequency is below the Nyquist frequency, a distortion known as *aliasing distortion* occurs, because of which the continuous-time signal cannot be recovered from its samples. The frequencies above the *folding frequency* are aliased as lower frequencies. The folding frequency corresponds to half the sampling frequency. For instance, a frequency $f_1 + \frac{F_s}{2}$ present in the continuous-time signal will appear as a frequency $\frac{F_s}{2} - f_1$, which is lower than the frequency f_1. Thus, a frequency higher than the folding frequency if present will alias itself as a lower frequency. This is the aliasing distortion. This is depicted in Fig. 2.10. In Fig. 2.10a the spectrum of a band-limited continuous-time signal with a maximum frequency f_c is shown. Figure 2.10b shows the spectrum of the sampled signal, where the sampling frequency is much higher than $2f_c$. There is no overlap between the replicas of the spectra. Therefore, the continuous-time signal can be recovered by filtering the sampled signal by an ideal lowpass filter with a cutoff frequency f_c. Figure 2.10c depicts the case where the sampling frequency is less than twice the maximum frequency of the continuous-time signal. Because of the overlap of adjacent spectra, the spectrum in the frequency range $-f_c \leq f \leq f_c$ is distorted and is the cause for the aliasing distortion.

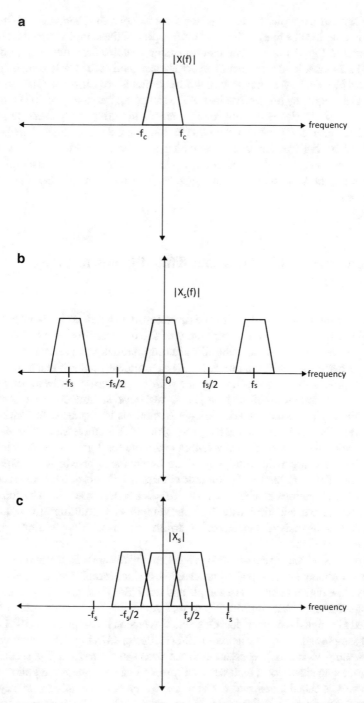

Fig. 2.10 An illustration of aliasing distortion: (**a**) spectrum of a continuous-time signal. (**b**) Oversampling case. (**c**) Undersampling case

As a second example, let us consider a continuous-time sinusoid of frequency 5 Hz. It is sampled at a rate of 20 per second. A plot of the sampled signal is shown in the top plot in Fig. 2.11a as a line plot for easier visualization. In the bottom plot of Fig. 2.11a a sinusoid at a frequency of 15 Hz sampled at 20 Hz is shown. The two plots look identical even though the two frequencies are different. The sinusoid at 15 Hz has a frequency higher than half of the sampling frequency of 20 Hz. It aliases itself as $15-10 = 5$ Hz signal component. This aliasing frequency is the same as that of the sinusoid at 5 Hz. Hence, the two sampled signals look alike. This is further ascertained by the frequency spectra, which are shown in Fig. 2.11b as top and bottom plots, respectively. Similarly, a third example of undersampling and oversampling of a continuous-time signal is illustrated in Fig. 2.12 and is self-explanatory.

2.7   Conversion of Continuous-Time Signals to Digital Signals

So far in our discussion we have treated discrete-time signals as having a continuum of amplitudes, meaning that the amplitudes of the discrete-time signals have infinite accuracy in amplitude. As a result all our computations such as convolution sum, etc., were performed with infinite accuracy. This ideal scenario changes when we deal with processing discrete-time signals with computers – hardware or software. There are two issues involved here. First, we have to convert the discrete-time signals into digital signals, which are approximations to signals with continuum of amplitudes. This is known as analog-to-digital (A/D) conversion. The degree of approximation depends on the *word length* available for digital representation of the amplitudes. The larger the word length, the better the approximation. The second issue deals with the accuracy of arithmetic operations. Errors due to limited accuracy of arithmetic operations manifest as noise. This is also the case with A/D conversion. In this chapter we will deal with A/D conversion and resulting errors. In a later chapter we will analyze the effect of arithmetic errors due to finite precision arithmetic operations.

The process of converting an analog signal to digital signal is depicted in Fig. 2.13. The input continuous-time signal is first sampled, and the sampled value is held until it is converted to a digital number. The sample and hold (S/H) functional block samples the input signal at a predetermined uniform rate and holds the sample value until it is converted to a digital representation. Once the conversion is completed, the S/H acquires the next sample and so on. The process of S/H is illustrated in Fig. 2.14, where an analog signal is sampled and held constant until the next sample arrives. The second block, namely, the quantizer block, represents the sampled value to the nearest allowed level. This process is called *quantization*. There are two types of quantizers, namely, *scalar* and *vector* quantizers. A scalar quantizer accepts a single analog sample and outputs a quantized value that approximates the input analog sample. A vector quantizer, on the

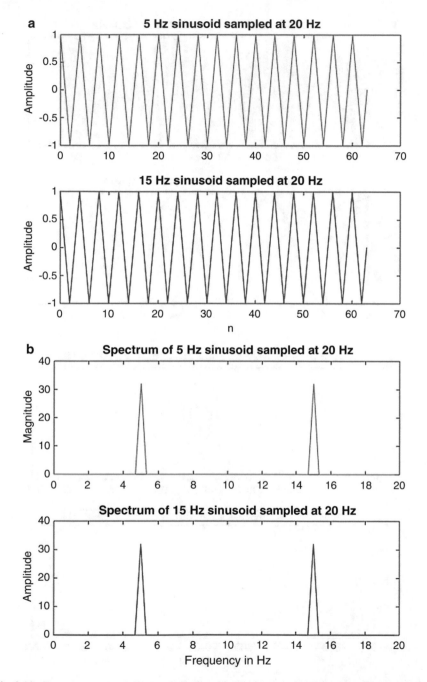

Fig. 2.11 Second example of aliasing distortion: (**a**) top plot, sampled 5 Hz sinusoid at a sampling frequency of 20 Hz; bottom plot, sampled 15 Hz sinusoid at a sampling frequency of 20 Hz. (**b**) Top plot, spectrum of 5 Hz signal at 20 Hz sampling rate; bottom plot, spectrum of 15 Hz signal sampled at 20 Hz rate. The 15 Hz signal appears as a 5 Hz signal

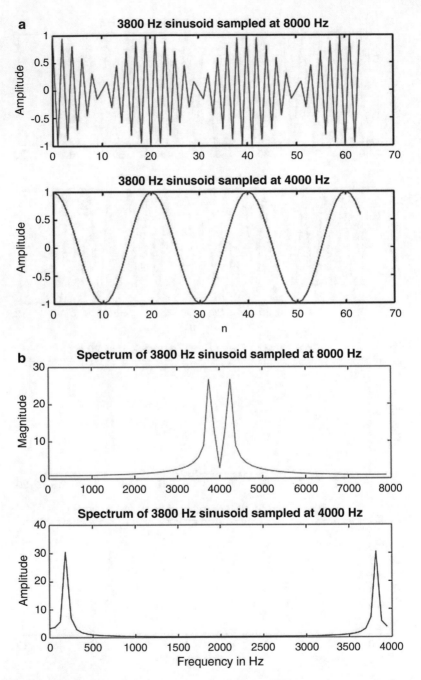

Fig. 2.12 A third example of aliasing distortion: (**a**) top plot, 3800 Hz signal at a sampling rate of 8000 Hz; bottom plot, same 3800 Hz signal sampled at a rate of 4000 Hz. (**b**) Top plot, spectrum of the signal at 8000 Hz sampling rate; bottom plot, spectrum of the same signal at 4000 Hz sampling rate. The 3800 Hz appears as 200 Hz signal

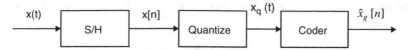

Fig. 2.13 A practical A/D converter model

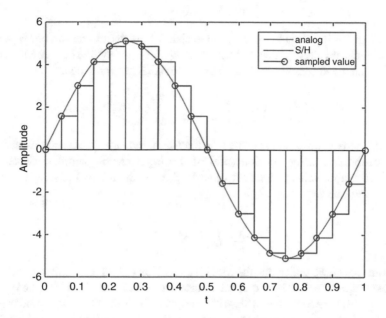

Fig. 2.14 Example of S/H function

other hand, accepts a vector of analog samples and outputs a vector of discrete samples that is close to the input vector. We will only deal with scalar quantizers here.

The design of a scalar quantizer amounts to dividing the input range of the analog signal amplitude into L + 1 levels and determining the corresponding L reconstruction or output levels. The relationship between the number of reconstruction levels L and the number of bits B of the quantizer is $L = 2^B$. The A/D converter has a fixed voltage or current amplitude limits for its input. For instance, it can have an input voltage limited to ± 1 V or 0 to 1 volt. If the number of binary digits (bits) allowed in an A/D converter is B bits, then there are $L = 2^B$ number of levels between the minimum and maximum input amplitude range. The quantizer assigns the current sample value a level that is closest to the analog sample value. Denote the L + 1 input decision intervals by $\{D_j, 1 \leq j \leq L + 1\}$ and the corresponding L output levels by $\{R_k, 1 \leq k \leq L\}$. The quantizer maps an input analog sample x to its nearest neighbor and is formally expressed as

$$Q(x) = \widehat{x} = R_k \tag{2.57}$$

In general, the lower and upper boundaries of the input decision intervals are defined by

$$D_1 = x_{min} \& D_{L+1} = x_{max} \tag{2.58}$$

Let us assume that the output levels are chosen so that the following is satisfied.

$$R_1 < R_2 < R_3 \cdots\cdots R_L \tag{2.59}$$

For a given input analog signal, the decision boundaries and the corresponding output levels are chosen such that the mean square error (MSE) between the analog and the quantized samples is a minimum. The MSE is expressed as

$$MSE = E\left\{ (x - \hat{x})^2 \right\} = \int_{D_1}^{D_{L+1}} (x - \hat{x})^2 p_x(x) dx \tag{2.60}$$

In the above equation, E denotes the statistical average or expectation and $p_x(x)$ the probability density function (pdf) of the input analog samples. Because the quantized output value is constant equal to R_k over the interval $[D_k, D_{k+1})$, (2.60) can be rewritten as

$$MSE = \sum_{m=1}^{L} \int_{D_m}^{D_{m+1}} (x - R_m)^2 p_x(x) dx \tag{2.61}$$

From (2.61), we notice that the MSE is a function of both the decision boundaries and the output levels. Therefore, the minimum value of the MSE in (2.61) can be found by differentiating the MSE with respect to both the decision boundaries and the output levels and setting them to zero and then solving the two equations. Thus,

$$\frac{\partial MSE}{\partial D_i} = (D_i - R_{i-1})^2 p_x(D_i) - (D_i - R_i)^2 p_x(D_i) = 0 \tag{2.62a}$$

$$\frac{\partial MSE}{\partial R_i} = 2 \int_{D_i}^{D_{i+1}} (x - R_i) p_x(x) dx = 0, 1 \leq i \leq L \tag{2.62b}$$

From (2.62a), we have

$$(D_i - R_{i-1})^2 = (D_i - R_i)^2 \tag{2.63}$$

Due to the fact that $R_i > R_{i-1}$, we determine the decision boundaries after simplifying (2.63) as

$$D_i = \frac{R_i + R_{i-1}}{2} \tag{2.64}$$

The output levels are obtained from (2.62b) as

$$R_i = \frac{\int_{D_i}^{D_{i+1}} x p_x(x) dx}{\int_{D_i}^{D_{i+1}} p_x(x) dx} \qquad (2.65)$$

The implications of the optimal quantizer are that the decision boundaries lie at the midpoints of the corresponding output levels and the optimal output levels are the centroids of the decision intervals. Since these two quantities are interdependent, there is no closed-form solution to the two Eqs. (2.64) and (2.65). The solution is obtained by iteration. Also, note that the design of an optimal quantizer requires a priori knowledge of the pdf of the input analog samples. In other words, the optimal scalar quantizer is a function of the pdf of the input analog samples. This type of quantizer is known as the *Lloyd-Max* quantizer. The decision intervals and the corresponding output levels of a Lloyd-Max quantizer are nonuniform. Moreover, the Lloyd-Max quantizer is dependent on the input signal. Therefore, each new signal must have its own quantizer for optimal performance. However, a closed-form solution exists for a uniform quantizer, which we will describe next. The design of a uniform quantizer is simple and is the reason for its widespread use in image and video compression standards.

Uniform Quantizer If the pdf of the input analog samples is uniform, there exists a closed-form solution to the decision boundaries and the output levels for the Lloyd-Max quantizer. Under this condition, we will find the decision intervals and the output levels will all be equal. Hence the quantizer is known as the *uniform quantizer*. Obviously, a uniform scalar quantizer is optimum for analog samples that have uniform pdf. A uniform pdf implies

$$p_x(x) = \frac{1}{x_{max} - x_{min}} = \frac{1}{D_{L+1} - D_1} \qquad (2.66)$$

Substituting (2.66) for the pdf in (2.65), the output levels of a scalar uniform quantizer are found to be

$$R_i = \frac{D_{i+1} + D_i}{2} \qquad (2.67)$$

Using (2.67) in (2.64), the following relationship is found

$$D_{i+1} - D_i = D_i - D_{i-1} = \Delta, 2 \leq i \leq L \qquad (2.68)$$

From the above equation, it is clear that the interval between two consecutive decision boundaries is the same and equals the quantization step size Δ. For instance, if the input amplitude range is between x_{min} and x_{max}, then for a B-bit quantizer, the step size is $\Delta = \frac{x_{max} - x_{min}}{2^B}$. The value assigned by the quantizer to an analog sample then lies half way between two consecutive levels. Therefore, the error or difference between a sample and its quantized value ranges between $\pm \frac{\Delta}{2}$. Larger the value of B smaller is the quantization error. Finally, the coder block assigns a B-bit binary code to the quantized level. Thus, a sample is converted to a digital number.

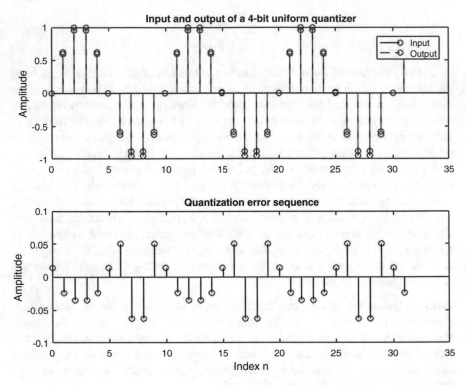

Fig. 2.15 Actual and quantized values of the sinusoid in Example 2.10 using a 4-bit uniform quantizer: top, input sinusoid in black stems and the quantized samples in red dashed stems; bottom, corresponding error sequence

Example 2.10 This example shows how to design a B-bit uniform quantizer based on an input sinusoid of specified amplitude, frequency, and sampling frequency. It then quantizes the sinusoid and calculates the resulting SNR in dB. The MATLAB M-file for this example is named *Example2_10.m*. The number of bits of quantization used in this example is 4. The amplitude range is ±1. The resulting SNR is found to be 23.07 dB. The actual input and the corresponding quantized values are shown in the top plot in Fig. 2.15. The quantization error sequence is shown in the bottom plot of Fig. 2.15. The input-output characteristic of the 4-bit uniform quantizer for the sinusoid in this example is shown in Fig. 2.16. As required, there are 16 steps between the amplitude range of ±1.

Coding the Quantized Values There are two ways to code the assigned level into a binary number. One way is to use what is called the *sign-magnitude* representation. In this method the magnitude of the sample value is represented by a b-bit binary number and a sign bit as the most significant bit (MSB) to indicate its sign. Thus, there are B = b + 1 bits in this digital representation. In the second method called *two's complement*, positive fractions are represented as in the sign-magnitude form. A negative fraction is represented in two's complement form as follows: first the

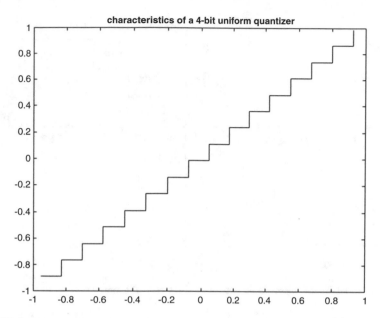

Fig. 2.16 Input-output characteristic of the 4-bit uniform quantizer of Example 2.10

magnitude is represented in binary number, the bits are complemented, and then a 1 is added to the least significant bit (LSB) to get the two's complement representation. For instance, let us represent the decimal fraction -0.875 in the two abovementioned formats. In sign-magnitude format, we first represent the magnitude of the decimal fraction in binary fraction, which is $0.875_{10} = 0.1110_2$. Then a "1" bit is inserted in the MSB position to get the sign-magnitude representation: $-0.875_{10} = 1.1110_2$. The "1" in the MSB corresponds to a negative value, and a "0" in the MSB corresponds to a positive value. The same decimal value in two's complement form is obtained by complementing each bit of the magnitude and then adding a "1" to the LSB. So, the complement of 0.875_{10} is 1.0001_2. Adding a "1" to the LSB gives the number $1.0001_2 + 0.0001_2 = 1.0010_2$. Thus, the two's complement representation of -0.875_{10} is 1.0010_2. As an example of a 3-bit A/D converter, Fig. 2.17 shows the input-output characteristics using two's complement representation. As can be seen from the figure, all input values greater than or equal to $\frac{7\Delta}{2}$ are assigned the same value of 3Δ. Similarly, all input values less than or equal to $-\frac{9\Delta}{2}$ are assigned the same value -4Δ.

2.8 Performance of A/D Converters

A/D converters come with different bit widths. Some are 8-bit converters, some are 12-bit converters, and others are 14- or 16-bit converters. We mentioned earlier that the output of an A/D converter is an approximation to the input samples. The degree

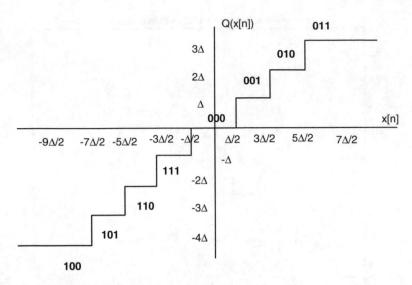

Fig. 2.17 Input-output characteristic of a 3-bit analog-to-digital converter

of approximation is a function of the bit width of the A/D converter. Because of the approximation carried out by an A/D converter, errors occur between the analog samples and the digital samples. This type of error is random and so is considered as noise. That is, there is no definite analytical expression to describe the errors sample by sample. It is, therefore, convenient and proper to describe the error due to quantization in terms of its averages. The most commonly used measure to describe noise is the variance. In order to estimate the variance of the noise due to quantization, one has to know its distribution. Here, by distribution we mean the probability of occurrence of the amplitudes of the noise due to quantization. In practice it is found that the quantization error is uniformly distributed between the range $\left[-\frac{\Delta}{2}, \frac{\Delta}{2}\right]$. That is to say that all amplitudes in this interval have the same probability of occurrence. Mathematically speaking, the uniform distribution is expressed as

$$p(e) = \frac{1}{\Delta}, \quad -\frac{\Delta}{2} \le e \le \frac{\Delta}{2} \tag{2.69}$$

with e corresponding to the possible amplitude of the quantization error. The mean or average value of the quantization error μ_e is obtained from

$$\mu_e = \int_{-\frac{\Delta}{2}}^{\frac{\Delta}{2}} e p(e) de = \frac{1}{\Delta} \int_{-\frac{\Delta}{2}}^{\frac{\Delta}{2}} e de = \frac{1}{2\Delta} e^2 \Big|_{-\frac{\Delta}{2}}^{\frac{\Delta}{2}} = 0 \tag{2.70}$$

So, the mean value of the quantization error is zero. The variance σ_e^2 of the quantization error is obtained from

$$\sigma_e^2 = \int_{-\frac{\Delta}{2}}^{\frac{\Delta}{2}} (e - \mu_e)^2 p(e) de = \frac{1}{\Delta} \int_{\frac{\Delta}{2}}^{\frac{\Delta}{2}} e^2 de = \frac{1}{3\Delta} e^3 \Big|_{-\frac{\Delta}{2}}^{\frac{\Delta}{2}} = \frac{\Delta^2}{12} \tag{2.71}$$

From Eq. (2.71), we notice that the variance of the noise due to quantization is proportional to the square of the step size. In terms of the word length B of the A/D converter, Eq. (2.71) amounts to

$$\sigma_e^2 = \frac{(x_{max} - x_{min})^2 2^{-2B}}{12} \tag{2.72}$$

The quantization noise variance of the A/D converter decreases exponentially with increasing bit width. The quantization noise variance alone is not enough to judge its effect on the processed signal. It depends on the power or variance of the analog signal being processed digitally. In other words, the effect of the noise due to quantization depends its variance relative to the signal variance. This is defined by the signal-to-noise ratio (SNR) and is usually expressed in decibel or dB for short. If the amplitude of the analog signal is assumed to be *uniformly distributed* in the range $\{x_{min}, x_{max}\}$, then its variance is found to be

$$\sigma_x^2 = \frac{(x_{max} - x_{min})^2}{12} \tag{2.73}$$

Then the SNR in dB of the A/D converter with B-bit bit width is defined as

$$SNR = 10 log_{10}\left(\frac{\sigma_x^2}{\sigma_e^2}\right) = 10 log_{10}\left(\frac{\frac{(x_{max}-x_{min})^2}{12}}{\frac{(x_{max}-x_{min})^2}{12}2^{-2B}}\right) \approx 6.02B, dB \tag{2.74}$$

From Eq. (2.74), we observe that the SNR in dB increases linearly with B. If B is increased by 1 bit, then the SNR increases by approximately 6 dB. That is to say that each additional bit in the A/D converter yields an improvement of about 6 dB in the resulting SNR.

MATLAB Examples   We can simulate A/D converters using the MATLAB Simulink system. Before building a hardware system, it is wise to first simulate it to assess its performance. If the system does not meet the target performance, one can fine-tune the design parameters and rerun the simulation to verify its performance. Thus, simulation not only saves time and energy but also guarantees performance. Simulink is a very useful tool in simulating algorithms and hardware systems. We will first demonstrate the sample and hold operation, using Simulink.

S/H Example Using Simulink   Figure 2.18 shows the block diagram simulating the S/H function. It consists of a continuous-time signal source, a pulse generator as the sampler, S/H block, and a time scope to display the signals. In order to draw the block diagram, first start the Simulink by clicking the *Simulink Library* on the toolbar menu on the MATLAB window or type *simulink* in the workspace. A Simulink Library Browser appears. Click the arrow pointing downward next to the icon showing simulink diagram on the toolbar, and select *New Model*. A new window appears. Select the *DSP System Toolbox* and then *Sources*. A list of sources appears on the right side as shown in Fig. 2.19. Let us choose the *Sine Wave* source and drag it to new untitled window. Double-click the Sine Wave block, and specify

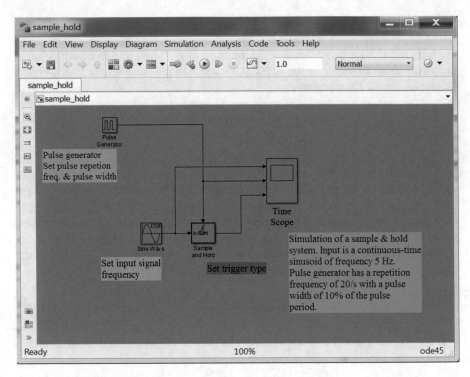

Fig. 2.18 Block diagram of sample and hold function using Simulink

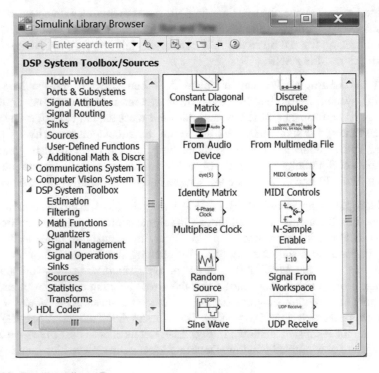

Fig. 2.19 Simulink Library Browser

Fig. 2.20 Block parameters
for sine wave source

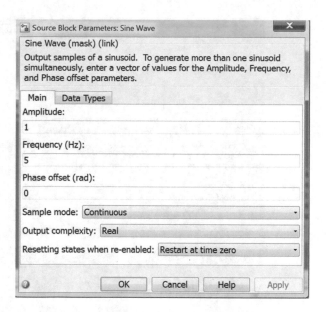

the parameters as shown in Fig. 2.20. Next, under *Simulink*, choose *Sources* and then
Pulse Generator and drag it to the untitled window. Double-click the *Pulse Generator* and fill in the blanks as shown in Fig. 2.21. Next, choose *DSP System Toolbox*
and then *Signal Operations*. From the list of blocks that appear on the right side,
choose the *Sample and Hold* block, and drag it to the untitled window. Double-click
the block, and fill in the blanks as shown in Fig. 2.22. Finally, choose *Sinks* under
DSP System Toolbox. From the list of sinks, choose *Time Scope*, and drag it to the
untitled window. Double-click the *Time Scope*. A time scope appears as shown in
Fig. 2.23. Figure 2.23 also shows the three time displays. This window was captured
after simulation. To display three signals in three rows, click *View*, and choose
Layout from the time scope window. By choosing *Configuration Properties* under
View in the toolbar, we can label the three displays. Now we have all the sub-blocks
and the corresponding parameters. We need to connect them in the order shown in
Fig. 2.18. To connect the output of one block to the input of a second block, first
click the output node, and then move the mouse to the input node of the second block
while pressing the left mouse button. Thus, we have the complete system. Next, save
the diagram with a name. To start the simulation, click the green button with an
arrow pointing to the right on the toolbar. If everything is syntactically correct,
MATLAB performs the simulation, and the various signals are displayed on the time
scope, as indicated in Fig. 2.23. In this example the simulation time is chosen to be
1 s. One can change the simulation time to suit the needs. In this example the input
continuous-time signal is a sine wave with a frequency of 5 Hz and amplitude unity.
Since the simulation time is chosen to be 1 s, there are five cycles of the sine wave, as
displayed on the topmost display. The middle display displays the pulse train of the
pulse generator. The pulse period is 0.05 s. Therefore, there are 20 pulses in 1 s, as
can be seen in the display. The bottommost display is the sample and hold signal.

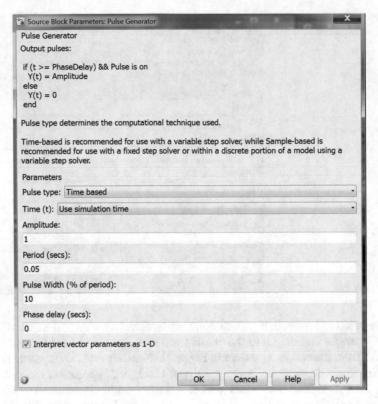

Fig. 2.21 Block parameters for pulse generator source

Fig. 2.22 Block parameters for sample and hold function

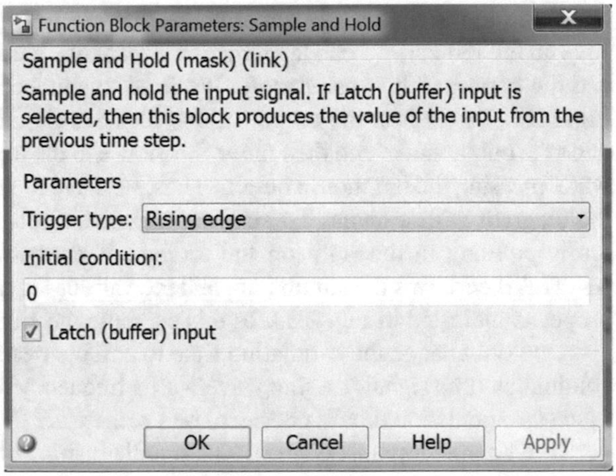

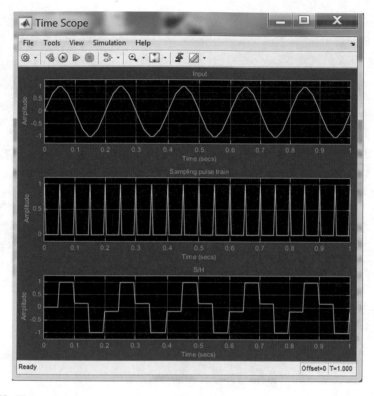

Fig. 2.23 Time scope

The analog signal is sampled at the rising edge of the pulse and held for about 45 ms duration. The pulse period is 50 ms, and the pulse width is 5 ms. So the S/H block holds the sample value for approximately 45 ms. Since there are 20 pulses in 1 s, there are 20 samples in that duration, as seen in Fig. 2.23.

Simulation of A/D Converter Using Simulink An A/D converter as shown in Fig. 2.13 first samples the input continuous-time signal and holds until the conversion is completed. We have simulated this sample and hold process as described above. Next, we will expand on this and include a quantizer to complete the A/D conversion process. Figure 2.24 shows the block diagram of an A/D converter using Simulink. As can be seen from the figure, we have included two input signal sources: a sine wave signal generator and a random signal generator. The sine wave function generates a continuous-time signal with an amplitude unity and a frequency of 5 Hz as shown in Fig. 2.25. The random signal generator produces a uniformly distributed random signal with amplitudes between −1 and +1 and is also a continuous function of time as seen in Fig. 2.26, which lists the parameters of the random signal generator. A manual switch is used to switch the input source between sine wave and random signal. Before starting the simulation, we have to double-click the switch to change the input

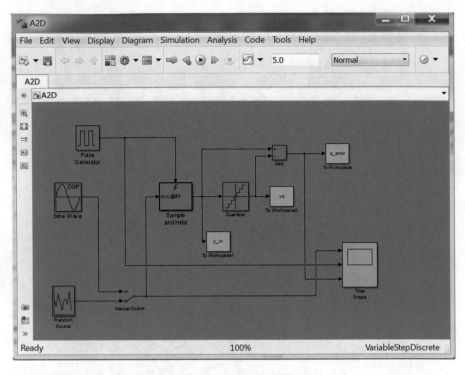

Fig. 2.24   Block diagram of an A/D converter using Simulink

Fig. 2.25   Parameters of the sine wave generator of Fig. 2.24

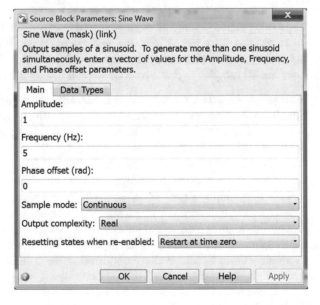

Fig. 2.26 Parameters of the random signal generator of Fig. 2.24

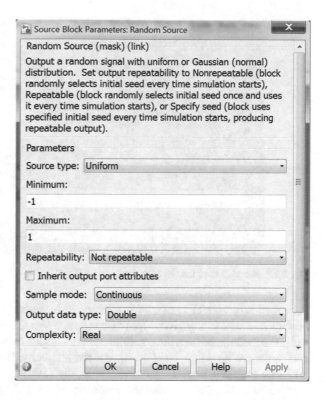

source. The S/H block has the same parameters as that used in the S/H example. The quantizer has 6 bits of quantization, and its parameters are shown in Fig. 2.27. Three outputs are generated in the simulation, namely, the output of S/H, the output of the quantizer, and the difference between these two signals. They are stored as vectors in the workspace with names as indicated in Fig. 2.24. The time scope displays the chosen input signal on the top plot, the pulse generator output on the middle plot, and the quantization error on the bottom plot, as shown in Fig. 2.28a. In this figure the input source is the sine wave. In Fig. 2.28b the random input signal is shown. With 1 min of simulation, the SNR due to quantization is found to be 36.36 dB for the sine wave and 36.11 dB for the random signal. These numbers agree with the SNRs obtained from analysis when the A/D bit width is 6 bits.

2.9 Summary

In this chapter we started with the mathematical description of discrete-time signals in general and some of the useful discrete-time signals in particular. Next we described linear time-invariant (LTI) discrete-time systems. Specifically, we

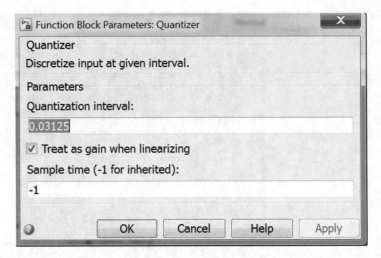

Fig. 2.27 Parameters of the uniform quantizer of Fig. 2.24

described how an LTI system is characterized in terms of its impulse response and how its response to any given input discrete-time signal is computed via convolution sum. We then established the condition that the impulse response must satisfy for the LTI system to be stable. Next we introduced an alternative method of describing an LTI discrete-time system, namely, the linear difference equation with constant coefficients. We showed how the response of an LTI discrete-time system to a specified input sequence could be obtained in closed-form solution. Several examples were included to nail the concept. Since discrete-time signals are mostly obtained from continuous-time or analog signals, we stated the sampling theorem also called Nyquist's theorem and showed how the analog signals can be recovered or reconstructed from their discrete-time counterparts. If the Nyquist sampling criterion is not satisfied, aliasing distortion will occur, and it cannot be removed. We exemplified this notion using a few examples. The sampling theorem that we talked about pertains to lowpass signals. Often, bandpass signals are encountered, especially in the field of communications. These signals are centered at a very high frequency with a narrow bandwidth. The sampling rate of these bandpass signals will be very high if Nyquist's condition is used. Instead, one can sample a bandpass signal at a much lower rate without incurring aliasing distortion. We verified this statement using an example. The next logical thing to do is to describe the process of converting an analog signal to digital signal. We described the A/D converter function by function with plots to illustrate the results. Since A/D conversion involves the approximation of analog samples using fixed number bits of representation, errors occur between the analog and digital values. These errors propagate through the discrete-time system and manifest as noise in the output. We, therefore, derived mathematical formula to measure the performance of an A/D converter. This formula expresses the signal-to-noise ratio of an A/D converter in terms of the number of bits used in the A/D converter. Finally we simulated the S/H operation

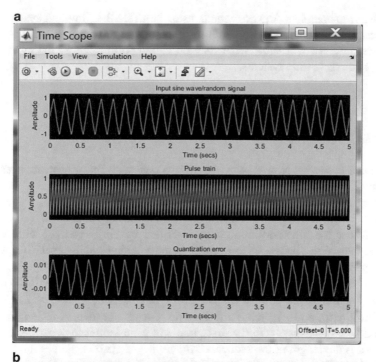

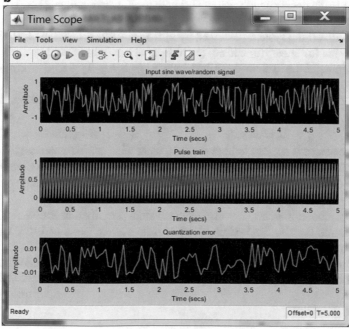

Fig. 2.28 Time scope display of signals used in Fig. 2.24: (**a**) sine wave, (**b**) random signal

as well as the complete A/D conversion using MATLAB's Simulink system. In the next chapter, we will deal with Z-transform and its use in describing LTI discrete-time systems. We will show what role the Z-transform plays in the analysis and design of discrete-time systems.

2.10   Problems

1. Give a few good reasons why we always deal with linear systems even if the actual systems are nonlinear.
2. Is the sequence, $x[n] = \alpha^n u[n]$, $|\alpha| \prec 1$ absolutely summable?
3. Is the sequence $x[n] = n^2 \alpha^n u[n]$, $|\alpha| \prec 1$ absolutely summable?
4. Consider the discrete-time system described by $y[n] = \alpha x^2[n]$ with α a real constant. Will you use this system to amplify a sinusoidal signal? If not, why? Explain.
5. Find the fundamental period of the sequence $x[n] = \cos(0.8n\pi + 0.25\pi)$.
6. What is the period of the sequence $x[n] = \sin\left[\frac{2\pi x 100}{1000}n\right] + \sin\left[\frac{2\pi x 150}{1000}n\right]$?
7. If $x[n]$, $y[n]$, and $g[n]$ represent three finite-length sequences of lengths N, M, and L, respectively, with the first sample of each sequence occurring at $n = 0$, what is the length of the sequence $x[n] \otimes y[n] \otimes g[n]$?
8. Evaluate the linear convolution of $x[n]$ with itself, where $x[n] = \{1, -1, 0, 1, -1\}$, $0 \leq n \leq 4$.
9. Determine if the system described by $y[n] = \alpha + x[n+1] + x[n] + x[n-1] + x[n-2]$ is (a) linear, (b) causal, (c) shift-invariant, and (d) stable.
10. Determine if the system described by $y[n] = x[n+1] + x[n] + x[n-1] + x[n-2]$ is causal.
11. Consider the two discrete-time LSI systems whose impulse responses are described by $h_1[n] = \begin{cases} 1, 0 \leq n \leq N-1 \\ 0, otherwise \end{cases}$ and $h_2[n] = \begin{cases} 1, -(N-1) \leq n \leq 0 \\ 0, otherwise \end{cases}$. If a unit step sequence is applied to both systems, what will be their responses?
12. Obtain the total solution for $n \geq 0$ of the discrete-time system described by $y[n] - 0.4y[n-1] - 0.05y[n-2] = 1.5 \cos(10\pi n)u[n]$, with initial conditions $y[-1] = 1$, $y[-2] = 0$.
13. Determine the impulse response of the system described in Problem 12 above.
14. Determine the rise time of the discrete-time system described in Problem 12. You may use MATLAB to solve the problem.
15. Find the impulse response and step response of the discrete-time system described by $y[n] - 0.7y[n-1] + 0.1y[n-2] = x[n] - 0.7x[n-1]$.

References

1. McGillem CD, Cooper GR (1974) Continuous and discrete signal and system analysis. Holt, Rhinehart and Winston, New York
2. Mitra SK (2011) Digital signal processing: a computer-based approach, 4th edn. McGraw Hill, New York
3. Oppenheim AV, Schafer RW (1975) Digital signal processing. Prentice-Hall, Englewood Cliffs
4. Papamichalis P (1990) Digital signal processing applications with the TMS320 family: theory, algorithms, and implementations, vol 3. Texas Instruments, Dallas
5. Proakis JG, Manolakis DG (1996) Digital signal processing: principles, Algorithms and Applications, 3rd edn. Prentice-Hall, Upper Saddle River

Chapter 3
Z-Transform

3.1 Z-Transform Definition

You may be asking what is the need to know Z-transform. What is Z-transform, anyway? Before we define it, let us know its use. Z-transform is very useful in the analysis of discrete-time signals and systems. We will see later that it is extremely powerful in the design of digital filters. It tells us everything about an LTI discrete-time system – its stability, its frequency response, the filter structure, etc. Therefore, it is a must for us to know what Z-transform is. By the way, Z-transform plays a similar role for the discrete-time signals and systems as does Laplace transform for the continuous-time counterparts.

Z-transform is a mapping of discrete-time sequences from the time domain into a complex variable domain. This complex variable for certain specific values can be considered as a frequency variable. Specifically, the Z-transform of a sequence $x[n]$, denoted $X(z)$, is defined as

$$X(z) = Z\{x[n]\} = \sum_{n=-\infty}^{\infty} x[n]z^{-n} \tag{3.1}$$

So, it is an infinite sum of the product of the sequence and the complex variable z raised to negative n. The Z-transform of a sequence exists only if the absolute of the summation on the right-hand side of (3.1) converges to a finite value. That is,

$$\left| \sum_{n=-\infty}^{\infty} x[n]z^{-n} \right| \le \sum_{n=-\infty}^{\infty} |x[n]z^{-n}| < \infty \tag{3.2}$$

Electronic supplementary material: The online version of this article (https://doi.org/10.1007/978-3-319-76029-2_3) contains supplementary material, which is available to authorized users.

© Springer International Publishing AG, part of Springer Nature 2019
K. S. Thyagarajan, *Introduction to Digital Signal Processing Using MATLAB with Application to Digital Communications*,
https://doi.org/10.1007/978-3-319-76029-2_3

For a given sequence $x[n]$, its Z-transform will exist, in general, for a certain range of values of the complex variable z. This range of values of z over which the Z-transform exists, i.e., Eq. (3.2) is satisfied, is called the *region of convergence* (ROC). As we will see later, the unit circle in the z-plane plays an important role in digital filters. In the complex z-plane, the unit circle is defined as those values of z satisfying the condition $|z| = 1$. Having defined the Z-transform, let us determine the Z-transform of some useful discrete-time signals before we go any further.

Z-Transform of an Impulse Sequence The Z-transform of a unit impulse sequence $\delta[n]$ is obtained from (3.1) as

$$Z\{\delta[n]\} = \sum_{n=-\infty}^{\infty} \delta[n]z^{-n} = 1 \tag{3.3}$$

since the unit impulse function is zero everywhere except at $n = 0$ at which it is unity. Thus, the Z-transform of a unit impulse sequence is a constant independent of z. Therefore, its ROC is the entire z-plane, meaning that it exists over the entire z-plane.

Z-Transform of a Unit Step Sequence Let us find the Z-transform of the unit step sequence $u[n]$ using the definition in (3.1):

$$Z\{u[n]\} = \sum_{n=-\infty}^{\infty} u[n]z^{-n} = \sum_{n=0}^{\infty} z^{-n} = \lim_{n\to\infty} \frac{1 - z^{-(n+1)}}{1 - z^{-1}} \tag{3.4}$$

The limiting value in (3.4) is finite only for values $|z| > 1$, in which case the Z-transform of the unit step sequence is

$$Z\{u[n]\} = \frac{1}{1 - z^{-1}}, |z| > 1 \tag{3.5}$$

From (3.5), we infer that the ROC of the Z-transform of the unit step sequence is outside the unit circle in the z-plane.

Z-Transform of a Real Causal Exponential Sequence Consider the exponential sequence $\alpha^n u[n]$, where α is a real constant. This sequence is causal because it is zero for n less than zero. Its Z-transform from the definition is

$$Z\{\alpha^n u[n]\} = \sum_{n=-\infty}^{\infty} \alpha^n u[n]z^{-n} = \sum_{n=0}^{\infty} \alpha^n z^{-n} = \sum_{n=0}^{\infty} \left(\alpha z^{-1}\right)^n \tag{3.6}$$

Observe that the summation in the above equation is a geometric series. If $|\alpha z^{-1}| < 1$ or, equivalently, if $|z| > |\alpha|$, then the summation in (3.6) converges, and so the Z-transform of an exponential sequence is

$$Z\{\alpha^n u[n]\} = \frac{1}{1 - \alpha z^{-1}}, |z| > |\alpha| \tag{3.7}$$

From (3.7), it is clear that the ROC of the Z-transform of a real causal exponential sequence is outside the circle of radius |a|.

Z-Transform of a Modulated Cosine Sequence We consider a modulated cosine sequence $x[n] = r^n \cos(n\omega_0)u[n]$, where r is a real positive constant. It is called a modulated cosine sequence because the amplitude of the cosine sequence is not constant but increases or decreases exponentially depending on whether r is greater than or less than unity. Using the definition in (3.1), we have

$$Z\{r^n \cos(n\omega_0)u[n]\} = \sum_{n=-\infty}^{\infty} r^n \cos(n\omega_0)u[n]z^{-n} = \sum_{n=0}^{\infty} r^n \cos(n\omega_0)z^{-n} \quad (3.8)$$

To obtain a closed-form expression for (3.8), we consider $\cos(n\omega_0) = Re\{e^{jn\omega_0}\}$. We can, therefore, write (3.8) as

$$Z\{r^n \cos(n\omega_0)u[n]\} = Re\{Z\{r^n e^{jn\omega_0}u[n]\}\} = Re\left\{\sum_{n=0}^{\infty} r^n e^{jn\omega_0}z^{-n}\right\} \quad (3.9)$$

Or,

$$Z\{r^n \cos(n\omega_0)u[n]\} = Re\left\{\sum_{n=0}^{\infty} \left(re^{j\omega_0}z^{-1}\right)^n\right\} \quad (3.9a)$$

Since $|e^{j\omega_0}| = 1$, for values of $|z| > |r|$, the right-hand side of (3.9a) converges to $\frac{1}{1-re^{j\omega_0}z^{-1}}$. Therefore,

$$Z\{r^n \cos(n\omega_0)u[n]\} = Re\left\{\frac{1}{1 - re^{j\omega_0}z^{-1}}\right\} \quad (3.10)$$

Since the denominator of Eq. (3.10) is complex, we multiply the numerator and denominator of (3.10) by the complex conjugate of the denominator. Then, the denominator becomes purely real because

$$\left(1 - re^{j\omega_0}z^{-1}\right)\left(1 - re^{-j\omega_0}z^{-1}\right) = 1 - 2rz^{-1}\cos(\omega_0) + r^2z^{-2} \quad (3.11)$$

The real part of the numerator is the real part of $1 - re^{-j\omega_0}z^{-1}$, which is $1 - rz^{-1}\cos(\omega_0)$. Hence,

$$Z\{r^n \cos(n\omega_0)u[n]\} = \frac{1 - rz^{-1}\cos(\omega_0)}{1 - 2rz^{-1}\cos(\omega_0) + r^2z^{-2}}, |z| > r \quad (3.12)$$

Again, the ROC of the Z-transform of the modulated cosine sequence is outside the circle of radius r in the z-plane.

Z-Transform of a Modulated Sine Sequence We are looking for the Z-transform of the sequence $r^n \sin(n\omega_0)u[n]$. We can use the same procedure used for the Z-transform of the modulated cosine sequence except that we take the imaginary part in Eq. (3.9) because $\sin(n\omega_0) = Im\{e^{jn\omega_0}\}$. After simplification, we arrive at

Table 3.1 Z-Transform of useful sequences

Sequence	Z-transform	ROC
$\delta[n]$	1	Everywhere
$u[n]$	$\frac{1}{1-z^{-1}}$	$|z| > 1$
$a^n u[n]$	$\frac{1}{1-az^{-1}}$	$|z| > |a|$
$r^n \cos(n\omega_0)u[n]$	$\frac{1-rz^{-1}\cos(\omega_0)}{1-2rz^{-1}\cos(\omega_0)+r^2z^{-2}}$	$|z| > r$
$r^n \sin(n\omega_0)u[n]$	$\frac{rz^{-1}\sin(\omega_0)}{1-2rz^{-1}\cos(\omega_0)+r^2z^{-2}}$	$|z| > r$

$$Z\{r^n \sin(n\omega_0)u[n]\} = \frac{rz^{-1}\sin(\omega_0)}{1-2rz^{-1}\cos(\omega_0)+r^2z^{-2}}, |z| > r \qquad (3.13)$$

From (3.13), the ROC of the Z-transform of a modulated sine sequence is found to be outside a circle of radius r in z-plane. Table 3.1 lists the Z-transforms of the sequences discussed above along with the corresponding ROC.

Example 3.1: Z-Transform of an Anticausal Exponential Sequence We want to find the Z-transform of the sequence $x[n] = -a^n u[-n-1]$. This sequence is anticausal because it exists for n less than 0. Using the definition of the Z-transform, we have

$$X(z) = \sum_{n=-\infty}^{\infty} x[n]z^{-n} = -\sum_{n=-\infty}^{-1} a^n z^{-n} = -\sum_{n=1}^{\infty} a^{-n}z^n$$

Note that

$$-\sum_{n=1}^{\infty} a^{-n}z^n = -\{a^{-1}z + a^{-2}z^2 + \cdots + a^{-n}z^n + \cdots\}$$

which can be rewritten as

$$-\sum_{n=1}^{\infty} a^{-n}z^n = -a^{-1}z\{1 + a^{-1}z + a^{-2}z^2 + \cdots + a^{-n}z^n + \cdots\}$$

Therefore, the Z-transform of the anticausal exponential sequence is

$$X(z) = -a^{-1}z\sum_{n=0}^{\infty} (a^{-1}z)^n = -\frac{a^{-1}z}{1-a^{-1}z}, |z| < |a|$$

The above equation can also be expressed as

$$X(z) = \frac{1}{1-az^{-1}}, |z| < |a|$$

Thus, the ROC of an anticausal exponential sequence is inside a circle of radius $|a|$, whereas the ROC of a causal exponential sequence is outside the circle of radius $|a|$.

Example 3.2: Z-Transform of a Finite-Length Exponential Sequence We want to find the Z-transform of an exponential sequence that is of finite duration in a closed form. Specifically, let the sequence be

$$x[n] = \begin{cases} a^n, L \le n \le M - 1 \\ 0, otherwise \end{cases}$$

Then, its Z-transform is obtained from

$$X(z) = \sum_{n=L}^{M-1} a^n z^{-n} = a^L z^{-L} + a^{L+1} z^{-(L+1)} + \cdots + a^{M-1} z^{-(M-1)}$$

By extracting the term $a^L z^{-L}$, the above equation results in

$$X(z) = a^L z^{-L} \sum_{n=0}^{M-L-1} \left(az^{-1}\right)^n = a^L z^{-L} \frac{1 - a^{M-L} z^{-(M-L)}}{1 - az^{-1}} = \frac{a^L z^L - a^M z^{-M}}{1 - az^{-1}}$$

As the sequence is of finite length, the summation is always finite everywhere in the z-plane except at $z = 0$ and $z = \infty$, provided |a| is finite. If $M > L \ge 0$, the ROC is the entire z-plane except $z = 0$. On the other hand, if L is negative and M is positive, the ROC is also the entire z-plane except at $z = 0$ and $z = \infty$. For the third possibility of both L and M negative, the region of convergence is again the entire z-plane except $z = \infty$.

3.2 Properties of Z-Transform

By using the properties of the Z-transform, it will be much more elegant and simple to analyze discrete-time signals and systems. We will list a few properties of the Z-transform with proofs.

Scaling by a Constant What happens to the Z-transform of a sequence if it is multiplied by a constant? More specifically, we are looking for the Z-transform of $ax[n]$ in terms of the Z-transform of $x[n]$. Thus,

$$Z\{ax[n]\} = \sum_{n=-\infty}^{\infty} ax[n]z^{-n} = a \sum_{n=-\infty}^{\infty} x[n]z^{-n} = aX(z) \tag{3.14}$$

In other words, the Z-transform of a sequence scaled by a constant equals the Z-transform of the original sequence scaled by that constant. Here the constant may be real or complex. We also observe that the ROC of the Z-transform of the scaled sequence is the same as that of the original sequence.

Linearity The Z-transform is linear, meaning that it satisfies the superposition rule. In other words, the Z-transform of a linear combination of sequences is the same linear combination of the individual Z-transforms. In mathematical terms, the following equation expresses the linearity:

$$Z\{ax_1[n] + bx_2[n]\} = \sum_{n=-\infty}^{\infty} \{ax_1[n] + bx_2[n]\}z^{-n} \qquad (3.15)$$

Since the summation operation is linear, Eq. (3.15) in conjunction with (3.14), we have

$$\begin{aligned} Z\{ax_1[n] + bx_2[n]\} &= a\sum_{n=-\infty}^{\infty} x_1[n]z^{-n} + b\sum_{n=-\infty}^{\infty} x_2[n]z^{-n} \\ &= aX_1(z) + bX_2(z) \end{aligned} \qquad (3.16)$$

Of course, the above equation holds in the ROC common to both Z-transforms.

Time Shifting What happens to the Z-transform of a sequence when the sequence is shifted in time by an integer number of samples? So,

$$Z\{x[n-m]\} = z^{-m}X(z), m \in Z \qquad (3.17)$$

The proof is simple. Using the definition of the Z-transform, we have

$$Z\{x[n-m]\} = \sum_{n=-\infty}^{\infty} x[n-m]z^{-n} \qquad (3.18)$$

Let $k = n - m$ in Eq. (3.18). Therefore, $n = k + m$. Eq. (3.18) becomes

$$Z\{x[n-m]\} = \sum_{k=-\infty}^{\infty} x[k]z^{-(m+k)} = z^{-m}\sum_{k=-\infty}^{\infty} x[k]z^{-k} = z^{-m}X(z) \qquad (3.19)$$

From the time-shifting property, we observe that a shift in time is equivalent to multiplying the corresponding Z-transform by the complex variable z raised to the negative power of the time shift.

Conjugation We may not have explicitly stated that the discrete-time sequence can be real or complex. The conjugation property states that the Z-transform of the complex conjugate of a sequence is the complex conjugate of the Z-transform as a function of the complex conjugate of z, that is,

$$Z\{x^*[n]\} = X^*(z^*) \qquad (3.20)$$

In the above equation, the symbol * denotes complex conjugation.

Proof Using the definition of Z-transform, we have

$$Z\{x^*[n]\} = \sum_{n=-\infty}^{\infty} x^*[n]z^{-n} \qquad (3.21)$$

Since $\{x^*\}^* = x$, we can rewrite (3.21) as

$$Z\{x^*[n]\} = \left\{ \sum_{n=-\infty}^{\infty} x[n](z^*)^{-n} \right\}^* = X^*(z^*) \tag{3.22}$$

Time Reversal The Z-transform of a time-reversed sequence is the Z-transform of the original sequence as a function of z^{-1}. That is,

$$Z\{x[-n]\} = X\left(\frac{1}{z}\right) \tag{3.23}$$

Proof Notice that time reversal corresponds to negative n as the time index. Now from the definition of the Z-transform, we have

$$Z\{x[-n]\} = \sum_{n=-\infty}^{\infty} x[-n]z^{-n} \tag{3.24}$$

In Eq. (3.24) let $m = -n$. Then, the Z-transform of the time-reversed sequence becomes

$$Z\{x[-n]\} = \sum_{m=\infty}^{-\infty} x[m]z^{m} = \sum_{m=-\infty}^{\infty} x[m](z^{-1})^{-m} = X\left(\frac{1}{z}\right) \tag{3.25}$$

Hence, the result.

Multiplication by an Exponential Sequence If a sequence $x[n]$ is multiplied by an exponential sequence α^n, the resulting Z-transform can be obtained by simply replacing the variable z in $X(z)$ by $\frac{z}{\alpha}$. That is,

$$Z\{\alpha^n x[n]\} = X\left(\frac{z}{\alpha}\right) \tag{3.26}$$

Proof Using the definition of the Z-transform, we obtain

$$Z\{\alpha^n x[n]\} = \sum_{n=-\infty}^{\infty} \alpha^n x[n]z^{-n} = \sum_{n=-\infty}^{\infty} x[n]\left(\frac{z}{\alpha}\right)^{-n} = X\left(\frac{z}{\alpha}\right) \tag{3.27}$$

Differentiation Rule The Z-transform of a sequence multiplied by the time index n equals the negative of the differential with respect to z of the Z-transform multiplied by z:

$$Z\{nx[n]\} = -z\frac{dX(z)}{dz} \tag{3.28}$$

Proof First, from the definition of the Z-transform, we have

$$Z\{nx[n]\} = \sum_{n=-\infty}^{\infty} nx[n]z^{-n} \tag{3.29}$$

Observe that

$$\frac{d}{dz}(z^{-n}) = -nz^{-(n+1)} \Rightarrow nz^{-n} = -z\frac{d}{dz}(z^{-n}) \tag{3.30}$$

Using (3.30) in (3.29), we find that

$$Z\{nx[n]\} = \sum_{n=-\infty}^{\infty} x[n]\left\{-z\frac{d}{dz}(z^{-n})\right\} \tag{3.31}$$

However, since differentiation operation is linear and here it is with respect to z, we can rewrite (3.31) as

$$Z\{nx[n]\} = -z\frac{d}{dz}\left\{\sum_{n=-\infty}^{\infty} x[n]z^{-n}\right\} = -z\frac{dX(z)}{dz} \tag{3.32}$$

Convolution The Z-transform of the convolution of two sequences is the product of their Z-transforms:

$$Z\{x[n] \otimes h[n]\} = X(z)H(z) \tag{3.33}$$

Proof From the definition of the Z-transform, we have

$$Z\{x[n] \otimes h[n]\} = \sum_{n=-\infty}^{\infty} \left\{\sum_{k=-\infty}^{\infty} x[k]h[n-k]\right\}z^{-n} \tag{3.34}$$

Replace $n - k$ by m in the above equation and interchange the order of summation. Then,

$$Z\{x[n] \otimes h[n]\} = \sum_{k=-\infty}^{\infty} x[k]\left\{\sum_{m=-\infty}^{\infty} h[m]z^{-(m+k)}\right\}$$

$$= \left\{\sum_{k=-\infty}^{\infty} x[k]z^{-k}\right\}\left\{\sum_{m=-\infty}^{\infty} h[m]z^{-m}\right\} = X(z)H(z) \tag{3.35}$$

In Table 3.2 we summarize the properties of the Z-transform for convenience.

Having defined the properties of the Z-transform, let us apply them to solve a few problems to demonstrate their use and elegance.

Table 3.2 Properties of the Z-transform

Property	Sequence	Z-Transform
Scaling by a constant	$ax[n]$	$aX(z)$
Linearity	$ax_1[n] + bx_2[n]$	$aX_1(z) + bX_2(z)$
Time shifting	$x[n - m], m \in Z$	$z^{-m}X(z)$
Conjugation	$x^*[n]$	$X^*(z^*)$
Time reversal	$x[-n]$	$X\left(\frac{1}{z}\right)$
Multiplication By an exponential sequence	$a^n x[n]$	$X\left(\frac{z}{a}\right)$
Differentiation	$nx[n]$	$-z\frac{dX(z)}{dz}$
Convolution	$x[n] \otimes h[n]$	$X(z)H(z)$

Example 3.3 Find the Z-transform and the corresponding ROC of the sequence $nr^n \cos (n\omega_0)u[n]$.

Solution Since the factor n occurs in the sequence, we can use the differentiation property. Therefore,

$$Z\{nr^n \cos (n\omega_0)u[n]\} = -z\frac{d}{dz}\{Z\{r^n \cos (n\omega_0)u[n]\}\} \qquad (3.36)$$

Using the Eq. (3.12) in (3.36), we have

$$Z\{nr^n \cos (n\omega_0)u[n]\} = -z\frac{d}{dz}\left\{\frac{1 - rz^{-1}\cos (\omega_0)}{1 - 2rz^{-1}\cos (\omega_0) + r^2z^{-2}}\right\} \qquad (3.37)$$

After differentiation and simplification of Eq. (3.37), we arrive at

$$\begin{aligned} &Z\{nr^n \cos (n\omega_0)u[n]\} \\ &= \frac{r^3z^{-3}\cos (\omega_0) - 2r^2z^{-2}(1 + \cos^2(\omega_0)) + 5rz^{-1}\cos (\omega_0) - 2}{(1 - 2rz^{-1}\cos (\omega_0) + r^2z^{-2})^2} \end{aligned} \qquad (3.38)$$

From Eq. (3.36), we infer that the ROC is outside the circle of radius r: $|z| > r$.

Example 3.4 Find the Z-transform and the ROC of the sequence $a^n u[-n]$.

Solution Let $X(z) = Z\{u[n]\}$. From the time reversal property of the Z-transform, we find that $Z\{u[-n]\} = X\left(\frac{1}{z}\right)$. Next, using the property, multiplication by an exponential sequence, we get $Z\{a^n u[-n]\} = X\left(\frac{1}{z/a}\right) = X\left(\frac{a}{z}\right)$. It is easy to see that the ROC is defined by $|z| < |a|$, that is, the ROC is inside a circle of radius $|a|$ in the z-plane.

Example 3.5 Find the Z-transform and the ROC of the sequence $n^2 \alpha^n u[n]$.

Solution Let us define the sequence $x[n] = \alpha^n u[n]$. Then, from the differentiation property of the Z-transform, we can write

$$Z\{n^2 x[n]\} = -z \frac{d}{dz}[Z\{nx[n]\}] = -z \frac{d}{dz}\left\{-z \frac{dX(z)}{dz}\right\} = z\left\{\frac{dX(z)}{dz} + z \frac{d^2 X(z)}{dz^2}\right\} \tag{3.39}$$

Since $X(z) = \frac{1}{1-\alpha z^{-1}}, |z| > |\alpha|$, Eq. (3.39) reduces to

$$Z\{n^2 \alpha^n u[n]\} = \frac{\alpha z^{-1}(1 + \alpha z^{-1})}{(1 - \alpha z^{-1})^3} \tag{3.40}$$

The ROC is |z| > |α|.

3.3   Z-Transform and Difference Equation

An LTI discrete-time system is governed by a constant coefficient linear difference equation, as seen in the previous chapter. We obtained the response of an LTI discrete-time system to a given input sequence in closed form by solving the corresponding difference equation. In this section we will describe how Z-transform can be used for the same purpose. To this end, consider a pth-order LTI discrete-time system described by a constant coefficient linear difference equation

$$y[n] = \sum_{i=0}^{q} a_i x[n-i] - \sum_{k=1}^{p} b_k y[n-k] \tag{3.41}$$

By taking the Z-transform on both sides of Eq. (3.41), we have

$$Z\{y[n]\} = Z\left\{\sum_{i=0}^{q} a_i x[n-i] - \sum_{k=1}^{p} b_k y[n-k]\right\} \tag{3.42}$$

We make use of the scaling and linearity properties of the Z-transform to rewrite (3.42) as

$$Z\{y[n]\} = \sum_{i=0}^{q} a_i Z\{x[n-i]\} - \sum_{k=1}^{p} b_k Z\{y[n-k]\} \tag{3.43}$$

Denoting $Y(z) = Z\{y[n]\}$ and $X(z) = Z\{x[n]\}$ and using the time-shifting property of the Z-transform, the LTI discrete-time system in the Z-domain is described by

$$Y(z) = \sum_{i=0}^{q} a_i z^{-i} X(z) - \sum_{k=1}^{p} b_k z^{-k} Y(z) \tag{3.44}$$

By collecting the like terms on the same side, we finally transform the difference Eq. (3.41) into

$$Y(z)\left\{\sum_{k=0}^{p} b_k z^{-k}\right\} = X(z)\left\{\sum_{i=0}^{q} a_i z^{-i}\right\} \tag{3.45}$$

In systems theory, the ratio of the Z-transform of the output of an LTI discrete-time system to its input is called the *transfer* or *system* function. Thus, the transfer function of the LTI discrete-time system described by (3.41) is found from (3.45) to be

$$H(z) = \frac{Y(z)}{X(z)} = \frac{a_0 + a_1 z^{-1} + a_2 z^{-2} + \cdots + a_q z^{-q}}{b_0 + b_1 z^{-1} + b_2 z^{-2} + \cdots + b_p z^{-p}}, b_0 = 1 \tag{3.46}$$

The system or transfer function is seen to be a rational polynomial in the complex variable z^{-1}. The order of the system is the maximum of p and q. A *proper transfer function* is defined as that for which $p > q$. Let us familiarize ourselves with transfer functions by solving a couple of examples.

Example 3.6 Find the transfer function in the Z-domain of the LTI discrete-time system described by the following difference equation:

$$\begin{aligned} y[n] = {}& 2x[n] + 2.6x[n-1] + 0.78x[n-2] + 0.05x[n-3] \\ & + 1.5y[n-1] - 0.47y[n-2] - 0.063y[n-3] \end{aligned} \tag{3.47}$$

Solution Apply Z-transform on both sides of (3.47) and we get

$$\begin{aligned} Y(z) = {}& 2X(z) + 2.6z^{-1}X(z) + 0.78z^{-2}X(z) + 0.05z^{-3}X(z) \\ & + 1.5z^{-1}Y(z) - 0.47z^{-2}Y(z) - 0.063z^{-3}Y(z) \end{aligned} \tag{3.48}$$

Then, by collecting the like terms and taking the ratio of the output to input transforms, we arrive at the system function as given in Eq. (3.49):

$$H(z) = \frac{Y(z)}{X(z)} = \frac{2 + 2.6z^{-1} + 0.78z^{-2} + 0.05z^{-3}}{1 - 1.5z^{-1} + 0.47z^{-2} + 0.063z^{-3}} \tag{3.49}$$

It is clear that the transfer function is a rational polynomial in z^{-1} of order 3. The difference equation given in this example involves current and previous samples of both the input and output and so is a recursive equation. The corresponding transfer function is seen from Eq. (3.49) to be a rational polynomial, that is, both the numerator and denominator are polynomials.

Example 3.7 Determine the transfer function in the Z-domain of the LTI system described by the following difference equation:

$$y[n] = \sum_{k=0}^{N} a_k x[n-k] \tag{3.50}$$

Solution We notice from the above equation that no previous output samples are used in computing the current output sample but only the current and previous input samples. So, what happens to the system function? By taking the Z-transform on both sides of (3.50) and then taking the ratio of output to input transforms, we get

$$H(z) = \frac{Y(z)}{X(z)} = \sum_{k=0}^{N} a_k z^{-k} \tag{3.51}$$

In this example we find that the transfer function is just an Nth-order polynomial. Equivalently, the transfer function in (3.51) is also a rational polynomial with the denominator equal to 1. Incidentally the system in Eq. (3.50) is called a *non-recursive* system whose transfer function is a polynomial of order N.

3.4   Poles and Zeros

We saw from previous discussions and examples that the transfer function of an LTI discrete-time system in the Z-domain is a rational polynomial in z or z^{-1}. Therefore, we can express the numerator and denominator polynomials of the transfer function in terms of their respective factors. Consider a pth-order transfer function of an LTI discrete-time system as given in Eq. (3.46). In terms of the factors of the polynomials, (3.46) can be rewritten as

$$H(z) = \frac{\prod_{i=1}^{q} \left(1 - \alpha_i z^{-1}\right)}{\prod_{j=1}^{p} \left(1 - \beta_j z^{-1}\right)}, p > q \tag{3.52}$$

From Eq. (3.52), we notice that the transfer function is zero whenever $z = \alpha_i$, $1 \le i \le q$. These are the roots of the numerator polynomial and are called the *zeros* of the transfer function. Corresponding to the degree q of the numerator of the transfer function, there are q zeros. On the other hand, when a factor in the denominator of (3.52) approaches zero, the transfer function approaches infinity, and the corresponding roots of the denominator are called the *poles* of the transfer function. That is, the poles are defined by the condition

$$H(z)|_{z \to \beta_j} \to \infty, 1 \le j \le p \tag{3.53}$$

Fig. 3.1 Pole-zero plot for Example 3.1

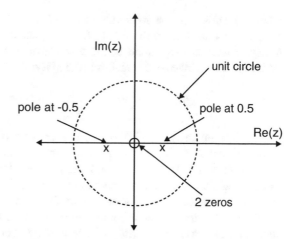

So, what is the use of knowing the poles and zeros of a transfer function of an LTI discrete-time system? Indeed it is a valid question. The stability of an LTI discrete-time system can be determined from its poles. Each pole factor $(1 - \beta_i z^{-1})$ gives rise to an exponential sequence in the discrete-time domain of the type $\beta_i^n u[n]$. We have seen this result earlier. Therefore, for this sequence to converge, the poles must satisfy $|\beta_i| < 1$. In other words, all poles must be less than unity in magnitude. The unit circle in the Z-domain is the border between stable and unstable regions. Apart from stability, poles and zeros effect in the design of digital filters, as we will see later.

Let us work out a few more examples to get familiar with poles and zeros.

Example 3.8 Find and plot the zeros and poles of the transfer function $H(z) = \frac{1}{1-0.25z^{-2}}$.

Solution Rewrite the transfer function in terms of z, which is $H(z) = \frac{z^2}{z^2-0.25}$. Since the numerator is a second-degree polynomial, it has two zeros; in this case both zeros are located at the origin. The denominator factors into $(z - 0.5)(z + 0.5)$. Therefore, the poles are at $z = 0.5$ and $z = -0.5$. The poles and zeros are plotted as shown in Fig. 3.1. It is customary to denote the zeros by open circles and the poles by crosses. The unit circle is also shown in the figure. Since the poles are situated inside the unit circle, the given system is stable in the BIBO sense. This establishes the fact that Z-transform is very useful in determining the stability of an LTI discrete-time system.

3.5 Inverse Z-Transform

The forward Z-transform maps a given discrete-time sequence into the Z-domain. More specifically, the Z-transform maps a discrete-time sequence into complex poles and zeros in the Z-domain. If we cannot recover the sequence from the Z-domain

back to the discrete-time domain, then there is no use for the Z-transform. The effort
we put in so far to learn the Z-transform is a shear waste. Fortunately, the inverse
mapping from the Z-domain to the discrete-time domain does exist. The formal
definition of the inverse Z-transform of a given Z-transform $X(z)$ is given below:

$$x[n] = Z^{-1}\{X(z)\} = \frac{1}{2\pi j} \oint X(z)z^{n-1}dz, j = \sqrt{-1} \qquad (3.54)$$

The integral in Eq. (3.54) is called a *contour integral*, and the contour C is
traversed counterclockwise. What have we gotten into? Who knows how to evaluate
contour integrals? Are we opening Pandora's box here? There is a way out of this
mess though. Indeed, we can compute the inverse Z-transform using partial fraction
expansion. The idea is to express a given function of z in terms of a sum of simple
fractions. Each fraction consists of one pole. Then, the inverse Z-transform is the
sum of the inverse Z-transforms of the individual fractions. Of course, there are cases
where the poles are not *simple* and multiple poles may exist. How do we then find the
inverse Z-transform? Let us cross the bridge when we come to it.

Partial Fraction Expansion for Simple Poles The Z-transform of a sequence $x[n]$
is a rational polynomial in z^{-1} and can be expressed as

$$X(z) = \frac{N(z)}{D(z)} \qquad (3.55)$$

where $N(z)$ and $D(z)$ are polynomials of degree q and p, respectively. Let us assume
the case of a proper function, meaning that $p > q$. By factoring $D(z)$ into its poles, we
can express (3.55) as

$$X(z) = \sum_{i=1}^{p} \frac{R_i}{1 - p_i z^{-1}} \qquad (3.56)$$

In the above equation, R_i are called the *residues* and p_i are, of course, the poles.
Assuming for the time being that the residues are known, then the inverse Z-transform is given by

$$x[n] = \sum_{i=1}^{p} R_i p_i^n u[n] \qquad (3.57)$$

Note that each pole factor corresponds to a sequence $p_i^n u[n] \leftrightarrow \frac{1}{1 - p_i z^{-1}}$, hence
the inverse Z-transform given in (3.57). It looks nice and dandy. How do we
determine the residues? Let us look into it. A residue at a particular pole is
determined by first multiplying the given function by that pole factor and then
evaluating the remaining function at the value of the pole in question. More
precisely, the residues are given by

$$R_i = \left(1 - p_i z^{-1}\right) X(z)\big|_{z=p_i}, 1 \leq i \leq p \tag{3.58}$$

After finding the residues, the discrete-time sequence is obtained from (3.57). This method of partial fraction expansion is easier than the method indicated in (3.54)!

Example 3.9 Find the inverse Z-transform of the function in Example 3.6.

Solution The given function is $X(z) = \frac{1}{1 - 0.25z^{-2}}$. From Example 3.8, the given function in terms of its pole factors is $X(z) = \frac{1}{(1 - 0.5z^{-1})(1 + 0.5z^{-1})}$. Since the given function is proper and has two simple poles, its partial fraction expansion takes the form:

$$X(z) = \frac{R_1}{1 - 0.5z^{-1}} + \frac{R_2}{1 + 0.5z^{-1}} \tag{3.59}$$

The residue corresponding to the pole at 0.5 is found from

$$R_1 = \left(1 - 0.5z^{-1}\right) X(z)\big|_{z=0.5} = \frac{1}{1 + 0.5z^{-1}}\bigg|_{z=0.5} = 0.5 \tag{3.60}$$

Note that we must first cancel the pole factor corresponding to the residue in question and then substitute the pole for z in the remaining function. Similarly, the second residue is obtained from

$$R_2 = \left(1 + 0.5z^{-1}\right) X(z)\big|_{z=-0.5} = \frac{1}{1 - 0.5z^{-1}}\bigg|_{z=-0.5} = 0.5 \tag{3.61}$$

Having found the two residues, the sequence corresponding to the given function is

$$x[n] = 0.5(0.5)^n u[n] + 0.5(-0.5)^n u[n] \tag{3.62}$$

It can be easily verified that the Z-transform of the sequence in Eq. (3.62) is the same as that given in the problem statement.

Partial Fraction Expansion of an Improper Function with Simple Poles Consider an improper function where the degree of the numerator polynomial is greater than or equal to that of the denominator. More specifically, let

$$X(z) = \frac{N(z)}{D(z)} \tag{3.63}$$

where, the degree of the numerator polynomial q is greater than or equal to the degree of the denominator polynomial p. First we have to divide $N(z)$ by $D(z)$ until the remainder fraction becomes a proper function as given in Eq. (3.64a):

$$X(z) = \sum_{i=0}^{q-p} a_i z^{-i} + \frac{N'(z)}{D(z)} \tag{3.64a}$$

After division, the degree of $N'(z)$ is less than p: $deg(N'(z)) < p$. Next, we have to express the remainder function in partial fraction and then obtain the inverse as before. From (3.64a), we find the inverse Z-transform of $X(z)$ to be

$$x[n] = \sum_{i=0}^{q-p} a_i \delta[n-i] + \sum_{j=1}^{p} R_j p_j^n u[n], \tag{3.64b}$$

where R_j are the residues of the poles p_j. It will be clearer if we work out an example.

Example 3.10 Find the inverse transform of the function given below:

$$X(z) = \frac{2 + 0.75z^{-1} + 0.625z^{-2} - 0.375z^{-3} - 0.125z^{-4}}{1 - 0.25z^{-1} - 0.125z^{-2}} \tag{3.65}$$

Solution Since the degree of the numerator is greater than that of the denominator, the given function is an improper function. So, we have to divide the numerator by the denominator until the remainder has a degree less than that of the denominator. Then we get

$$X(z) = 1 + z^{-1} + z^{-2} + \frac{1}{1 - 0.25z^{-1} - 0.125z^{-2}} \tag{3.66}$$

The remainder function can be expanded in partial fraction and is obtained as follows:

$$\frac{1}{(1 - 0.5z^{-1})(1 + 0.25z^{-1})} = \frac{A}{(1 - 0.5z^{-1})} + \frac{B}{(1 + 0.25z^{-1})} \tag{3.67a}$$

$$A = \left. \frac{1}{(1 + 0.25z^{-1})} \right|_{z=0.5} = \frac{2}{3} \tag{3.67b}$$

$$B = \left. \frac{1}{(1 - 0.5z^{-1})} \right|_{z=-0.25} = \frac{1}{3} \tag{3.67c}$$

Therefore, the given function is rewritten in terms of its partial fractions as

$$X(z) = 1 + z^{-1} + z^{-2} + \frac{2/3}{1 - 0.5z^{-1}} + \frac{1/3}{1 + 0.25z^{-1}} \tag{3.68}$$

The inverse Z-transform of the function in (3.68) is the sum of the inverse transforms. Thus,

$$x[n] = Z^{-1}\{X(z)\} = \delta[n] + \delta[n-1] + \delta[n-2] + \frac{2}{3}(0.5)^n u[n] + \frac{1}{3}(-0.25)^n u[n]$$

Partial Fraction Expansion of a Function with Complex Poles There is another scenario wherein there are poles that are complex (not purely real). How do we handle this situation? Remember that the Z-transform of a real sequence is a real rational function in the variable z^{-1}, meaning that the coefficients of the polynomials are real. Therefore, if there is a complex pole, it must occur with its conjugate. Otherwise, the coefficients of the polynomials cannot be real. Similarly, the residues of the pair of complex poles must also be complex conjugates. As an example, consider the following function with a pair of complex conjugate poles. Then it can be expressed in partial fractions as

$$X(z) = \frac{A}{1 - re^{j\theta}z^{-1}} + \frac{A^*}{1 - re^{-j\theta}z^{-1}} \tag{3.69}$$

The residue in Eq. (3.69) is complex and can be expressed in magnitude-phase form as $A = |A|e^{j\varphi}$. The inverse transform is obtained by adding the inverse transforms of the two terms on the right-hand side of (3.69), which is

$$x[n] = 2|A|r^n \cos(n\theta + \varphi)u[n] \tag{3.70}$$

Example 3.11 Find the inverse Z-transform of the function given below:

$$X(z) = \frac{1 - 0.433z^{-1}}{1 - 0.866z^{-1} + 0.25z^{-2}}$$

Solution The poles are found to be complex conjugates, and one of them is $p = 0.433 + j0.25 = 0.5e^{j\frac{\pi}{6}}$. Since the given function is a proper function, we can express it as

$$X(z) = \frac{A}{1 - 0.5e^{j\frac{\pi}{6}}z^{-1}} + \frac{A^*}{1 - 0.5e^{-j\frac{\pi}{6}}z^{-1}} \tag{3.71}$$

The residue A is found from

$$A = \frac{1 - 0.433z^{-1}}{1 - 0.5e^{-j\frac{\pi}{6}}z^{-1}}\bigg|_{z=0.5e^{j\frac{\pi}{6}}} = 0.5 \tag{3.72}$$

Since A is real, its complex conjugate is itself. After adding the two inverse transforms, we arrive at

$$x[n] = (0.5)^n \cos\left(n\frac{\pi}{6}\right)u[n] \tag{3.73}$$

Partial Fraction Expansion of a Function with Multiple-Order Poles So far we have described how to determine the inverse Z-transform of a function with either simple or complex poles. There is a third possibility in that the poles of multiple-order can occur. To make the statement clear, if there is a pole at $z = p$ of order k, then the corresponding pole factor in the denominator of the Z-function will be of the type $(1 - pz^{-1})^k$. The partial fraction expansion corresponding to this multiple-order pole will take the form

$$X(z) = \sum_{j=1}^{k} \frac{R_j}{(1 - pz^{-1})^j} \tag{3.74}$$

Note that not just the kth-order pole but poles of order one up to k occur in the expansion. The k residues are obtained using what is known as L'Hospital rule. According to this rule, the residues in (3.74) are described by

$$R_j = \frac{1}{(k-j)!(-p)^{k-j}} \frac{d^{k-j}}{d(z^{-1})^{k-j}} \left(1 - pz^{-1}\right)^k X(z) \Big|_{z=p} \tag{3.75}$$

Though the formula looks rather formidable, we will see that it is not so in practice. To prove, let us work out an example.

Example 3.12 Find the partial fraction expansion of the function

$$X(z) = \frac{1}{(1 - 0.5z^{-1})(1 - 0.75z^{-1})^2} \tag{3.76}$$

Solution Because there is a pole of order two at $z = 0.75$, we can write the partial fraction expansion of the given function as

$$X(z) = \frac{R_1}{(1 - 0.5z^{-1})} + \frac{R_2}{(1 - 0.75z^{-1})} + \frac{R_3}{(1 - 0.75z^{-1})^2} \tag{3.77}$$

The residue corresponding to the simple pole is found as before and is given by

$$R_1 = \left(1 - 0.5z^{-1}\right) X(z) \Big|_{z=0.5} = \frac{1}{(1 - 0.75z^{-1})^2} \Big|_{z=0.5} = 4 \tag{3.78}$$

The other two residues are given by

$$R_2 = \frac{1}{(2-1)!(-0.75)^{2-1}} \frac{d}{dz^{-1}} \left(\frac{1}{1 - 0.5z^{-1}}\right) \Big|_{z=0.75}$$
$$= \left(\frac{-4}{3}\right) \frac{0.5}{(1-0.5z^{-1})^2} \Big|_{z=0.75} = -6 \tag{3.79}$$

$$R_3 = \left(1 - 0.75z^{-1}\right)^2 X(z) \Big|_{z=0.75} = 3 \tag{3.80}$$

From the partial fraction expansion, we can find the inverse Z-transform as follows: the inverse Z-transform of the first partial fraction is $4(0.5)^n u[n]$. Similarly, the second partial fraction gives rise to $-6(0.75)^n u[n]$. It can be shown using the differentiation rule that the terms $(n + 1)(0.75)^n u[n]$ and $\frac{1}{(1-0.75z^{-1})^2}$ form a Z-transform pair. Therefore, we have

$$x[n] = 4(0.5)^n u[n] - 6(0.75)^n u[n] + 3(n+1)(0.75)^n u[n] \qquad (3.81)$$

3.6 MATLAB Examples

Let us work out some more examples to capture the essence of the discussions we have had so far. Though the MATLAB tool is very useful in solving problems related to discrete-time signals and systems, it is a good idea to strengthen our analytical ability as well. We are going to solve a few more problems relating to the Z-transform using both analytical method and MATLAB tools.

Example 3.13 Given the function $H(z) = \frac{z+1.7}{(z+0.3)(z-0.5)}$, find its inverse Z-transform.

Solution First we will rewrite the given function in terms of the complex variable z^{-1}, which results in

$$H(z) = \frac{z^{-1} + 1.7z^{-2}}{(1 + 0.3z^{-1})(1 - 0.5z^{-1})} = \frac{z^{-1} + 1.7z^{-2}}{1 - 0.2z^{-1} - 0.15z^{-2}} \qquad (3.82)$$

Since the degree of the numerator is equal to the degree of the denominator, $H(z)$ is an improper function. Therefore, we have to divide the numerator by the denominator until the remainder function is a proper function. This gives us the following function:

$$H(z) = -11.333 + \frac{11.33 - 1.26667z^{-1}}{(1 + 0.3z^{-1})(1 - 0.5z^{-1})} \qquad (3.83)$$

Next, we express the remainder function in partial fraction expansion. The result is

$$H(z) = -11.333 + \frac{5.83}{1 + 0.3z^{-1}} + \frac{5.5}{1 - 0.5z^{-1}} \qquad (3.84)$$

The inverse Z-transform of (3.84) is the sum of the inverse Z-transforms of the individual terms. The first term is a constant, and its inverse Z-transform is that constant times the impulse sequence. The second and the third terms correspond to exponential sequences. Thus,

$$h[n] = Z^{-1}\{H(z)\} = -11.333\delta[n] + 5.83(-0.3)^n u[n] + 5.5(0.5)^n u[n] \qquad (3.85)$$

In order to solve this problem using MATLAB, we must have the Signal Processing Toolbox. Assuming we have the DSP System Toolbox installed, the function *residuez* is used to calculate the residues, poles, and the quotients of a given

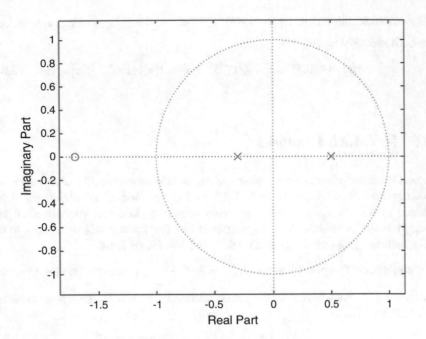

Fig. 3.2 Pole-zero plot of $H(z)$ of Example 3.13

Z-transform. The proper way to call this function in the MATLAB workspace is as follows: *[R,P,K] = residuez(B,A)*. R denotes the residues, P the poles, and K the quotients, all as 1D vectors. The arguments of the *residuez* function are B and A. B is a vector of coefficients of the numerator polynomial of H(z) in ascending powers of z^{-1}, and A is a vector of coefficients of the denominator polynomial, also in ascending powers of z^{-1}. So, corresponding to (3.82), the vector $B = [0\ 1\ 1.7]$. The first element is zero because the constant term is zero. The next element 1 is the coefficient of z^{-1}, and the element 1.7 is the coefficient of z^{-2}. In the same manner, the vector $A = [1 - 0.2 - 0.15]$. In MATLAB syntax, the elements of a vector are separated by at least one space. Next, we type this equation "[R,P, K] = residuez(B,A)" and press return. We get the results that tally with our analytical results. One more thing about using MATLAB is that it has a function named *isstable*, which can be used to test if a given LTI discrete-time system is stable or not. The arguments of this function are the coefficients of the polynomials B and A. The function returns a "1" if the system is stable and a "0" if unstable. The MATLAB file to solve this problem is named *Example3_11.m*. Figure 3.2 shows the pole-zero plots of H(z) in Example 3.13.

Example 3.14 An LTI discrete-time system is described by the following difference equation: $y[n] = 3x[n] + 0.4y[n-1] + 0.05y[n-2]$. Determine its (a) transfer function, (b) impulse response, and (c) step response.

Solution The Z-transform of the above difference equation is found to be

$$Y(z) = 3X(z) + 0.4z^{-1}Y(z) + 0.05z^{-2}Y(z) \qquad (3.86)$$

Then the transfer function of the said LTI system is the ratio of $Y(z)$ to $X(z)$:

$$H(z) = \frac{Y(z)}{X(z)} = \frac{3}{1 - 0.4z^{-1} - 0.05z^{-2}} \qquad (3.87)$$

The impulse response is the inverse Z-transform of $H(z)$. Since the degree of the numerator polynomial of $H(z)$ in z^{-1} is less than that of its denominator, $H(z)$ is a proper function. Therefore, its partial fraction expansion consists of two terms corresponding to the two poles. The poles of $H(z)$ are the roots of its denominator polynomial and are found to be at 0.5 and -0.1. Thus, the partial fraction expansion of $H(z)$ is

$$H(z) = \frac{A}{1 - 0.5z^{-1}} + \frac{B}{1 + 0.1z^{-1}} \qquad (3.88)$$

The residues at the poles are $A = (1 - 0.5z^{-1})X(z)|_z = {}_{0.5} = 2.5$ and $B = (1 + 0.1z^{-1})X(z)|_z = {}_{-0.1} = 0.5$. The impulse response is the inverse Z-transform of $H(z)$, which is

$$h[n] = 2.5(0.5)^n u[n] + 0.5(-0.1)^n u[n] \qquad (3.89)$$

Next, we have to compute the step response of the system. In this case, the input is a unit step function and the output is called the step response. To determine the step response using the Z-transform method, we simply multiply the transfer function H(z) by the Z-transform of the input. This arises from the property of the convolution in the discrete-time domain of the Z-transform. The Z-transform of a unit step sequence was found to be $\frac{1}{1-z^{-1}}$. Therefore,

$$Y(z) = H(z)X(z) = \frac{3}{(1 - 0.5z^{-1})(1 + 0.1z^{-1})(1 - z^{-1})}$$
$$= \frac{R_1}{(1 - z^{-1})} + \frac{R_2}{(1 - 0.5z^{-1})} + \frac{R_3}{(1 + 0.1z^{-1})} \qquad (3.90)$$

The residues are found to be $R_1 = 5.454$, $R_2 = -2.5$, and $R_3 = 0.0455$. Therefore, the unit step response of the given system is

$$y[n] = 5.454u[n] - 2.5(0.5)^n u[n] + 0.0455(-0.1)^n u[n] \qquad (3.91)$$

We have already seen the use of the MATLAB function *residuez* to calculate the residues, poles, and the quotients of a function of z. MATLAB has two more useful functions to calculate the impulse response and step response of an LTI discrete-time system. These functions are, respectively, *impz(B,A)* and *stepz(B,A)*, where B and A are the vectors of coefficients of the numerator and denominator polynomials of a given function of z as explained before. In this problem, we use both analytical solutions and MATLAB functions to compare the results. Figure 3.3a is a pole-zero plot of the transfer function of the discrete-time system in Example 3.14. The pole-zero plot of the system response $Y(z)$ is shown in Fig. 3.3b. The impulse response using the closed-form solution is compared with that obtained by solving the given difference equation, and the results are plotted as shown in Fig. 3.4. The impulse response obtained from using the MATLAB function *impz* is shown in the top plot of Fig. 3.5. It seems to agree with both response sequences shown in Fig. 3.4. Finally, Fig. 3.6 plots the step response of the system: the top plot is using the closed-form solution and the bottom plot using the difference equation. As a comparison, the bottom plot of Fig. 3.6 is obtained using the MATLAB function *stepz*. Again, all three plots are in agreement. The MATLAB file named *Example3_12*.m is used to solve the problem.

Example 3.15 Let us consider the LTI discrete-time system of Example 3.14 and calculate its response to an input sinusoid $x[n] = \cos(0.4n)u[n]$ using MATLAB.

Solution The Z-transform of the input sinusoidal sequence is a second-order function as given by

$$X(z) = Z\{\cos(0.4n)u[n]\} = \frac{1 - \cos(0.4)z^{-1}}{1 - 2\cos(0.4)z^{-1} + z^{-2}} \qquad (3.92)$$

The Z-transform of the output sequence is the product of $X(z)$ and $H(z)$:

$$Y(z) = \left(\frac{1 - \cos(0.4)z^{-1}}{1 - 2\cos(0.4)z^{-1} + z^{-2}}\right)\left(\frac{3}{(1 - 0.4z^{-1} - 0.05z^{-2})}\right) \qquad (3.93)$$

In MATLAB, we can convert the product $X(z)H(z)$ into a single function using the built-in function *sos2tf*. The arguments for this function are (1) the coefficients of the numerator and denominator polynomials of the two second-order sections as a matrix and (2) the gain. For this example, the coefficient matrix is a 2×6 matrix. The first row contains the coefficients of the numerator polynomial of $X(z)$ followed by the coefficients of the denominator polynomial of $X(z)$. The second row consists of the coefficients corresponding to $H(z)$. Thus:

$$C = \begin{bmatrix} 1 & -\cos(0.4) & 0 & 1 & -2\cos(0.4) & 1 \\ 1 & 0 & 0 & 1 & -0.4 & -0.05 \end{bmatrix} \text{ and the gain } G = 3. \qquad (3.94a)$$

The MATLAB function returns the coefficients of the numerator polynomial of $Y(z)$ as Nr = [3 −2.7633 0 0] and that of the denominator polynomial as Dr = [1 −2.2422 1.6869 −0.3079 −0.05]. Using these coefficients, we can express the output Z-transform as

$$Y(z) = \frac{3 - 2.7633z^{-1}}{1 - 2.2422z^{-1} + 1.6869z^{-2} - 0.3079z^{-3} - 0.05z^{-4}} \qquad (3.94b)$$

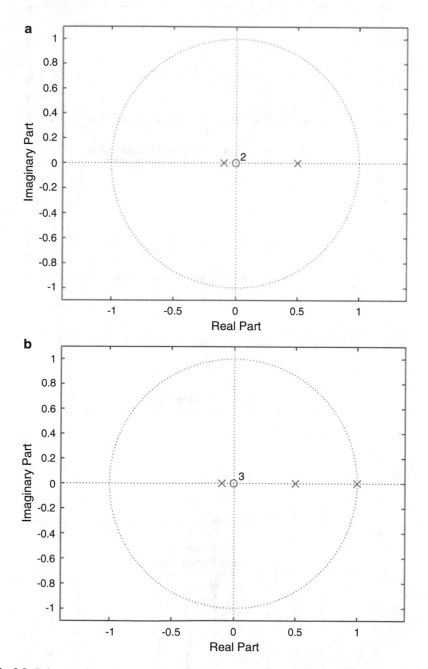

Fig. 3.3 Pole-zero plots of H(z) and Y(z) of Example 3.12: (**a**) pole-zero plot of H(z), (**b**) pole-zero plot of Y(z)

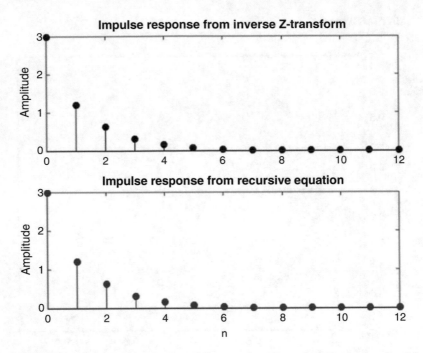

Fig. 3.4 Impulse response of the LTI discrete-time system of Example 3.12: top plot, closed-form solution; bottom plot, impulse response by recursively solving the difference equation of Example 3.12

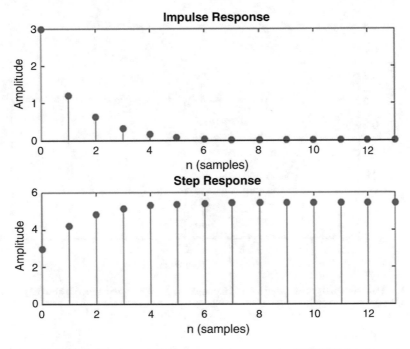

Fig. 3.5 Calculation of impulse and step responses using respective MATLAB functions: top plot, impulse response; bottom plot, step response

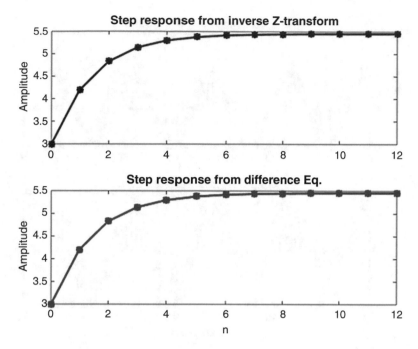

Fig. 3.6 Step response of the discrete-time system of Example 3.12: top plot, closed-form solution; bottom plot, solution obtained by solving recursively the difference equation of Example 3.12

Then we use the function *residuez* to calculate the residues of the output Z-transform. The partial fraction expansion takes the form:

$$Y(z) = \frac{0.2788 + j0.7317}{1 - (0.9211 + j0.3893)z^{-1}} + \frac{0.2788 - j0.7317}{1 - (0.9211 - j0.3893)z^{-1}}$$
$$+ \frac{-1.6004}{1 - 0.5z^{-1}} + \frac{0.0428}{1 + 0.1z^{-1}} \tag{3.95}$$

Once the residues are determined, the sum of the inverse Z-transform of each partial fraction gives the response in the discrete-time domain. Denote the two complex conjugate poles by $p_{1,2} = |p|e^{\pm j\theta}$ and the corresponding residues by $R_{1,2} = |R|e^{\pm j\varphi}$. Then the response of the system to the sinusoidal input is written as

$$y[n] = 2|R| \cos(n\theta + \varphi)u[n] + R_3(0.5)^n u[n] + R_4(-0.1)^n u[n] \tag{3.96}$$

where $R_3 = -1.6004$ and $R_4 = 0.0428$. The system response is shown in Fig. 3.7. As a comparison, the response obtained by recursively solving the difference equation is plotted in the bottom plot of Fig. 3.7. The M-file to solve this problem is named *Example3_13.m*.

Example 3.16 As another example of the use of MATLAB, let us specify the zeros and poles of an LTI discrete-time system and then obtain the transfer function. We will also calculate the impulse and step responses of the system in question. For this

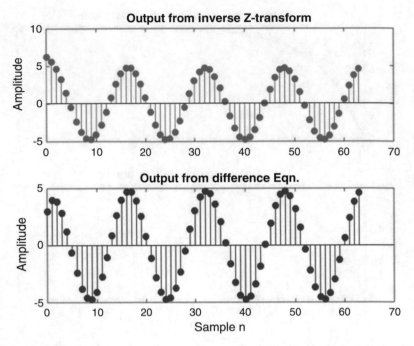

Fig. 3.7 Plot of the response of the system in Example 3.13 to a sinusoidal input. Top plot: response using MATLAB tool. Bottom plot: response by recursively solving the difference equation

system there are five zeros, all at 0. There is a real pole at 0.85 and two pairs of complex conjugate poles at $0.5e^{\pm j\frac{\pi}{3}}$, $0.5e^{\pm j\frac{\pi}{12}}$. The gain K = 1.

Solution The MATLAB function *zp2tf* accepts the zeros, poles, and the gain and returns the coefficients of the numerator and denominator polynomials of the transfer function. The zeros and poles must be column vectors. Thus, the function call is [Nr, Dr] = zp2tf(Z,P,K). For the specified poles and zeros, we have a fifth-order transfer function as given by

$$H(z) = \frac{1}{1 - 2.3159z^{-1} + 2.229z^{-2} - 1.202z^{-3} + 0.374z^{-4} - 0.0531z^{-5}} \quad (3.97)$$

Having found the transfer function, next we call the function *impz* with Nr and Dr as its arguments. This will plot the impulse response of the transfer function. Similarly, we call the function *stepz* with the same arguments Nr and Dr, which will plot the step response. If we want the transfer function in cascades of first- and/or second-order sections, we can use the function *zp2sos* with arguments Z, P, and K. It will return a matrix of coefficients of the transfer function in second-order sections. The actual function call is [SOS, G] = zp2sos(Z,P,K). For this example, we will get two second-order sections and one first-order section. The matrix of coefficients SOS is of size 3×6 with each row corresponding to the numerator and denominator polynomials of the respective section, and G is the overall gain. The three sections are cascaded. Let the sections be denoted by $H_i(z)$, $i = 1, 2, 3$. Using the matrix of coefficients SOS obtained, we have

$$H_1(z) = \frac{1}{1 - 0.85z^{-1}} \tag{3.98a}$$

$$H_2(z) = \frac{1}{1 - 0.9569z^{-1} + 0.25z^{-2}} \tag{3.98b}$$

$$H_3(z) = \frac{1}{1 - 0.5z^{-1} + 0.25z^{-2}} \tag{3.98c}$$

$$G = 1 \tag{3.98d}$$

Thus, the overall system as a cascade is described by

$$H(z) = GH_1(z)H_2(z)H_3(z) \tag{3.99}$$

Note that the order in which the sections are cascaded is not important. The pole-zero plot is shown in Fig. 3.8. It shows five zeros at $z = 0$, two pairs of complex conjugate poles, and a real pole, as specified in the problem. The impulse and step

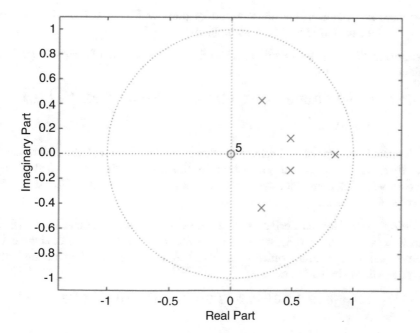

Fig. 3.8 Pole-zero plot of the transfer function of Example 3.16

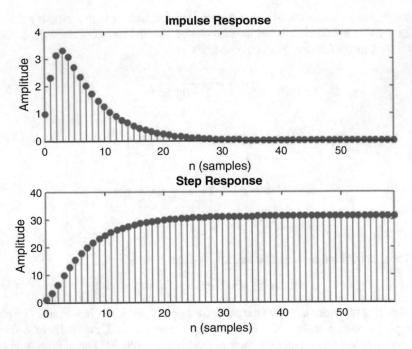

Fig. 3.9 Impulse and step responses of the system in Example 3.16

responses are shown in the top and bottom plots of Fig. 3.9. The M-file named
Example3_14.m is used to solve this problem.

Example 3.17 Consider a sixth-order FIR filter, whose transfer function is specified
below

$$H(z) = 0.0033 + 0.059z^{-1} + 0.2492z^{-2} + 0.377z^{-3} + 0.2492z^{-4} + 0.059z^{-5}$$
$$+ 0.0033z^{-6}$$

Compute the impulse and step responses of the FIR filter and also determine its
rise time. Then express the given transfer function in cascades of first- and/or
second-order functions. Plot the poles and zeros of the FIR filter. Use MATLAB
to solve the problem.

Solution As we saw earlier, to calculate the step response, we call the MATLAB
function *[h, n] = stepz(B,A)*, where B is the vector of coefficients of the numerator
polynomial in ascending order of z^{-1} and A the corresponding denominator
polynomial, as shown below.

$$B = [0.0033, 0.059, 0.2492, 0.377, 0.2492, 0.059, 0.0033]$$

$$A = [1, 0, 0, 0, 0, 0, 0]$$

In MATLAB, the elements of the vectors will be separated by a space and not comma. For the FIR filter, A is the vector of the same size of B with all elements equal to zero except the first element. By calling the function *stepz(B,A)* without output arguments, MATLAB will plot the step response.

To express the FIR transfer function in a cascaded form, we will have to call the function *[sos,g] = tf2sos(B,A)*. Again, the vectors B and A are defined as above. For this example, when the above function is invoked, it returns the matrix

$$sos = \begin{bmatrix} 1 & 15.6764 & 39.2434 & 1 & 0 & 0 \\ 1 & 0.3995 & 0.0255 & 1 & 0 & 0 \\ 1 & 1.803 & 1 & 1 & 0 & 0 \end{bmatrix}$$

and the gain $g = 0.0033$. In the above matrix, each row corresponds to a first- or second-order transfer function. The first three elements are the coefficients of the numerator polynomial, and the second three elements correspond to the denominator polynomial, both in ascending powers of z^{-1}. Because the given transfer function corresponds to an FIR filter, the denominator polynomials have the first element equal to unity and the rest to zero. Thus, the three second-order transfer functions are

$$H_1(z) = 1 + 15.6764z^{-1} + 39.2434z^{-2}$$

$$H_2(z) = 1 + 0.3995z^{-1} + 0.0255z^{-2}$$

$$H_3(z) = 1 + 1.803z^{-1} + z^{-2}$$

The transfer function of the cascaded FIR filter will then take the form

$$H(z) = gH_1(z)H_2(z)H_3(z)$$

Note that the order in which the individual sections are cascaded does not affect the overall response. However, when this cascaded filter is implemented with finite precision hardware, then the ordering of the sections matters.

To obtain the pole-zero plot of the FIR filter in a single section, we can invoke the MATLAB function *zplane(B,A)*, which will plot the zeros and poles of the transfer function with numerator polynomial B and denominator polynomial A. The zeros will be shown as open circles and the poles as crosses along with the unit circle in the z-plane. Similarly, the function call *zplane(sos)* will plot the poles and zeros of the cascaded filter. Last but not the least, the step response of the cascaded FIR filter is obtained by calling the function with input and output arguments *[s,n] = stepz(sos)*, where sos is the matrix of coefficients of the polynomials of the second-order sections. Note that since the input argument in the above function call does not have the gain factor, the step response will be 1/gain times the response of the single-section FIR filter. To obtain the same amplitude range, the step response of the

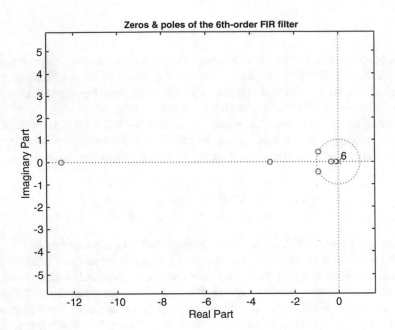

Fig. 3.10 Pole-zero plot of the sixth-order FIR filter of Example 3.17 as a single section

cascaded FIR filter must be scaled by the gain factor. Finally, the rise time of the FIR filter is obtained by calling the function $R = risetime(step\_response, sample\_instants)$, where *step_response* is the step response of the FIR filter and *sample_instants* is the set of time indices over which the step response was calculated. By calling simply *risetime(step_response)*, MATLAB plots the step response and shows the rise time graphically. The pole-zero plots of the FIR filter with single and cascaded section are shown in Figs. 3.10 and 3.11, respectively. For the single section, all the poles are located at the origin. The step response of the FIR filter with single section is shown in Fig. 3.12 as a stem plot. Figure 3.13 shows the step response of the cascaded FIR filter scaled by the gain factor. The two step responses seem identical. The rise time plot of the FIR filter is shown in Fig. 3.14. Note that MATLAB marks the two points corresponding to the 10% and 90% of the final value (steady state) of the step response. The rise time is found to be 2.6548 samples. The M-file *Example3_17.m* is used to solve this problem.

Example 3.18 In this example we will deal with an IIR digital filter, whose transfer function is described by

$$H(z) = \frac{N(z)}{D(z)}$$

$$= \frac{0.0162 + 0.808z^{-1} + 0.1615z^{-2} + 0.1615z^{-3} + 0.808z^{-4} + 0.0162z^{-5}}{1 - 1.506z^{-1} + 1.8316z^{-2} - 1.2374z^{-3} + 0.5447z^{-4} - 0.1156z^{-5}}$$

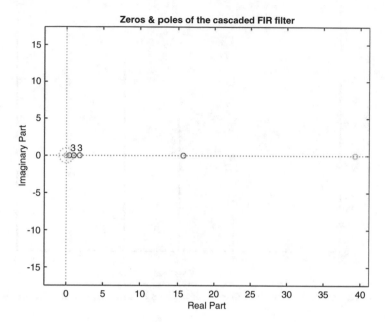

Fig. 3.11 Pole-zero plot of the sixth-order FIR filter of Example 3.17 as a cascade of three second-order sections

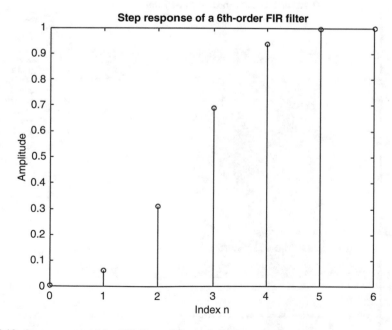

Fig. 3.12 Step response of the FIR filter of Example 3.17 as a single section

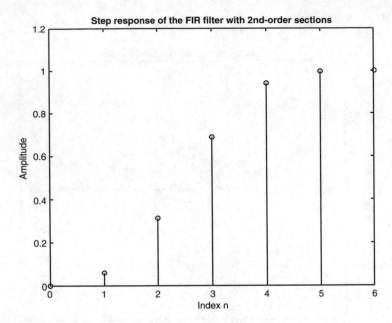

Fig. 3.13 Step response of the FIR filter of Example 3.17 as a cascade of three second-order sections

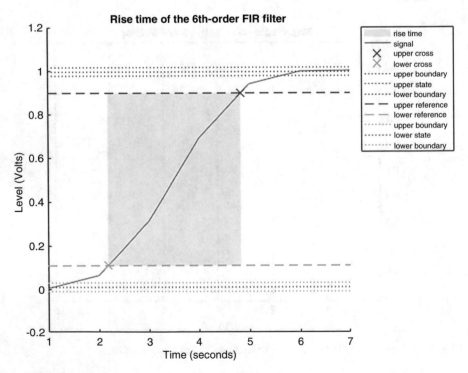

Fig. 3.14 Plot of the step response of the single-section FIR filter of Example 3.17 showing the rise time

As we did in the previous example, we will use MATLAB to plot the poles and zeros of H(z). We will then compute the impulse and step responses of the IIR filter followed by finding the cascade and parallel forms of the given transfer function.

Solution By calling the MATLAB function *zplane(B,A)*, we obtain the pole-zero plot of the transfer function of the IIR filter, where the two vectors are defined as

$$B = [0.0162, 0.808, 0.1615, 0.1615, 0.808, 0.0162]$$

$$A = [1, -1.506, 1.8316, -1.2374, 0.5447, -0.1156]$$

Note that in MATLAB, there will be no commas but only space(s) between the elements of the vectors B and A. The commas are included here for clarity. One can verify in MATLAB if the given IIR filter is stable or not using the statement *isstable (B,A)*. If the filter is stable, the function returns a 1 and if it is unstable, it will return a 0. In this example, the filter is stable. The impulse and step responses are calculated in the same manner as indicated in the previous example. To obtain the cascade realization, we call the function *[sos,g] = tf2sos(B,A)*. The function returns the coefficients of the numerator and denominator polynomials of the first- and/or second-order functions and the overall gain. For this example, the following second-order sections are obtained.

$$sos = \begin{bmatrix} 1 & 49.6799 & 0 & 1 & -0.437 & 0 \\ 1 & 0.4792 & 0.0789 & 1 & -0.6747 & 0.3566 \\ 1 & -0.2825 & 0.2552 & 1 & -0.3943 & 0.7418 \end{bmatrix}$$

From the above equation, we can write the three second-order sections as

$$H_1(z) = \frac{1 + 49.6799z^{-1}}{1 - 0.437z^{-1}}$$

$$H_2(z) = \frac{1 + 0.4792z^{-1} + 0.0789z^{-2}}{1 - 0.6747z^{-1} + 0.3556z^{-2}}$$

$$H_3(z) = \frac{1 - 0.2825z^{-1} + 0.2552z^{-2}}{1 - 0.3943z^{-1} + 0.7418z^{-2}}$$

and the gain is $g = 0.0162$. The impulse response of the IIR filter in a single section is computed by invoking the MATLAB function *[h,n] = impz(B,A)*, where n is a vector of the sample index and h is the impulse response of the filter defined at the points in n. Similarly, the impulse response of the cascaded IIR filter can be found using the function call *[h,n] = impz(sos)*. However, due to the gain factor, the actual impulse response of the cascaded IIR filter is g times the impulse response h obtained through the above function call. The step response of the single-section and cascaded IIR filters are obtained in the manner as indicated in the previous example.

To obtain the parallel form, we use the function call $[R,P,K] = residuez(B,A)$, where B and A are the vectors as defined above. The vector R contains the residues corresponding to the poles P, and K is a vector of quotients. If the given transfer function is an improper function, then the vector K contains the quotients. Otherwise, it is an empty vector. The poles of the transfer function of the given IIR filter are found to be

$$p_{1,2} = 0.1971 \pm j0.8384$$

$$p_{3,4} = 0.3374 \pm j0.4927$$

$$p_5 = 0.4370$$

Since the filter order is 5, there are two sets of complex conjugate poles and one real pole. The corresponding residues are given by

$r_{1,2} = -0.5324 \pm j0.0695, r_{3,4} = -0.1370 \pm j0.6067, r_5 = 1.4951$

Since the transfer function is improper, the quotient $K = -0.1401$. Thus, the parallel form of the IIR filter has two second-order sections corresponding to the two complex conjugate poles and one first-order section corresponding to the real pole. In addition, the IIR filter in parallel form has a constant corresponding to the quotient K. In order to convert the complex conjugate pole factor into a second-order transfer function, we invoke the same *residuez* function but with different arguments. For instance, to convert the first pair of complex conjugate poles to a second-order transfer function, we use the function call $[b,a] = residuez(R(1:2), P(1:2), 0)$. The *residuez* function takes the first pair of the complex conjugate residues and poles and returns the coefficients of the numerator and denominator polynomials of the second-order transfer function in the vectors b and a, respectively. Here, we have to use zero for K. To obtain the transfer function corresponding to the real pole, we have to use the function call $[b,a] = residuez([R(5) 0], [P(5) 0], 0)$. Since there is only one real pole, the pole vector P has the first element corresponding to the real pole, and the second element is zero. Similar vector is used for the residue. Of course, K corresponds to zero. The details can be found in the M-file *Example3_18.m*. Using these call functions, we obtain the following individual transfer functions of the IIR filter in parallel form.

$$H_1(z) = \frac{-1.0647 + 0.0934z^{-1}}{1 - 0.3943z^{-1} + 0.7418z^{-2}}$$

$$H_2(z) = \frac{-0.2740 + 0.6903z^{-1}}{1 - 0.6747z^{-1} + 0.3566z^{-2}}$$

$$H_3(z) = \frac{1.4951}{1 - 0.4370z^{-1}}$$

The fourth section is a constant equal to K, which is -0.1401 (Figs. 3.15 and 3.16).

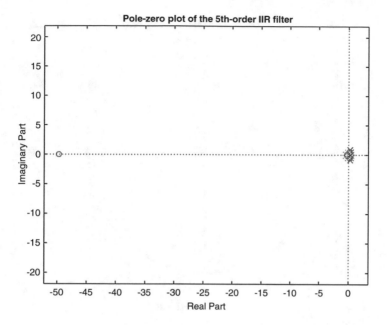

Fig. 3.15 Pole-zero plot of the IIR filter in Example 3.18

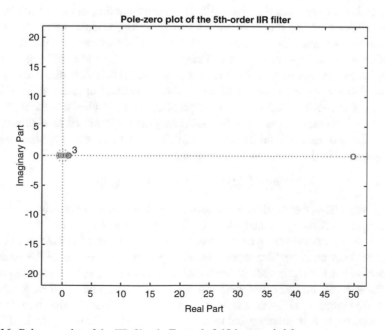

Fig. 3.16 Pole-zero plot of the IIR filter in Example 3.18 in cascaded form

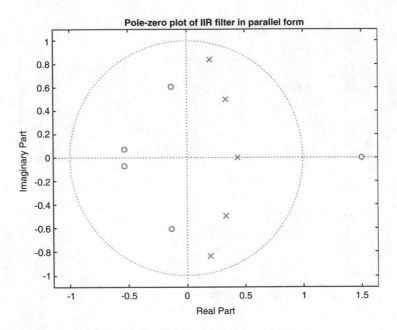

Fig. 3.17   Pole-zero plot of the IIR filter in Example 3.18 in parallel form

Having determined the individual transfer functions of the parallel form of the IIR filter, we then draw the pole-zero plot of all the sections. This is shown in Fig. 3.17. The impulse responses of the given IIR filter corresponding to the single, cascade, and parallel forms are shown in Fig. 3.18 in the top, middle, and bottom plots, respectively. As seen from the figure, the three plots are in agreement. For the cascade form, the impulse response obtained from the function call *impz* must be multiplied by the gain factor g. In the case of the parallel form, the impulse responses of the individual sections must be added to obtain the true impulse response. This is due to the fact that the sections operate in parallel, which implies that the input is the same for all the sections and the individual outputs are summed to obtain the true output. In the case of an improper transfer function, the impulse response corresponding to the quotients is expressed as

$$K(1)\delta[n] + K(2)\delta[n-1] + \cdots + K(L)\delta[n-L]$$

In this example, the quotient K is a scalar, and therefore there is only the first term and is equal to K for $n = 0$ and zero for all other sample indices n.

Next, we compute the step response of the individual sections of the parallel form using the function *stepz*. The input arguments are the coefficients of the numerator and denominator polynomials of the individual sections. The true step response is then the sum of the individual step responses. For the scalar quotient K, the step response is simply K times the unit step function. That is, the step response corresponding to K is simply equal to K for all sample indices *n*. Figure 3.19 depicts

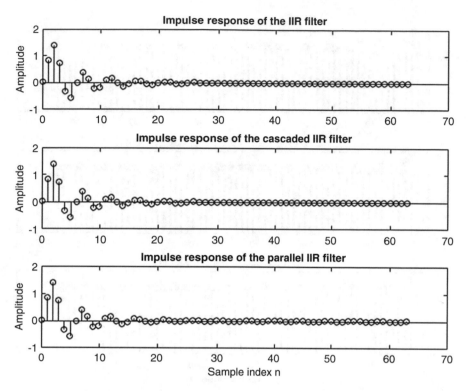

Fig. 3.18 Impulse responses of the IIR filter in Example 3.18: top, single section; middle, cascade form; bottom: parallel form

the step responses for the three cases. Again, all three plots seem to be identical. Again, the step response obtained using the function *stepz* must be multiplied by the gain factor g in the case of the cascade form. The rise time plots for all the three cases are shown in Figs. 3.20, 3.21, and 3.22, which are obtained by simply calling the function *risetime* with step response as its input argument. The rise time is found to be 1.6464 samples for all the three cases.

3.7 Summary

As mentioned earlier, Z-transform plays an important role in the analysis of discrete-time signals and systems. We defined the Z-transform of a sequence and its ROC. Z-transform is a mapping of a discrete-time sequence onto a complex plane. The Z-transform of several important sequences such as the impulse, unit step, real, and complex exponential sequences was determined along with their ROCs. Next we introduced poles and zeros of the Z-transform of sequences and illustrated how the ROC is related to the poles. We also dealt with LTI discrete-time systems in the

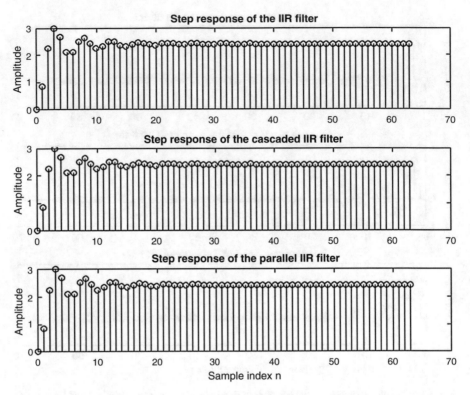

Fig. 3.19 Step responses of the IIR filter in Example 3.18: top, single section; middle, cascade form; bottom, parallel form

Z-domain and defined what is known as the system or transfer function. The transfer function in the Z-domain tells everything about an LTI discrete-time system. Z-transform without the ability to recover a sequence in the discrete-time domain is of no value. The inverse Z-transform enables us to reconstruct a sequence from its Z-transform. Though the inverse Z-transform involves contour integration in the Z-domain, simpler methods, such as partial fraction expansion, exist. We worked out several examples of calculating the response of an LTI discrete-time system to a given input using partial fraction expansion method. Depending on the nature of the poles, partial fraction expansion method differs. Again, a few examples are included in this chapter to clarify the methods. Finally, the use of MATLAB in calculating the zeros and poles of a transfer function was established by way of examples. We also mentioned that Z-transform is very useful in the design of LTI discrete-time systems. We will defer to a later chapter to elucidate this statement. The next chapter describes the discrete-time Fourier transform (DTFT), which is a frequency domain representation of discrete-time signals and systems. We will establish the connection between the Z-transform and the DTFT as well.

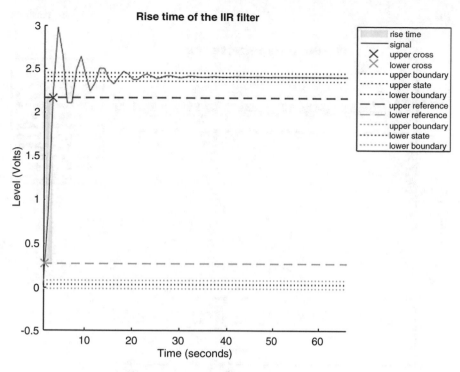

Fig. 3.20 Plot of the step response of the single-section IIR filter of Example 3.18 showing the rise time

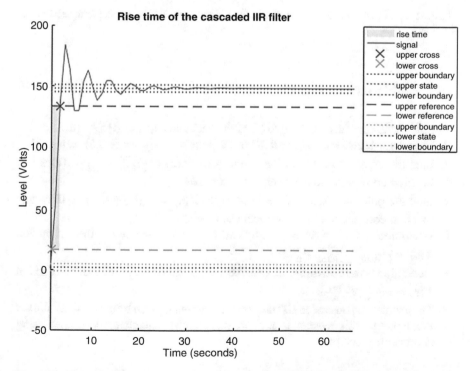

Fig. 3.21 Plot of the step response of the IIR filter of Example 3.18 in cascaded form showing the rise time

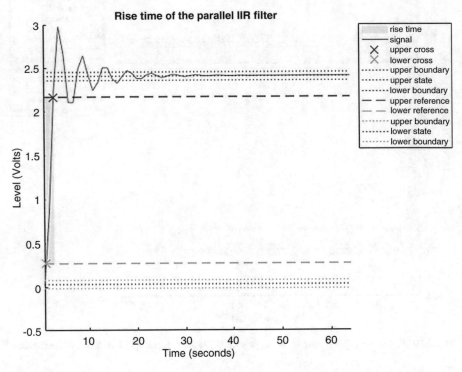

Fig. 3.22 Plot of the step response of the IIR filter of Example 3.18 in parallel form showing the rise time

3.8 Problems

1. Find the Z-transform and the ROC of the sequence $x[n] = (0.5)^n u[n]$.
2. Find the Z-transform and the ROC of the sequence $x[n] = (-0.5)^n u[n-2]$.
3. Find the poles and zeros of the Z-transform $X(z) = \frac{7+3.6z^{-1}}{1+0.9z^{-1}+0.18z^{-2}}$. Does this function correspond to a convergent sequence?
4. Find the poles and zeros of the Z-transform $X(z) = \frac{1-z^{-2}}{1-0.25z^{-1}-0.125z^{-2}}$. Does this function correspond to a convergent sequence?
5. Determine the inverse Z-transform corresponding the function $X(z) = \frac{4.5-1.3z^{-1}+1.4z^{-2}}{(1-0.3z^{-1})(1+0.5z^{-1}+0.9z^{-2})}$.
6. Determine the inverse Z-transform corresponding to the function $X(z) = \frac{1.5-0.125z^{-1}}{1+0.375z^{-1}-0.0625z^{-2}}$.
7. The impulse response of an LTI discrete-time system is given by $h[n] = (0.2)^n u[n]$. If the input to this system is $x[n] = (0.3)^n u[n]$, calculate the output using Z-transform method.

8. The impulse response of an LTI discrete-time system is given by $h[n] = (-0.2)^n$ $u[n]$. If the input to this system is $x[n] = (0.3)^n u[n]$, calculate the output using Z-transform method.

9. Consider the causal LTI discrete-time system described by $y[n] = 3x[n] + 0.4y[n-1] + 0.05y[n-2]$. (a) Determine its transfer function and (b) compute the response of the system to a unit step input.

10. Find the impulse response and step response of the LTI discrete-time system whose transfer function is given by $H(z) = \frac{1+z^{-1}}{1-0.4z^{-1}-0.05z^{-2}}$.

11. Find the step response of the LTI discrete-time system whose transfer function is described by $H(z) = \frac{1-0.5z^{-1}}{1-0.7z^{-1}+0.1z^{-2}}$.

12. An LTI discrete-time system is described by $y[n] - 0.7y[n-1] + 0.1y[n-2] = x[n] - 0.6x[n-1]$. Determine its (a) transfer function in the Z-domain, (b) its impulse response, and (c) step response.

13. The Z-transform of a causal sequence $x[n]$, $n \geq 0$ is denoted by $X(z)$. Show that $x[0] = \lim_{z \to \infty} X(z)$. This is known as the initial value theorem.

14. Determine the DC value of the LTI system whose transfer function is described by $H(z) = \frac{1+6z^{-1}+15z^{-2}+20z^{-3}+15z^{-4}+6z^{-5}+z^{-6}}{1-2.38z^{-1}+2.91z^{-2}-2.06z^{-3}+0.88z^{-4}-0.21z^{-5}+0.022z^{-6}}$. Hint: The DC value corresponds to $z = 1$.

15. For the same transfer function in Problem 14, find the initial and final values. The final value is given by $x[\infty] = \lim_{z \to 1} (z-1)X(z)$.

References

1. Churchill RV, Brown JW (1990) Introduction to complex variables and applications, 5th edn. McGraw-Hill, New York
2. Jury EI (1973) Theory and application of the z-transform method. Robert E. Krieger, Huntington
3. Milne-Thomson LM (1951) Calculus of finite differences. Macmillan, London
4. Mitra SK (2011) Digital signal processing: a computer-based approach, 4th edn. McGraw Hill, New York
5. Oppenheim AV, Shafer RW (1989) Discrete-time signal processing. Prentice-Hall, Englewood Cliffs
6. Oppenheim AV, Willsky AS, Young IT (1983) Signals and systems. Prentice-Hall, Englewood Cliffs
7. Rabiner LR, Gold B (1975) Theory and application of digital signal processing. Prentice-Hall, Englewood Cliffs

Chapter 4
Frequency Domain Representation of Discrete-Time Signals and Systems

4.1 Introduction

Signals, continuous-time, or discrete-time occur in the time domain. We, therefore, described rather elaborately discrete-time signals and systems in the time domain. For easier and more efficient ways to analyze such signals and systems, we next introduced the Z-transform, which is an alternative representation of discrete-time signals and systems. The Z-transform maps a discrete-time signal or an LTI discrete-time system from the discrete-time domain into a complex plane. In this plane, the discrete-time signals and systems are represented by their poles and zeros. There is another domain in which a discrete-time signal or equivalently an LTI discrete-time system can be represented. This domain is the frequency domain. We can visualize a signal more easily in the frequency domain than in the time domain. For instance, a sum of sinusoidal signals with differing frequencies is hard to identify in the time domain individually. On the other hand, such a signal can be easily identified individually in the frequency domain. This is illustrated in Figs. 4.1a and b. The discrete-time signal is shown in Fig. 4.1a. It consists of three sinusoids at frequencies 13, 57, and 93 Hz with amplitudes 1, 1.5, and 2, respectively, at a sampling frequency of 500 Hz. It is hard to discern the individual sinusoids from the figure. Figure 4.1b shows the discrete-time Fourier transform (DTFT) representation of the signal in Fig. 4.1a. One can clearly distinguish the three components in frequency and relative amplitude. The DTFT also greatly aids in the design of LTI discrete-time systems. This chapter deals with the representation of discrete-time signals and systems in the frequency domain. More specifically, we will define a mapping known as the discrete-time Fourier transform that characterizes discrete-time signals and systems in the frequency domain. As a consequence, we will show the

Electronic supplementary material: The online version of this article (https://doi.org/10.1007/978-3-319-76029-2_4) contains supplementary material, which is available to authorized users.

© Springer International Publishing AG, part of Springer Nature 2019
K. S. Thyagarajan, *Introduction to Digital Signal Processing Using MATLAB with Application to Digital Communications*,
https://doi.org/10.1007/978-3-319-76029-2_4

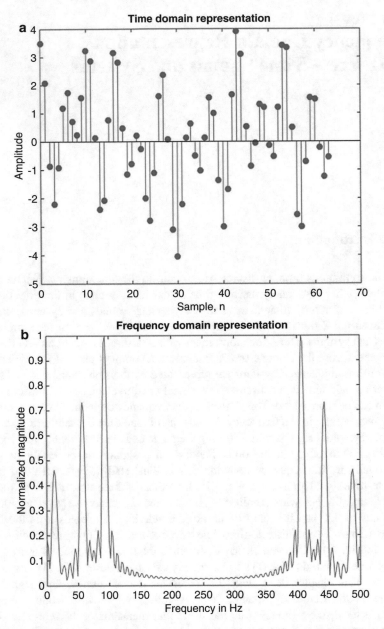

Fig. 4.1 Visualization of a discrete-time sequence that consists three frequencies 13, 57, and 93 Hz with a sampling frequency of 500 Hz: (**a**) sequence in the discrete-time domain, (b) the same sequence in the frequency domain. The three frequencies are clearly distinguishable

relationship between the Z-transform and DTFT. We will also observe that the DTFT characterizes an LTI discrete-time system in the frequency domain. Because of this property, we can define the filtering operations. Filters such as lowpass, highpass, bandpass, bandstop, etc., can be characterized more efficiently in the frequency domain. It further leads us to the design of such filters (Tables 4.1, 4.2 and 4.3).

Table 4.1 Typical discrete-time sequences and corresponding DTFTs

Sequence	Discrete-time Fourier transform
$\delta[n]$	1
$1, \quad -\infty < n < \infty$	$\sum\limits_{k=-\infty}^{\infty} 2\pi\delta(\Omega + 2\pi k)$
$u[n]$	$\dfrac{1}{1 - e^{-j\Omega}} + \sum\limits_{k=-\infty}^{\infty} \pi\delta(\Omega + 2\pi k)$
$e^{jn\Omega_0}$	$\sum\limits_{k=-\infty}^{\infty} 2\pi\delta(\Omega - \Omega_0 + 2\pi k)$
$\alpha^n u[n], \quad |\alpha| < 1$	$\dfrac{1}{1-ae^{-j\Omega}}$

Table 4.2 Properties of discrete-time Fourier transform

Property	Sequence	DTFT
Linearity	$ax_1[n] + bx_2[n]$	$aX_1(e^{j\Omega}) + bX_2(e^{j\Omega})$
Time reversal	$x[-n]$	$X(e^{-j\Omega})$
Time shifting	$x[n - N], N \in Z$	$e^{-jN\Omega}X(e^{j\Omega})$
Modulation	$e^{jn\Omega_0}x[n]$	$X\left(e^{j(\Omega-\Omega_0)}\right)$
Differentiation	$nx[n]$	$j\dfrac{dX\left(e^{j\Omega}\right)}{d\Omega}$
Convolution in the time domain	$x[n] \otimes h[n]$	$X(e^{j\Omega})H(e^{j\Omega})$
Convolution in freq. domain	$x[n]h[n]$	$X(e^{j\Omega}) \otimes H(e^{j\Omega})$
Energy conservation	$\sum\limits_{n=-\infty}^{\infty} |x[n]|^2$	$\dfrac{1}{2\pi} \int_{-\pi}^{\pi} \left|X\left(e^{j\Omega}\right)\right|^2 d\Omega$

Table 4.3 DTFT of ideal filters

Filter type	Cutoff frequencies	DTFT, $H(e^{j\Omega})$
Lowpass	Ω_c	$\dfrac{\sin(n\Omega_c)}{n\pi}, \quad -\infty < n < \infty$
Highpass	Ω_c	$\begin{cases} 1 - \dfrac{\Omega_c}{\pi}, n = 0 \\ -\dfrac{\sin(n\Omega_c)}{n\pi}, otherwise \end{cases}$
Bandpass	Ω_1 and Ω_2	$\dfrac{\sin(n\Omega_2)}{n\pi} - \dfrac{\sin(n\Omega_1)}{n\pi}, \quad -\infty < n < \infty$
Bandstop	Ω_1 and Ω_2	$\begin{cases} 1 - \dfrac{\Omega_2 - \Omega_1}{\pi}, n = 0 \\ \dfrac{\sin(n\Omega_1)}{n\pi} - \dfrac{\sin(n\Omega_2)}{n\pi}, otherwise \end{cases}$

4.2   Discrete-Time Fourier Transform

The DTFT of a sequence $x[n]$ is defined as

$$X\left(e^{j\Omega}\right) = \sum_{n=-\infty}^{\infty} x[n]e^{-jn\Omega} \tag{4.1}$$

In Eq. (4.1), the frequency variable Ω is a normalized frequency expressed in radians and is continuous. Actually, $\Omega = \frac{2\pi f}{F_s}$, where f is the frequency in Hz and F_s is the sampling frequency in Hz. Also notice that the DTFT of a discrete-time signal is, in general, complex. From the definition of the DTFT, it is implicit that the signal in the time domain is discrete, while its DTFT in the frequency domain is continuous. One other important fact about the DTFT is that it is a periodic function with period 2π. Why? Because

$$X\left(e^{j(\Omega+2\pi k)}\right) = \sum_{n=-\infty}^{\infty} x[n]e^{-jn(\Omega+2\pi k)} = \sum_{n=-\infty}^{\infty} x[n]e^{-jn\Omega} = X\left(e^{j\Omega}\right), k \in Z \tag{4.2}$$

The DTFT in Eq. (4.1) exists only if the summation on the right-hand side is finite. This implies that the sequence $x[n]$ must be absolutely summable. Otherwise the DTFT of the sequence does not exist, meaning that there is no representation of the sequence in the frequency domain. As we noted from the definition, the DTFT of a sequence is complex. Therefore, it can be expressed in magnitude – phase form:

$$X\left(e^{j\Omega}\right) = \left|X\left(e^{j\Omega}\right)\right|e^{j\theta(\Omega)} \tag{4.3}$$

4.2.1   DTFT and Z-Transform

Since both DTFT and Z-transform are complex functions, it is easy to relate the two as

$$X\left(e^{j\Omega}\right) = X(z)\big|_{z=e^{j\Omega}} \tag{4.4}$$

In other words, the DTFT of a sequence is its Z-transform evaluated on the *unit circle* in the Z-domain. Since we have already discussed the Z-transform, we can easily determine the DTFT of discrete-time sequences and LTI discrete-time systems from the corresponding Z-transforms.

Example 4.1 Find the DTFT of the sequence $(-0.5)^n u[n]$.

Solution First let us use the definition of DTFT. Then,

$$X(e^{j\Omega}) = \sum_{n=-\infty}^{\infty} x[n]e^{-jn\Omega} = \sum_{n=0}^{\infty} (-0.5)^n e^{-jn\Omega} = \sum_{n=0}^{\infty} \left(-0.5e^{-j\Omega}\right)^n$$
$$= \frac{1}{1+0.5e^{-j\Omega}} \tag{4.5}$$

The Z-transform of the given sequence has been found to be $X(z) = \frac{1}{1+0.5z^{-1}}$. Therefore, its DTFT is

$$X(e^{j\Omega}) = \frac{1}{1+0.5z^{-1}}\bigg|_{z=e^{j\Omega}} = \frac{1}{1+0.5e^{-j\Omega}} \tag{4.6}$$

Because the DTFT is complex, we can express it in magnitude-phase form. Multiply the numerator and denominator of (4.5) by the complex conjugate of the denominator to obtain

$$X(e^{j\Omega}) = \frac{1+0.5e^{j\Omega}}{(1+0.5e^{-j\Omega})(1+0.5e^{j\Omega})} = \frac{(1+0.5\cos(\Omega)) + j0.5\sin(\Omega)}{1.25 + \cos(\Omega)} \tag{4.7}$$

Therefore,

$$\left| X(e^{j\Omega}) \right| = \frac{1}{\sqrt{1.25 + \cos(\Omega)}}, \quad -\pi \le \Omega \le \pi \quad \text{and} \tag{4.8a}$$

$$\theta(\Omega) = \tan^{-1}\left(\frac{0.5\sin(\Omega)}{1+0.5\cos(\Omega)} \right), \quad -\pi \le \Omega \le \pi \tag{4.8b}$$

Figure 4.2 shows the plots of the magnitude in dB and phase in degrees of the DTFT in Eqs. 4.8a and 4.8b. The magnitude in dB is $20log_{10}(|X(e^{j\Omega})|)$.

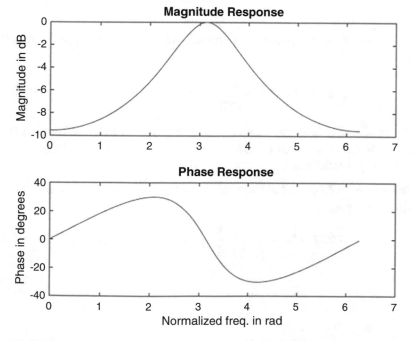

Fig. 4.2 DTFT of the sequence in Example 4.1: Top plot: magnitude in dB of the DTFT, Bottom plot: phase in degrees

4.2.2   DTFT of Some Typical Sequences

Let us find the DTFT of some useful sequences that we encounter often in the discrete-time signals and systems.

DTFT of the Unit Impulse Using the definition of the DTFT, we have

$$DTFT\{\delta[n]\} = \sum_{n=-\infty}^{\infty} \delta[n]e^{-jn\Omega} = 1, \forall\Omega \tag{4.9}$$

As expected, the DTFT of the unit impulse sequence is unity for all frequencies.

DTFT of a Constant The DTFT of a sequence with unit value for all discrete-time instants is

$$DTFT\{1\} = \sum_{n=-\infty}^{\infty} e^{-jn\Omega} = \sum_{k=-\infty}^{\infty} 2\pi\delta(\Omega + 2\pi k), k\epsilon Z \tag{4.10}$$

The DTFT of 1 consists of a sequence of impulses of strength 2π, one impulse in each period.

DTFT of the Unit Step Sequence The DTFT of the unit step sequence can be shown to be

$$DTFT\{u[n]\} = \frac{1}{1 - e^{-j\Omega}} + \sum_{k=-\infty}^{\infty} \pi\delta(\Omega + 2\pi k), k\epsilon Z \tag{4.11}$$

DTFT of a Complex Exponential Sequence The DTFT of the complex exponential sequence $e^{jn\Omega_0}$ is given by

$$DTFT\{e^{jn\Omega_0}\} = \sum_{k=-\infty}^{\infty} 2\pi\delta(\Omega - \Omega_0 + 2\pi k), k\epsilon Z \tag{4.12}$$

As can be seen from the above equation, the DTFT of a complex exponential sequence is a sequence of impulses of strength 2π at multiples of the frequency of the complex exponential sequence.

DTFT of a Real Exponential Sequence The DTFT of a real exponential sequence $\alpha^n u[n]$ is found to be

$$DTFT\{\alpha^n u[n]\} = \frac{1}{1 - \alpha e^{-j\Omega}}, |\alpha| < 1, -\pi \leq \Omega \leq \pi \tag{4.13}$$

4.3 Inverse Discrete-Time Fourier Transform

The discrete-time signal is recovered from its DTFT by the process of inverse DTFT. Unlike the inverse Z-transform, inverse DTFT is much simpler to perform. Since the DTFT of a sequence is periodic with period 2π, it can be expanded in a Fourier series. The coefficients of the Fourier series of the DTFT are the samples of the sequence in question. Therefore, the sequence in question is given by

$$x[n] = \frac{1}{2\pi} \int_{-\pi}^{\pi} X(e^{j\Omega}) e^{jn\Omega} d\Omega \tag{4.14}$$

In practice, it may be simpler to obtain the inverse DTFT (IDTFT) using partial fraction expansion and then identifying the individual terms with the corresponding IDTFT. In doing so, we will also use the properties of the DTFT to further simplify the process of IDTFT.

4.4 Properties of DTFT

Let us briefly list the properties of the DTFT. These properties will come in handy when we solve complex problems in the frequency domain.

Linearity DTFT is a linear transform, which implies that the DTFT of a linear combination of sequences is the same linear combination of the individual DTFTs. In other words,

$$DTFT\{ax_1[n] + bx_2[n]\} = aX_1(e^{j\Omega}) + bX_2(e^{j\Omega}) \tag{4.15}$$

The proof is simple. Since the DTFT involves a summation, which is a linear operation, the DTFT is linear.

Time Reversal The DTFT of a time-reversed sequence $x[-n]$ is a function of the negative frequency:

$$DTFT\{x[-n]\} = X(e^{-j\Omega}) \tag{4.16}$$

Proof: By definition,

$$DTFT\{x[-n]\} = \sum_{n=-\infty}^{\infty} x[-n]e^{-jn\Omega} = \sum_{m=-\infty}^{\infty} x[m]e^{-jm(-\Omega)} = X(e^{-j\Omega}) \tag{4.17}$$

The significance of the time-reversal property is that the unit circle is traversed in clockwise direction. Note that for the non-time-reversed sequence, the unit circle is traversed in the counterclockwise direction in the DTFT domain.

Time Shift The DTFT of a time-shifted sequence is the DTFT of the sequence with a phase shift. That is,

$$DTFT\{x[n-N]\} = e^{-jN\Omega}X\left(e^{j\Omega}\right), N\epsilon Z \qquad (4.18)$$

Proof: $DTFT\{x[n-N]\} = \sum_{n=-\infty}^{\infty} x[n-N]e^{-jn\Omega}$

Let $m = n - N$ in the above equation. Then, we can write (4.18) as

$$DTFT\{x[n-N]\} = \sum_{m=-\infty}^{\infty} x[m]e^{-j(m+N)\Omega} = e^{-jN\Omega}\sum_{m=-\infty}^{\infty} x[m]e^{-jm\Omega}$$
$$= e^{-jN\Omega}X\left(e^{j\Omega}\right) \qquad (4.19)$$

So, according to the time-shift property, a shift in time is equivalent to a phase shift in the frequency.

Frequency Shift The DTFT of a modulated sequence is equivalent to a frequency shift. More specifically,

$$DTFT\left\{e^{jn\Omega_0}x[n]\right\} = X\left(e^{j(\Omega-\Omega_0)}\right) \qquad (4.20)$$

Proof: Note that multiplying a sequence by a complex exponential sequence is called modulation. Using the definition of DTFT, we can write

$$DTFT\left\{e^{jn\Omega_0}x[n]\right\} = \sum_{n=-\infty}^{\infty} x[n]e^{jn\Omega_0}e^{-jn\Omega} = \sum_{n=-\infty}^{\infty} x[n]e^{-jn(\Omega-\Omega_0)}$$
$$= X\left(e^{j(\Omega-\Omega_0)}\right) \qquad (4.21)$$

Differentiation in the Frequency Domain This property is described by the following equation:

$$DTFT\{nx[n]\} = j\frac{dX(e^{j\Omega})}{d\Omega} \qquad (4.22)$$

Proof: By definition, $DTFT\{nx[n]\} = \sum_{n=-\infty}^{\infty} nx[n]e^{-jn\Omega}.$

We can write $ne^{-jn\Omega} = j\frac{d}{d\Omega}e^{-jn\Omega}$. Substituting this for $ne^{-jn\Omega}$ in the above equation and taking the derivative outside the summation, we get

$$DTFT\{nx[n]\} = j\frac{d}{d\Omega}\left\{\sum_{n=-\infty}^{\infty} x[n]e^{-jn\Omega}\right\} = j\frac{dX(e^{j\Omega})}{d\Omega} \qquad (4.23)$$

Convolution in the Time Domain The DTFT of the convolution of two sequences is the product of the individual DTFTs. That is,

$$DTFT\{x[n] \otimes h[n]\} = X(e^{j\Omega})H(e^{j\Omega}) \tag{4.24}$$

Proof: Using the definition of DTFT and convolution sum, we have

$$DTFT\{x[n] \otimes h[n]\} = \sum_{n=-\infty}^{\infty} \left\{ \sum_{k=-\infty}^{\infty} x[k]h[n-k] \right\} e^{-jn\Omega} \tag{4.25}$$

By interchanging the order of summation and replacing n − k by m, we can write (4.25) as

$$
\begin{aligned}
DTFT\{x[n] \otimes h[n]\} &= \sum_{k=-\infty}^{\infty} x[k] \left\{ \sum_{m=-\infty}^{\infty} h[m]e^{-j(m+k)\Omega} \right\} \\
&= \left\{ \sum_{k=-\infty}^{\infty} x[k]e^{-jk\Omega} \right\} \left\{ \sum_{m=-\infty}^{\infty} h[m]e^{-jm\Omega} \right\} \\
&= X(e^{j\Omega})H(e^{j\Omega})
\end{aligned}
\tag{4.26}
$$

Convolution in the Frequency Domain The duality of the convolution in the time domain is that the DTFT of the convolution of two DTFTs is the product of the two corresponding sequences. Another way of stating is that the DTFT of the product of two sequences is the convolution of the corresponding DTFTs. Thus,

$$DTFT\{x[n]h[n]\} = X(e^{j\Omega}) \otimes H(e^{j\Omega}) \tag{4.27}$$

Proof: By definition,

$$DTFT\{x[n]h[n]\} = \sum_{n=-\infty}^{\infty} x[n]h[n]e^{-j\Omega} \tag{4.28}$$

Now, substitute the IDTFT of h[n] for h[n] in Eq. (4.28), and change the order of summation and integration to obtain

$$
\begin{aligned}
DTFT\{x[n]h[n]\} &= \sum_{n=-\infty}^{\infty} x[n] \left\{ \frac{1}{2\pi} \int_{-\pi}^{\pi} H(e^{j\theta})e^{jn\theta} \right\} e^{-jn\Omega} \\
&= \frac{1}{2\pi} \int_{-\pi}^{\pi} H(e^{j\theta}) \left\{ \sum_{n=-\infty}^{\infty} x[n]e^{-jn(\Omega-\theta)} \right\} d\theta \\
&= \frac{1}{2\pi} \int_{-\pi}^{\pi} H(e^{j\theta})X(e^{j(\Omega-\theta)}) d\theta
\end{aligned}
\tag{4.29}
$$

Energy Conservation One other property of DTFT is that the energy in a sequence is conserved, meaning that one can compute the energy in a sequence either in the discrete-time domain or in the frequency domain. They are identical. Sometimes it is much easier to determine the signal energy in the frequency domain than in the discrete-time domain and vice versa. The statement implies that

$$\sum_{n=-\infty}^{\infty} |x[n]|^2 = \sum_{n=-\infty}^{\infty} x[n]x^*[n] = \frac{1}{2\pi} \int_{-\pi}^{\pi} |X(e^{j\Omega})|^2 d\Omega \qquad (4.30)$$

From the above equation, we observe that the sum of the magnitude squared of a sequence, which is its energy, is the same as the area under the magnitude squared of its DTFT in one period.

Enough of the properties of the DTFT; let us look at a few examples to familiarize ourselves with what we have learned so far.

Example 4.2 Find the DTFT of the sequence $\alpha^{|n|}$, $|\alpha| < 1$.

Solution From the definition of DTFT,

$$DTFT\{\alpha^{|n|}\} = \sum_{n=-\infty}^{\infty} \alpha^{|n|}e^{-jn\Omega} = \sum_{n=1}^{\infty} \left(\alpha e^{j\Omega}\right)^n + \sum_{n=0}^{\infty} \left(\alpha e^{-j\Omega}\right)^n \qquad (4.31)$$

Since $|\alpha| < 1$, the two summations on the right-hand side converge. So,

$$DTFT\{\alpha^{|n|}\} = \frac{\alpha e^{j\Omega}}{1 - \alpha e^{j\Omega}} + \frac{1}{1 - \alpha e^{-j\Omega}} = \frac{1 - \alpha^2}{1 + \alpha^2 - 2\alpha \cos(\Omega)} \qquad (4.32)$$

Since the given sequence is real and even, its DTFT is purely real.

Example 4.3 Find the DTFT of $\alpha^n u[n-1]$, $|\alpha| < 1$.

Solution We can rewrite the given sequence as $\alpha(\alpha^{n-1}u[n-1])$. The sequence within the parenthesis is the same sequence $\alpha^n u[n]$ shifted by one sample to the right. We can, therefore, use the time-shifting property to obtain

$$DTFT\{\alpha^n u[n-1]\} = \alpha DTFT\{\alpha^{n-1}u[n-1]\} = \frac{\alpha e^{-j\Omega}}{1 - \alpha e^{-j\Omega}} \qquad (4.33)$$

In magnitude-phase representation, the DTFT of the given sequence is

$$|DTFT\{\alpha^n u[n-1]\}| = \frac{|\alpha|}{\sqrt{1+\alpha^2 - 2\alpha \cos(\Omega)}}, \theta(\Omega) = -\Omega - \tan^{-1}\left(\frac{\alpha \sin\Omega}{1 - \alpha \cos\Omega}\right) \qquad (4.34)$$

Example 4.4 Find the DTFT of the sequence $n\alpha^n u[n]$, $|\alpha| < 1$.

Solution Because of the factor n appearing in the given sequence, we can use the differentiation property to obtain

$$DTFT\{n\alpha^n u[n]\} = j\frac{d}{d\Omega} DTFT\{\alpha^n u[n]\} = j\frac{d}{d\Omega}\left\{\frac{1}{1-\alpha e^{-j\Omega}}\right\} = \frac{\alpha e^{-j\Omega}}{(1-\alpha e^{-j\Omega})^2}$$

(4.35)

The magnitude of the DTFT is $\frac{|\alpha|}{1+\alpha^2-2\alpha\cos(\Omega)}$, and the phase is $-\Omega - 2\tan^{-1}\left(\frac{\alpha\sin\Omega}{1-\alpha\cos\Omega}\right)$

Example 4.5 Find the DTFT of the finite-length sequence

$$x[n] = \begin{cases} 1, 0 \leq n \leq N-1 \\ 0, \quad \text{otherwise} \end{cases}$$

(4.36)

Solution Using the definition of the DTFT, we can write

$$X(e^{j\Omega}) = \sum_{n=-\infty}^{\infty} x[n]e^{-j\Omega} = \sum_{n=0}^{N-1} e^{-j\Omega} = \frac{1-e^{-jN\Omega}}{1-e^{-j\Omega}}$$

(4.37)

We can simplify the above equation as

$$X(e^{j\Omega}) = \frac{e^{-jN\frac{\Omega}{2}}\left(e^{jN\frac{\Omega}{2}} - e^{-jN\frac{\Omega}{2}}\right)}{e^{-j\frac{\Omega}{2}}\left(e^{j\frac{\Omega}{2}} - e^{-j\frac{\Omega}{2}}\right)} = e^{-j(N-1)\frac{\Omega}{2}}\frac{\sin\left(N\frac{\Omega}{2}\right)}{\sin\left(\frac{\Omega}{2}\right)}$$

(4.38)

Observe that the given finite-length sequence is a rectangular window, and so its DTFT is a function with side lobes. One other thing we notice here is that the sequence is limited in the time domain, and so it is not limited in the frequency domain!

Example 4.6 Find the DTFT of the sequence $y[n] = x[n] \otimes x^*[-n]$.

Solution As a result of the convolution property, the DTFT of $y[n]$ equals the product of the DTFTs of $x[n]$ and $x^*[-n]$. First, let us find the DTFT of the sequence $x^*[-n]$, which by definition, is

$$DTFT\{x^*[-n]\} = \sum_{n=-\infty}^{\infty} x^*[-n]e^{-jn\Omega} = \left\{\sum_{n=-\infty}^{\infty} x[-n]e^{jn\Omega}\right\}^*$$

(4.39)

In (4.39), substitute $m = -n$ to get

$$DTFT\{x^*[-n]\} = \left\{\sum_{m=-\infty}^{\infty} x[m]e^{-jm\Omega}\right\}^* = X^*(e^{j\Omega})$$

(4.40)

Therefore, we have

$$DTFT\{x[n]\otimes x^*[-n]\} = X(e^{j\Omega})X^*(e^{j\Omega}) = |X(e^{j\Omega})|^2$$

(4.41)

Thus, the DTFT of the convolution of a sequence with its own time-reversed conjugate is the magnitude squared of the DTFT of the sequence. We also notice that the square of the magnitude function is real and even!

4.5   Frequency Domain Representation of LTI Discrete-Time Systems

It is customary to describe an LTI discrete-time system in terms of a linear difference equation with constant coefficients. In order to describe such a system in the frequency domain, we must determine its transfer function in the frequency domain. Consider an LTI discrete-time system described by the following difference equation, where the coefficients are constants:

$$y[n] = \sum_{j=0}^{q} a_j x[n-j] - \sum_{k=1}^{p} b_k y[n-k], p > q \qquad (4.42)$$

By applying the DTFT on both sides of (4.42) and making use of the time-shifting property of the DTFT, we get

$$Y\left(e^{j\Omega}\right) = X\left(e^{j\Omega}\right) \sum_{m=0}^{q} a_m e^{-jm\Omega} - Y\left(e^{j\Omega}\right) \sum_{n=1}^{p} b_n e^{-jn\Omega} \qquad (4.43)$$

We can then express the transfer function of the LTI discrete-time system in the frequency domain as

$$H\left(e^{j\Omega}\right) = \frac{Y(e^{j\Omega})}{X(e^{j\Omega})} = \frac{a_0 + a_1 e^{-j\Omega} + a_2 e^{-j2\Omega} + \cdots\cdots + a_q e^{-jq\Omega}}{1 + b_1 e^{-j\Omega} + b_2 e^{-j2\Omega} + \cdots\cdots + b_p e^{-jp\Omega}} \qquad (4.44)$$

As expected, the transfer function of an LTI discrete-time system in the frequency domain is a rational function in the variable $e^{j\Omega}$. Recall that the system described by (4.42) in the discrete-time domain or equivalently by (4.44) in the frequency domain is a recursive system. Hence its transfer function is a rational polynomial in the variable $e^{j\Omega}$. On the other hand, a non-recursive system will have a transfer function where the denominator is identically equal to 1.

Example 4.7 The impulse response of an LTI discrete-time system is given by $h[n] = 0.5^n u[n]$. If the input to this system is $x[n] = 0.75^n u[n]$, find the response of the system using the DTFT.

Solution Since the given system is LTI, its response is the convolution of its impulse response sequence and the input sequence. In the frequency domain, the DTFT of the system response is the product of the DTFT of its impulse response and the input. First, we find the DTFT of the impulse response and the input from our previous discussion. Therefore,

$$H(e^{j\Omega}) = DTFT\{h[n]\} = \frac{1}{1 - 0.5e^{-j\Omega}} \tag{4.45a}$$

$$X(e^{j\Omega}) = DTFT\{x[n]\} = \frac{1}{1 - 0.75e^{-j\Omega}} \tag{4.45b}$$

The DTFT of the system output is

$$Y(e^{j\Omega}) = DTFT\{y[n]\} = DTFT\{h[n] \otimes x[n]\} = H(e^{j\Omega})X(e^{j\Omega})$$

$$= \frac{1}{(1 - 0.5e^{-j\Omega})(1 - 0.75e^{-j\Omega})} \tag{4.46}$$

The time domain response is the IDTFT of (4.46). One way to find the IDTFT is to express (4.46) in partial fractions and then identify each fraction with a real exponential sequence. So,

$$Y(e^{j\Omega}) = \frac{A}{1 - 0.5e^{-j\Omega}} + \frac{B}{1 - 0.75e^{-j\Omega}} \tag{4.47}$$

The residues are given by

$$A = (1 - 0.5e^{-j\Omega})Y(e^{j\Omega})\big|_{e^{j\Omega}=0.5} = -2 \tag{4.48a}$$

$$B = (1 - 0.75e^{-j\Omega})Y(e^{j\Omega})\big|_{e^{j\Omega}=0.75} = 3 \tag{4.48b}$$

Note that $IDTFT\{\frac{1}{1-0.5e^{-j\Omega}}\} = 0.5^n u[n]$ and $IDTFT\{\frac{1}{1-0.75e^{-j\Omega}}\} = 0.75^n u[n]$. Therefore, the response of the given LTI discrete-time system is

$$y[n] = -2(0.5)^n u[n] + 3(0.75)^n u[n] \tag{4.49}$$

Example 4.8 An LTI discrete-time system consists of a cascade of two systems $h_1[n]$ and $h_2[n]$, where

$$h_1[n] = \delta[n] + \delta[n-1] \text{ and} \tag{4.50a}$$

$$h_2[n] = \beta^n u[n], |\beta| < 1 \tag{4.50b}$$

Determine the value of β such that the magnitude of the overall frequency response of the system is unity.

Solution In a cascade connection, the output of the first system is the input to the second system and so on. This implies that the overall impulse response is the convolution of the impulse response of the individual sections. Therefore, the DTFT of the overall system is the product of the individual DTFTs. Thus,

$$H(e^{j\Omega}) = H_1(e^{j\Omega})H_2(e^{j\Omega}) = \frac{(1 + e^{-j\Omega})}{1 - \beta e^{-j\Omega}} \qquad (4.51)$$

The magnitude of the overall frequency response is given by

$$|H(e^{j\Omega})| = \left| \frac{(1 + \cos\Omega) - j\sin\Omega}{(1 - \beta\cos\Omega) + j\beta\sin\Omega} \right| = \sqrt{\frac{2 + 2\cos\Omega}{1 + \beta^2 - 2\beta\cos\Omega}} \Rightarrow \beta = -1 \quad (4.52)$$

An LTI discrete-time system whose magnitude of the frequency response is a constant is known as an *allpass* system.

4.5.1 Steady State Response of LTI Discrete-Time Systems

We have seen earlier that the particular solution of a linear difference equation with constant coefficients is proportional to the input sequence. The complementary solution is, in general, a decaying function. So, when the complementary solution or the transient response disappears, only the particular solution remains. This is called the *steady state* response of the system. In particular, when the input to an LTI discrete-time system is a sinusoidal sequence of a specified frequency, its response in the steady state is the same input sinusoid except that its amplitude and phase are modified by the value of the transfer function at that frequency. We can, therefore, express the steady state response of an LTI discrete-time system to an input sinusoid $e^{jn\Omega_0}$ as

$$y_{ss}[n] = |H(e^{j\Omega_0})| e^{j\Omega_0(n-\tau)} \qquad (4.53)$$

From Eq. (4.53) we notice that the amplitude of the output sinusoid is the magnitude of the transfer function at the input frequency and the lagging phase angle equals τ times the input frequency. If the phase response is linear, then the phase delay equals the negative of the phase angle divided by the input frequency. If the phase angle represented by $\theta(\Omega)$ is linear, then the phase delay or simply the delay is expressed by

$$\tau = -\frac{\theta(\Omega)}{\Omega_0} \; samples \qquad (4.54)$$

Recall that the transfer function $H(e^{j\Omega})$ is the DTFT of the impulse response $h[n]$ of an LTI discrete-time system. Since the impulse response is unique to a given system, the corresponding frequency response is also unique to the system.

Example 4.9 The impulse response of an LTI discrete-time system is given by

$$h[n] = 0.5^n u[n] \tag{4.55}$$

Find the steady state response of the system if the input is $x[n] = \cos\left(\frac{n\pi}{6}\right) u[n]$.

Solution The frequency response or the transfer function of the given system is the DTFT of its impulse response:

$$H(e^{j\Omega}) = DTFT\{h[n]\} = DTFT\{0.5^n u[n]\} = \frac{1}{1 - 0.5e^{-j\Omega}} \tag{4.56}$$

The transfer function in magnitude-phase form of the given system is found to be

$$\left|H(e^{j\Omega})\right| = \frac{1}{\sqrt{1.25 - \cos\Omega}} \tag{4.57a}$$

$$\theta(\Omega) = -\tan^{-1}\left(\frac{0.5 \sin\Omega}{1 - 0.5 \cos\Omega}\right), rad \tag{4.57b}$$

Since the input is a unit amplitude sinusoid at a frequency $\frac{\pi}{6}$ rad, the steady state response of the given system is given by

$$y_{ss}[n] = \left|H(e^{j\frac{\pi}{6}})\right| \cos\left(\frac{n\pi}{6} - \theta\left(\frac{\pi}{6}\right)\right) u[n] \tag{4.58}$$

From Eqs. (4.57a) and (4.57b), we find that $\left|H(e^{j\frac{\pi}{6}})\right| \approx 1.6138$, and the phase angle is $\theta\left(\frac{\pi}{6}\right) \approx -0.415283^r$ or $-23.79°$. Therefore, the steady state response of the given system to the input sinusoid is

$$y_{ss}[n] = 1.6138 \cos\left(\frac{n\pi}{6} - 23.79°\right) u[n] \tag{4.59}$$

The input and the system response are plotted as a function of the sample index and shown in Fig. 4.3 as top and bottom plots, respectively, using MATLAB. The system response is obtained by calling the function *conv*. It accepts two sequences as vectors and returns a sequence that is of length equal to the sum of the lengths of the impulse response and input minus one. However, we plot the response in length equal to the input sequence. As can be seen from the figure, the system response is the same sinusoid as the input with a change in its amplitude. We also notice a delay of one sample in the output sequence due to the phase shift in the transfer function, which agrees with the analytical result. The M-file for this problem is named *Example4_9.m*.

Group Delay When the phase response of an LTI discrete-time system is not linear, then its phase delay is not constant but is a function of the input frequency. When a group of sinusoidal frequencies is present in the input, it is customary to find the phase delay over this group of frequencies. It is called the *group delay* and is defined as

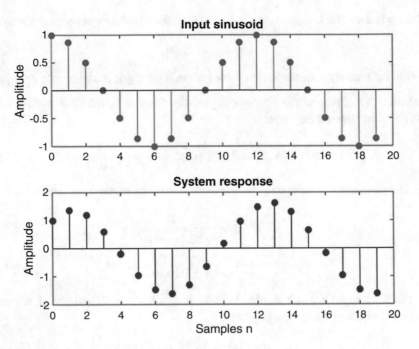

Fig. 4.3 Steady state response of the system in Example 4.9 due to a sinusoidal input. Top plot: input sequence. Bottom plot: output sequence

$$\tau_{gd} = -\frac{d\theta(\Omega)}{d\Omega} \tag{4.60}$$

Let us illustrate the calculation of group delay by way of an example.

Example 4.10 Calculate the group delay of the system in Example 4.9 and plot the result.

Solution The phase response of the given system is shown in Eq. (4.57b). Then, the group delay is given by

$$\tau_{gd} = -\frac{d}{d\Omega}\left\{-\tan^{-1}\left(\frac{0.5\sin\Omega}{1 - 0.5\cos\Omega}\right)\right\} \tag{4.61}$$

To obtain the derivative in Eq. (4.61), first let us rewrite (4.57b) as

$$\tan\theta(\Omega) = \frac{-0.5\sin\Omega}{1 - 0.5\cos\Omega} \tag{4.62}$$

Now differentiate (4.62) with respect to Ω. Therefore,

$$\frac{1}{\cos^2\theta}\frac{d\theta}{d\Omega} = \frac{-(1 - 0.5\cos\Omega)(0.5\cos\Omega) + (0.5\sin\Omega)(0.5\sin\Omega)}{(1 - 0.5\cos\Omega)^2} \quad (4.63)$$

From (4.62), we get

$$\cos\theta(\Omega) = \frac{1 - 0.5\cos\Omega}{\sqrt{1.25 - \cos\Omega}}. \quad (4.64)$$

Using (4.64) in (4.63) and after algebraic manipulation, we obtain

$$\frac{d\theta(\Omega)}{d\Omega} = \frac{0.25 - 0.5\cos\Omega}{1.25 - \cos\Omega} \quad (4.65)$$

Using Eq. (4.65) in (4.61), we finally obtain the expression for the group delay as

$$\tau_{gd} = \frac{-(0.25 - 0.5\cos\Omega)}{1.25 - \cos\Omega} \quad (4.66)$$

The group delay in (4.66) is shown in Fig. 4.4 top plot as a function of Ω in the interval $0 \le \Omega \le \pi$. The MATLAB function to evaluate the group delay is *grpdelay*. It accepts the coefficients of the numerator and denominator polynomials of the

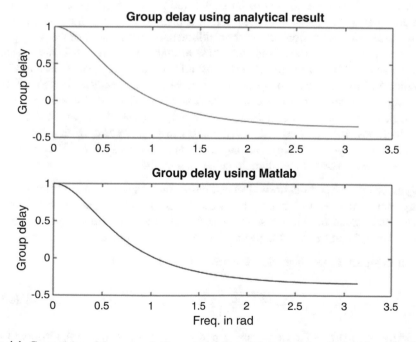

Fig. 4.4 Group delay of the system in Example 4.9. Top plot: group delay obtained from equation (4.66). Bottom plot: group delay obtained using the MATLAB function

transfer function, both in ascending powers of $e^{-j\Omega}$. The group delay calculated using the MATLAB function is shown in the bottom plot in Fig. 4.4. They seem to agree completely. The M-file named *Example4_10.m* is used to solve this problem.

4.5.2 Concept of Filtering

A filter is meant to remove unwanted frequency components and pass other frequency components of interest from an input signal. The effect of filtering is easily visualized in the frequency domain. We know from earlier discussion that the DTFT of the response of an LTI discrete-time system to an input sequence is the product of the DTFTs of the input and impulse response as given by

$$Y(e^{j\Omega}) = X(e^{j\Omega})H(e^{j\Omega}) \tag{4.67}$$

The shape of $X(e^{j\Omega})$ can be altered by a proper design of the shape of $H(e^{j\Omega})$. Since the transfer function $H(e^{j\Omega})$ of the filter is a rational polynomial in $e^{-j\Omega}$, the design of a filter amounts to determining the order and coefficients of the polynomials. Because the transfer function in the frequency domain is a complex function, there are two factors to be considered in the design of a filter, namely, the magnitude and phase. One can design a filter to have a certain shape of the magnitude without any concern for the resulting phase response. The resulting phase may be linear or nonlinear. Nonlinear phase response may not be tolerated in some applications, while other applications may tolerate nonlinear phase. For instance, nonlinear phase in filtering of speech or music will not have any noticeable effect in hearing. But nonlinear phase can cause a contour effect in processing images. Though a filter may be designed to satisfy only the magnitude of the frequency specifications, the resulting nonlinear phase can be corrected so that the overall phase is linear yet satisfying the magnitude specifications. We will consider the design of LTI discrete-time filters in detail in later chapters.

Types of Filter Specifications Depending on the shape of the magnitude of the frequency response, a filter may be classified as lowpass, highpass, bandpass, or bandstop filter. In the ideal case, a lowpass filter passes all frequencies of interest with unity gain and rejects the rest of the frequencies completely.

Ideal Lowpass Filter The ideal lowpass filter can be described by

$$H_{LP}(e^{j\Omega}) = \begin{cases} 1, |\Omega| \leq \Omega_c \\ 0, \Omega_c < |\Omega| \leq \pi \end{cases} \tag{4.68}$$

Figure 4.5a shows the frequency response specified in Eq. (4.68). Observe from (4.68) that the phase response is assumed to be zero. In equation (4.68), Ω_c is called the cutoff frequency. Since the frequency response is periodic with period 2π, the frequency response of the ideal lowpass filter is zero in the intervals $-\pi \leq \Omega < -\Omega_c$

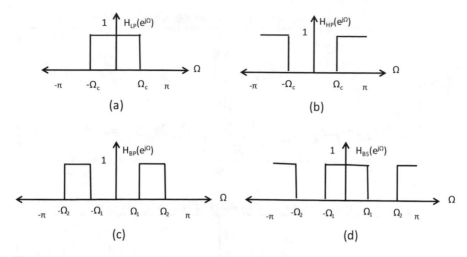

Fig. 4.5 Frequency responses of ideal LP, HP, BP, and BS filters. (**a**) Lowpass filter, (**b**) highpass filter, (**c**) bandpass filter, (**d**) bandstop filter

and $\Omega_c < \Omega \leq \pi$. Using the inverse DTFT, we can determine the corresponding impulse response of the ideal lowpass filter as

$$h_{LP}[n] = \frac{1}{2\pi} \int_{-\pi}^{\pi} H\left(e^{j\Omega}\right) e^{jn\Omega} d\Omega = \frac{1}{2\pi} \int_{-\Omega_c}^{\Omega_c} e^{jn\Omega} d\Omega = \frac{\sin\left(n\Omega_c\right)}{n\pi}, \quad -\infty < n < \infty$$

(4.69)

As can be seen from (4.69), the impulse response of the ideal lowpass filter is of infinite duration. It has a main lobe and side lobes with decreasing amplitudes. As an example, the impulse response of an ideal lowpass filter with a cutoff frequency of 0.4π is shown in the top plot of Fig. 4.6 over the index from -31 to 31. Its frequency response is shown in Fig. 4.7 in the top plot over the frequency range 0 to π, π being half the sampling frequency. Because the impulse response is truncated, the corresponding frequency response has ripples in both the passband and stopband regions. We will discuss this effect later in the book.

Ideal Highpass Filter A highpass filter rejects all frequencies up to the cutoff frequency and passes the rest of the frequencies as described by

$$H_{HP}\left(e^{j\Omega}\right) = \begin{cases} 0, |\Omega| < \Omega_c \\ 1, \Omega_c \leq |\Omega| \leq \pi \end{cases}$$

(4.70)

The ideal frequency response of the highpass filter is shown in Fig. 4.5b. The corresponding impulse response can be obtained using the inverse DTFT as shown below:

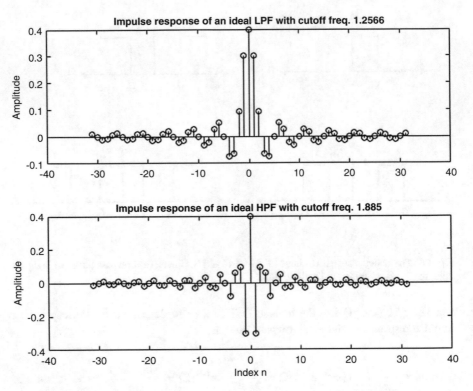

Fig. 4.6 Impulse responses. Top: lowpass filter with a cutoff frequency 0.4 π rad. Bottom: highpass filter with a cutoff frequency 0.6 π rad

$$h_{HP}[n] = \frac{1}{2\pi} \int_{-\pi}^{\pi} H(e^{j\Omega}) e^{jn\Omega} d\Omega = \frac{1}{2\pi} \left[\int_{-\pi}^{-\Omega_c} e^{jn\Omega} d\Omega + \int_{\Omega_c}^{\pi} e^{jn\Omega} d\Omega \right] \qquad (4.71)$$

After evaluating the integral, we find the impulse response to be

$$h_{HP}[n] = -\frac{\sin(n\Omega_c)}{n\pi} + \frac{\sin(n\pi)}{n\pi}, \quad -\infty < n < \infty \qquad (4.72)$$

The second term in (4.72) equals 1 at $n = 0$ and zero at all other values of n. The first term at $n = 0$ is $\frac{-\Omega_c}{\pi}$, and for all other values of n, it follows the expression given there. Therefore, the impulse response of the ideal highpass filter in the discrete-time domain takes the form

$$h_{HP}[n] = \begin{cases} 1 - \dfrac{\Omega_c}{\pi}, & n = 0 \\[2mm] -\dfrac{\sin(n\Omega_c)}{n\pi}, & otherwise \end{cases} \qquad (4.73)$$

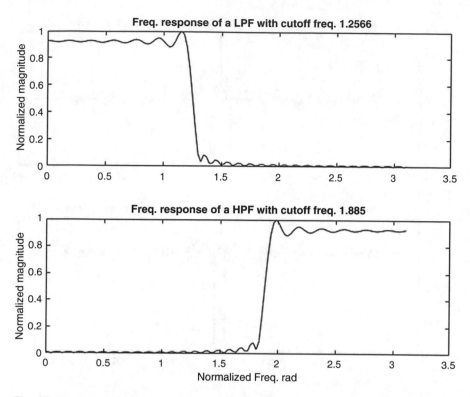

Fig. 4.7 Frequency responses. Top: lowpass filter with a cutoff frequency 0.4 π rad. Bottom: highpass filter with a cutoff frequency 0.6 π rad

The bottom plot in Fig. 4.6 shows the impulse response of an ideal highpass filter with a cutoff frequency of 0.6π, and the bottom plot in Fig. 4.7 shows the corresponding frequency response. In MATLAB, the frequency response is obtained using the *fft* function. It performs the discrete Fourier transform of the discrete-time sequence. The details of the function *fft* will be discussed in a later chapter.

Ideal Bandpass Filter   A bandpass filter passes all frequencies within a band of frequencies called the *passband* and rejects the frequencies outside the passband. The frequency characteristic of an ideal bandpass filter is described by

$$H_{BP}\left(e^{j\Omega}\right) = \begin{cases} 0, |\Omega| < \Omega_1 \\ 1, \Omega_1 \le |\Omega| \le \Omega_2 \\ 0, \Omega_2 < |\Omega| \le \pi \end{cases} \tag{4.74}$$

Figure 4.5c below illustrates the frequency response of the ideal bandpass filter described in (4.74). As before, we use the inverse DTFT to find the corresponding impulse response, which is

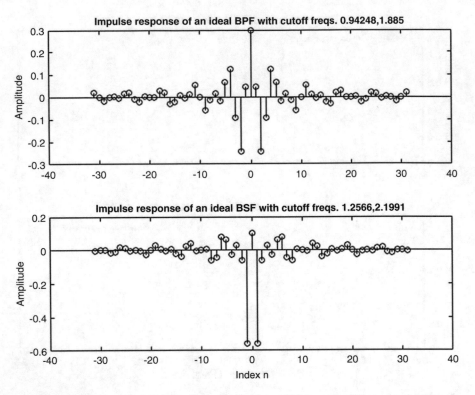

Fig. 4.8 Impulse responses. Top: bandpass filter with cutoff frequencies $0.3\,\pi$ rad and $0.6\,\pi$ rad. Bottom: bandstop filter with cutoff frequencies $0.4\,\pi$ rad and 0.7π

$$h_{BP}[n] = \frac{1}{2\pi} \int_{-\pi}^{\pi} H\left(e^{j\Omega}\right) e^{jn\Omega} d\Omega = \frac{1}{2\pi} \int_{-\Omega_2}^{-\Omega_1} e^{jn\Omega} d\Omega + \frac{1}{2\pi} \int_{\Omega_1}^{\Omega_2} e^{jn\Omega} d\Omega \quad (4.75)$$

After evaluating the integral in (4.75), we obtain the impulse response of the ideal bandpass filter as

$$h_{BP}[n] = \frac{\sin\left(n\Omega_2\right)}{n\pi} - \frac{\sin\left(n\Omega_1\right)}{n\pi}, \quad -\infty < n < \infty \quad (4.76)$$

Using (4.76), the impulse response of a bandpass filter with band edges at 0.3π and 0.6π is illustrated in the top plot in Fig. 4.8. The corresponding frequency response is depicted in the top plot in Fig. 4.9.

Ideal Bandstop Filter Contrary to the bandpass filter, a bandstop filter rejects a band of frequencies and passes the rest of the frequencies in an input signal. An ideal bandstop filter has the frequency response given by

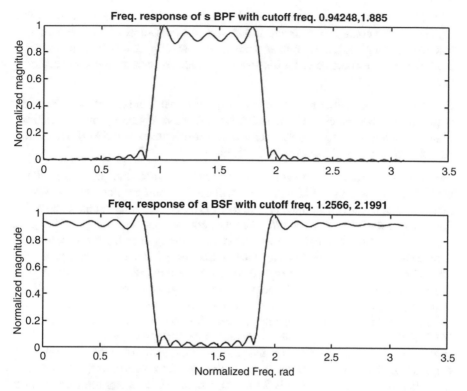

Fig. 4.9 Frequency responses. Top: bandpass filter with cutoff frequencies 0.3 π rad and 0.6 π rad. Bottom: bandstop filter with cutoff frequencies 0.4 π rad and 0.7 π

$$H_{BS}\left(e^{j\Omega}\right) = \begin{cases} 1, |\Omega| \leq \Omega_1 \\ 0, \Omega_1 < |\Omega| < \Omega_2 \\ 1, \Omega_2 \leq |\Omega| \leq \pi \end{cases} \tag{4.77}$$

Figure 4.5d illustrates the frequency response of the ideal bandstop filter defined in (4.77). The impulse response corresponding to (4.77) is obtained from the inverse DTFT and is given by

$$h_{BS}[n] = \frac{1}{2\pi}\left\{ \int_{-\pi}^{-\Omega_2} e^{jn\Omega} d\Omega + \int_{-\Omega_1}^{\Omega_1} e^{jn\Omega} d\Omega + \int_{\Omega_2}^{\pi} e^{jn\Omega} d\Omega\right\} \tag{4.78}$$

After evaluating the integral in (4.78) and with some algebraic manipulation, we obtain the impulse response of the ideal bandstop filter and is given by

$$h_{BS}[n] = \begin{cases} 1 - \dfrac{(\Omega_2 - \Omega_1)}{\pi}, n = 0 \\ \dfrac{\sin(n\Omega_1)}{n\pi} - \dfrac{\sin(n\Omega_2)}{n\pi}, otherwise \end{cases} \tag{4.79}$$

With this brief introduction to filtering, let us look at an example to further understand the concept of filtering. In Fig. 4.8, the bottom plot shows the impulse response of an ideal bandstop filter whose lower and upper cutoff frequencies are 0.4π and 0.7π, respectively. Its frequency response appears in the bottom plot in Fig. 4.9.

Example 4.11 Consider a discrete-time sequence that is a sum of two sinusoidal sequences at frequencies of 100 and 1200 Hz with a sampling frequency of 4000 Hz. Filter this sequence by a lowpass filter with a cutoff frequency of 500 Hz. Show that the 1200 Hz sinusoid is rejected by the filter.

Solution Since we have not yet learned to design a discrete-time filter, we will use the built-in function in MATLAB. As we will see in a later chapter, there are several types of filters available in the literature to choose. For the time being, let us pick a filter type called the *Butterworth* filter. Of course, one can design a lowpass, highpass, etc., filter belonging to the Butterworth type. Since the problem asks for a lowpass filter, we will design a lowpass Butterworth discrete-time or digital filter. The MATLAB function to design a lowpass Butterworth filter is *butter*. This function accepts the filter order and a cutoff frequency normalized to half the sampling frequency. It returns the coefficients of the numerator and denominator polynomials of the transfer function of the filter. So, the actual function call to MATLAB is [B,A] = butter(N,Wc), where N is the filter order, Wc is the normalized cutoff frequency, and B and A are the vectors of coefficients of the numerator and denominator polynomials, respectively, of the transfer function of the Butterworth filter. In this example, Wc equals 500/2000, which is 0.25. Once the lowpass filter has been designed, we next generate a sequence that is the sum of two sinusoids at 100 and 1200 Hz. The sinusoid at a frequency of 100 Hz is equivalent to a normalized frequency of 0.05. Remember that the normalization factor is half the sampling frequency. Since the normalized cutoff frequency is 0.25, the 100 Hz sinusoid is within the passband of the filter and so will be present in the output without any attenuation. On the other hand, the sinusoid at a frequency of 1200 Hz has a normalized frequency of 0.6, which is outside the passband of the lowpass filter. Therefore, this component must be absent in the output. So far we have designed a filter and generated the input sequence. Next we must filter the input sequence by the filter. The MATLAB function *filter* accepts the two polynomials B and A and the input sequence as arguments and returns the filtered sequence as a vector. We can then plot the input and the filtered sequences to verify if the 1200 Hz signal is absent in the filtered output. As we have seen earlier, it is easier to visualize the frequencies in the frequency domain than in the discrete-time domain. We will, therefore, use the MATLAB function *fft* to compute the frequency domain representation of the input and output sequences. The complete MATLAB code for this problem is in the M-file named *Filtering_example.m*. The magnitude of the frequency response of the lowpass digital filter is plotted as a function of the frequency in Hz and is shown in Fig. 4.10. The magnitude at the cutoff frequency of 500 Hz is 0.707. Next, the input and filtered sequences are plotted and shown in Fig. 4.11. The top plot is the input sequence and the bottom plot is the filtered sequence. As can be

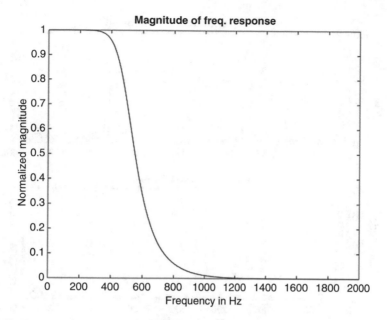

Fig. 4.10 Frequency response of the fifth-order Butterworth lowpass filter

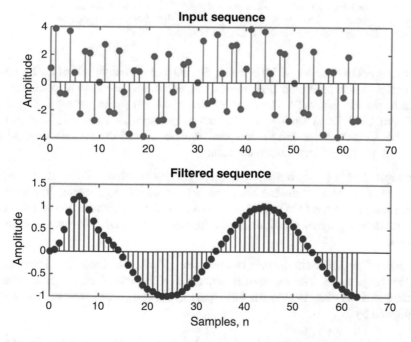

Fig. 4.11 Input and output sequences in Example 4.11. Top plot: input sequence consisting of the sum of 100 and 1200 Hz sinusoids. Bottom plot: output of the lowpass filter

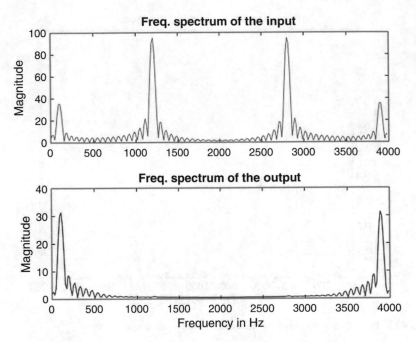

Fig. 4.12 Frequency spectra of the input and output sequences in Example 4.11. Top plot: frequency spectrum of the input sequence. Bottom plot: frequency spectrum of the filtered sequence. Both spectra are plotted as a function of frequency in Hz

seen, it is not easy to say which is which. So, we plot the magnitude of the frequency spectrum of the input and filter output sequences and are shown in the top and bottom plots in Fig. 4.12. The input spectrum has two peaks corresponding to the 100 and 1200 Hz sinusoids, whereas the output spectrum has only one peak at 100 Hz. This confirms that the lowpass filter has passed the 100 Hz sinusoid and rejected the other component in the input.

Example 4.12 Let us consider a communications problem. We want to design a bandpass filter to separate the higher of the two carrier frequencies present in an amplitude modulated (AM) signal. The two carrier frequencies are 540 and 850 kHz. The modulating signal is a sinusoid at a frequency of 5 kHz. We want to show the results.

Solution The AM radio frequencies range from 540 kHz to 1700 kHz. These are the carrier frequencies. The message is contained in the modulating signal, which is limited to 5 kHz bandwidth. An AM signal in the continuous-time domain can be expressed by

$$x_{AM}(t) = A_c(1 + km(t)) \cos(\omega_c t) \tag{4.80}$$

In (4.80), A_c is the amplitude of the carrier signal, $k = \frac{A_m}{A_c}$ is the modulation index, A_m is the amplitude of the modulating signal, and $\omega_c = 2\pi f_c$ is the carrier frequency in radian/s. The AM signal is a bandpass signal with a center frequency f_c and a bandwidth of 10 kHz. The frequencies above the carrier frequency are called the *upper sideband*, and those below the carrier frequency are called the *lower sideband*. In this example, there are two carrier frequencies present, and we want to filter out the lower carrier frequency component. We can express the sum of the two carriers as

$$x(t) = A_1(1 + k_1 m(t)) \cos(\omega_{c1} t) + A_2(1 + k_2 m(t)) \cos(\omega_{c2} t) \qquad (4.81)$$

For the sake of argument, we assume the two modulation indices to be different. First, we have to convert the analog signal to a discrete-time signal by sampling it. The highest frequency present in the signal in Eq. (4.81) is $f_{c2} + f_m$, where $f_{c2} = 850\ kHz$ and $f_m = 5\ kHz$. Let us choose a sampling frequency $F_s = 3\ (f_{c2} + f_m) = 2.565\ MHz$. The sampled AM signal is then written as

$$x[n] = A_1(1 + k_1 m(nT)) \cos(n\Omega_{c1}) + A_2(1 + k_2 m(nT)) \cos(n\Omega_{c2}), \qquad (4.82)$$

where $\Omega_{c1} = \frac{\omega_{c1}}{F_s}$ and $\Omega_{c2} = \frac{\omega_{c2}}{F_s}$. The next task is to design a bandpass digital filter with a center frequency f_{c2}, lower edge frequency $W_1 = f_{c2} - f_m$, and upper edge frequency of $W_2 = f_{c2} + f_m$, both edge frequencies normalized to half the sampling frequency. The MATLAB function *butter(N,W)* designs a Butterworth bandpass filter of order 2N and band edge frequencies as a vector W. Note that if the frequency argument is a scalar, it designs a lowpass filter as we did in the previous example. The function returns the coefficients of the numerator and denominator polynomials of the Butterworth bandpass filter of order 2N. Finally, we filter the signal in (4.82) through the designed BP filter to pass the higher carrier frequency and reject the lower frequency. The results are shown in Figs. 4.13, 4.14, and 4.15. Figure 4.13 shows the AM signal in (4.82) in the discrete-time and frequency domains. For the sake of clarity, the signal in the discrete-time is plotted as a continuous function. In the bottom plot of Fig. 4.13, the spectrum is plotted between zero and half the sampling frequency. As can be seen from the figure, there are two carrier frequencies present in the AM signal. The magnitude of frequency response of the bandpass filter is shown in Fig. 4.14. The bandpass filtered AM signal in discrete-time and frequency domains is shown in Fig. 4.15. From the bottom plot of Fig. 4.15, we see that only the higher carrier frequency component is present at the output of the bandpass filter. It is important to mention that a Butterworth bandpass filter is chosen as an example. One can choose a different type of filter such as Chebyshev or elliptic to achieve the same result with a lower order.

Example 4.13 Image filtering example As another example, we will consider filtering an image to see its effect on the image. Specifically, we will use a *finite impulse response* (FIR) lowpass filter to filter a gray-scale image. So, design a lowpass FIR filter of order 14 with a cutoff frequency of 0.1, and then filter a gray-scale image with the designed filter, and display the original and filtered images.

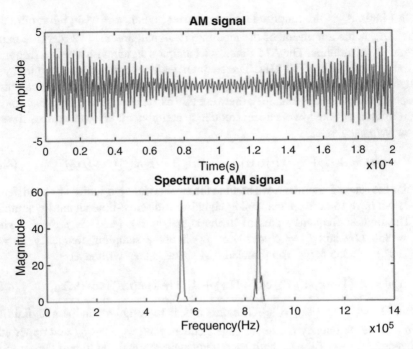

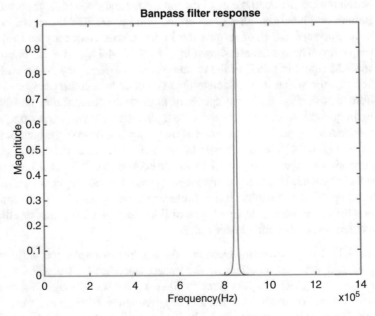

Fig. 4.13 AM signal of Example 4.12. Top plot: signal in the discrete-time domain. Bottom plot: same signal in the frequency domain

Fig. 4.14 Magnitude of the frequency response of a sixth-order Butterworth bandpass filter centered at 850 kHz with a bandwidth of 10 kHz

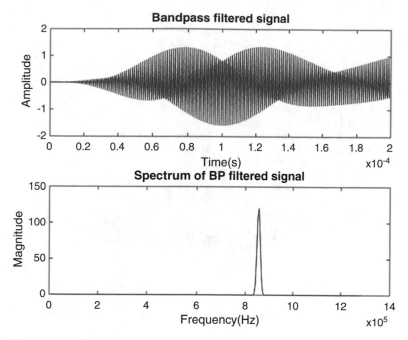

Fig. 4.15 Bandpass filtered AM signal in Eq. (4.74). Top plot: filtered signal in the discrete-time domain. Bottom plot: filtered signal in the frequency domain

Solution   An FIR filter has an impulse response, which is of finite length as opposed to an *infinite impulse response* (IIR) filter, which has an infinite duration. We can use the MATLAB function *fir1*, which accepts the filter order and cutoff frequency normalized to half the sampling frequency as arguments. It returns the impulse response of the filter as a vector. Thus, we use the MATLAB statement $h = fir1(N,w)$. As is, this function designs a lowpass FIR filter using Hamming window. We will later learn about the windowing techniques. For now, it is enough to know that we can design a lowpass FIR filter using the above statement. Once the filter has been designed, we read an image using the MATLAB function *imread*. This function can read an image with different formats, which can be found in the MATLAB. If the image read is an RGB (color) image, then we convert it to a gray-scale image using the MATLAB function *rgb2gray*. This gray-scale image is then filtered through the designed lowpass FIR filter via the MATLAB function *imfilter*, which accepts the input 2D image and the 2D filter impulse response and returns the filtered image as a 2D matrix. The details of this function can be obtained from MATLAB. Finally, the image can be displayed using the MATLAB function *imshow*. The MATLAB codes are listed in the M-file named *Image_filtering_example.m*. First, the magnitude of the frequency response of the designed lowpass FIR filter is plotted as a surface plot and shown in Fig. 4.16. The two frequencies range from zero to half the sampling frequency in radians. The input gray-scale image is shown in Fig. 4.17, and the lowpass filtered image is shown in

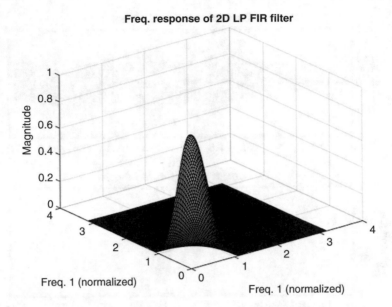

Fig. 4.16 Surface plot of the 2D lowpass FIR filter with a normalized cutoff frequency of 0.1

Fig. 4.17 Original gray-scale image

Lowpass filtered grayscale image

Fig. 4.18 Image in Fig. 4.12 filtered through the lowpass FIR filter of Fig. 4.11. The effect of lowpass filtering is to smudge the image as is evident from the figure

Fig. 4.18. The original image has finer details as seen in Fig. 4.17. The effect of lowpass filtering is to smooth or blur the image. The amount of blurring depends on the cutoff frequency. From Fig. 4.18, we notice the loss of details on the bikes, helmets, etc.

4.5.3   Calculation of DTFT Using MATLAB

So far we used analytical tools to determine the DTFT of discrete-time signals and systems. In the previous chapter, we learned to characterize LTI discrete-time systems in terms of their transfer functions in the Z-domain. In general, the transfer function of an LTI discrete-time system is a rational polynomial in the complex variable z. There are two types of LTI discrete-time systems or filters. If the denominator polynomial is not just a constant, then the corresponding filter is called an IIR (infinite impulse response) filter. An IIR filter has both numerator and denominator polynomials in z. On the other hand, if the denominator polynomial of the transfer function is identically equal to 1, then the filter is termed an FIR (finite impulse response) filter. We have also established the relationship between the Z-transform and the DTFT. More specifically, the DTFT of a discrete-time signal x[n] can be obtained from its Z-transform through

$$X\left(e^{j\Omega}\right) = X(z)\big|_{z=e^{j\Omega}} \tag{4.83}$$

The same rule applies to the LTI discrete-time systems. Once the Z-transform is given, the corresponding DTFT can be determined by simply using (4.83). If we want to compute the DTFT at a set of values of Ω, then we can use MATLAB to compute the DTFT at those points. More specifically, MATLAB has the function *freqz* to compute the DTFT. The arguments of the *freqz* function are the vectors of coefficients of the numerator and denominator polynomials of the Z-transform of a discrete-time signal or system. There are other parameters that are optional. For more information about the function *freqz*, one may obtain them by typing "help freqz" in the MATLAB workspace. Let us consider the following example.

Example 4.14 Calculate and plot the DTFT of the discrete-time sequence described by

$$x[n] = 0.75^n \cos\left(\frac{n\pi}{8}\right) u[n] \tag{4.84}$$

Solution Using the Z-transform property, we obtain

$$X(z) = \frac{1 - 0.75 z^{-1} \cos\left(\frac{\pi}{8}\right)}{1 - 1.5 z^{-1} \cos\left(\frac{\pi}{8}\right) + \frac{9}{16} z^{-2}} = \frac{B(z)}{A(z)} \tag{4.85}$$

From (4.85), we find the vectors to be $B = \left[1 - 0.75 \cos\left(\frac{\pi}{8}\right) \; 0\right]$ and $A = \left[1 - 1.5 \cos\left(\frac{\pi}{8}\right) \; \frac{9}{16}\right]$. Note that both vectors must be of the same size. Next, we call the function [X,W] = freqz(B,A,256). The function returns the DTFT of x [n] in X over 256-point vector W, uniformly distributed in the interval $[0, \pi]$. Note that π corresponds to half the sampling frequency. The normalized magnitude and phase of the DTFT are shown in the top and bottom plots in Fig. 4.19. The MATLAB code is listed in the M-file *Example4_14.m*.

Example 4.15 Calculation of the DTFT of a Finite-Length Sequence Using MATLAB In the previous example, we considered calculating the DTFT of an infinite-length sequence. Using MATLAB, we can also compute the DTFT of a finite-length sequence by calling the same function *freqz*. In this case, the Z-transform of a finite-length sequence has only the numerator polynomial in z^{-1} with the denominator polynomial being identically equal to 1. Therefore, the function call takes the form

$$[H, W] = freqz(h, N) \tag{4.86}$$

where h is the finite-length sequence, H is its DTFT at the N points W, uniformly distributed over the interval $[0, \pi]$. We will, therefore, compute the DTFTs of the impulse responses of the ideal filters that we discussed above using (4.86) in MATLAB and plot the normalized magnitudes and phase responses. The DTFTs of the ideal lowpass, highpass, bandpass, and bandstop filters are shown in Figs. 4.20, 4.21, 4.22, and 4.23, respectively. As can be seen from the plots, the phase functions are wrapped around 2π. However, the actual phase is a continuous

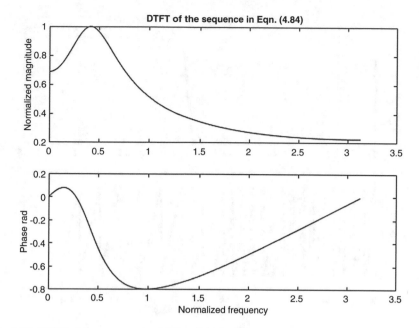

Fig. 4.19 DTFT of the sequence in (4.84) using MATLAB

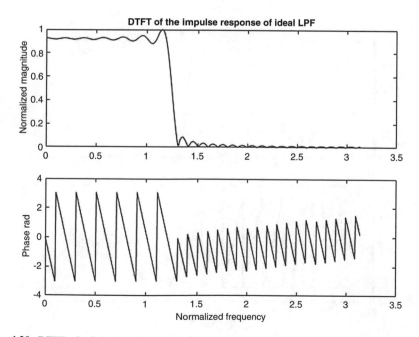

Fig. 4.20 DTFT of a finite-length lowpass filter

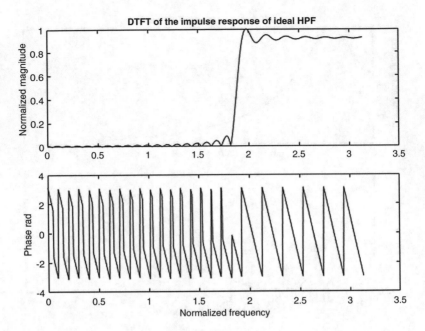

Fig. 4.21 DTFT of a finite-length highpass filter

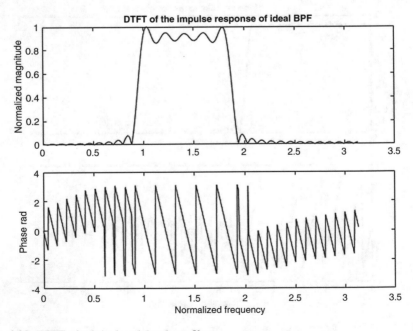

Fig. 4.22 DTFT of a finite-length bandpass filter

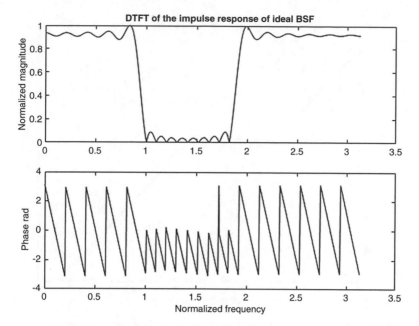

Fig. 4.23 DTFT of a finite-length bandstop filter

function of the frequency. We can unwrap the phase in MATLAB using the function *unwrap*. In Figure 4.24 is shown the unwrapped phase of the bandstop filter as an example. The M-file named *Example4_15.m* is used to solve this problem.

Example 4.16 MATLAB Example to Determine Cascade and Parallel Forms of IIR Filter Given the transfer function of an IIR filter in the frequency domain, determine the cascade and parallel forms using MATLAB. Compute the impulse and step responses of the three forms of the IIR filter. Also, generate a white Gaussian noise, and filter it through the IIR filter, and plot the DTFT of the filtered noise.

Solution The given frequency domain transfer function of the IIR filter is described by

$$H(e^{j\Omega})$$
$$= \frac{b_0 + b_1 e^{-j\Omega} + b_2 e^{-j2\Omega} + b_3 e^{-j3\Omega} + b_4 e^{-j4\Omega} + b_5 e^{-j5\Omega} + b_6 e^{-j6\Omega} + b_7 e^{-j7\Omega} + b_8 e^{-j8\Omega}}{a_0 + a_1 e^{-j\Omega} + a_2 e^{-j2\Omega} + a_3 e^{-j3\Omega} + a_4 e^{-j4\Omega} + a_5 e^{-j5\Omega} + a_6 e^{-j6\Omega} + a_7 e^{-j7\Omega} + a_8 e^{-j8\Omega}}$$

which is an eighth-order IIR filter, where the numerator and denominator polynomials are given by

$$B = [0.0789, -0.0457, -0.0418, -0.0172, 0.1319, -0.0172, -0.0418, -0.0457, 0.0789]$$

$$A = [1, -1.0003, 2.0927, -1.5789, 2.1993, -1.1056, 1.0185, -0.2907, 0.2016]$$

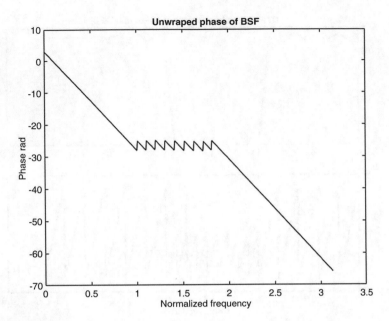

Fig. 4.24 Unwrapped phase response of the bandpass filter whose frequency response is shown in Fig. 4.23

To find the cascade form, we use the function call *[sos,g]* = *tf2sos(B,A)*, which returns the coefficients of the numerator and denominator polynomials of the first- and or second-order sections. The overall gain factor is $g = 0.0789$. For this example, since the filter order is 8, there are four second-order sections and are found to be

$$H_{c1}\left(e^{j\Omega}\right) = \frac{1 + 1.6647e^{-j\Omega} + e^{-j2\Omega}}{1 + 0.2247e^{-j\Omega} + 0.4951e^{-j2\Omega}}$$

$$H_{c2}\left(e^{j\Omega}\right) = \frac{1 - 1.8278e^{-j\Omega} + e^{-j2\Omega}}{1 - 0.7172e^{-j\Omega} + 0.5357e^{-j2\Omega}}$$

$$H_{c3}\left(e^{j\Omega}\right) = \frac{1 + 1.0562e^{-j\Omega} + e^{-j2\Omega}}{1 + 0.66e^{-j\Omega} + 0.8613e^{-j2\Omega}}$$

$$H_{c4}\left(e^{j\Omega}\right) = \frac{1 - 1.4722e^{-j\Omega} + e^{-j2\Omega}}{1 - 1.1677e^{-j\Omega} + 0.8825e^{-j2\Omega}}$$

To obtain the parallel form, we first have to convert the transfer function to the residues, poles, and quotients. This is achieved by calling *[R,P,K]* = *residuez(B,A)*, where B and A are the coefficients of the numerator and denominator polynomials of the given transfer function. Once the residues, poles, and the quotients are obtained, the transfer functions of the individual sections of the parallel form are determined using the same *residuez* function but with input arguments R and P and with K = 0. The details of the MATLAB commands can be found in the M-file *Example 4_16.m*. The four sections of the parallel form obtained are given by the following.

$$H_{p1}\left(e^{j\Omega}\right) = \frac{-0.0874 + 0.1263e^{-j\Omega}}{1 - 1.1677e^{-j\Omega} + 0.8825e^{-j2\Omega}}$$

$$H_{p2}\left(e^{j\Omega}\right) = \frac{-0.1069 - 0.1396e^{-j\Omega}}{1 + 0.66e^{-j\Omega} + 0.8613e^{-j2\Omega}}$$

$$H_{p3}\left(e^{j\Omega}\right) = \frac{-0.0991 - 0.4192e^{-j\Omega}}{1 - 0.7172e^{-j\Omega} + 0.5357e^{-j2\Omega}}$$

$$H_{p4}\left(e^{j\Omega}\right) = \frac{-0.0191 + 0.564e^{-j\Omega}}{1 + 0.2247e^{-j\Omega} + 0.4951e^{-j2\Omega}}$$

with K = 0.3914. Once the cascade and parallel forms have been determined, we can then calculate the impulse and step responses using the functions *impz* and *stepz*. Again, the details are found in the abovementioned M-file. Finally, we generate the white Gaussian noise using the function call *x = sigma x randn(256,1)*, where sigma is the standard deviation. This will generate a Gaussian random vector of dimension 256x1. In order to determine the filtered noise, we first calculate the DTFT of the noise and the filter transfer function and then multiply the two point by point. To determine the DTFT of a vector, we can use the function *freqz*. This product gives the DTFT of the filtered noise. Having determined all the required items, we then plot the various functions as follows.

The magnitude of the DTFT of the given IIR filter is shown in Fig. 4.25, which also includes the DTFTs of the cascade and parallel forms. All three are identical. From the figure, we notice that the given IIR filter is a bandpass filter. The phase

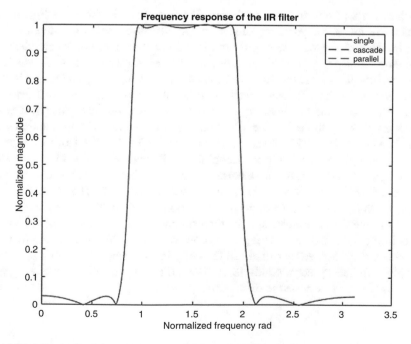

Fig. 4.25 Magnitude of the DTFT of the IIR filter of Example 4.16

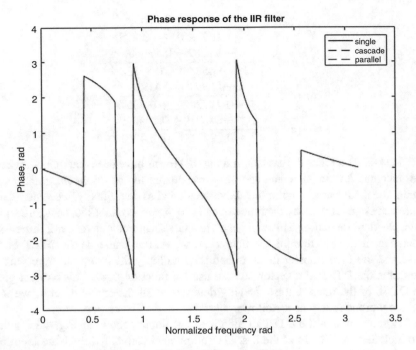

Fig. 4.26 Phase of the DTFT of the IIR filter of Example 4.16

responses of all the three forms of the IIR filter are shown in Fig. 4.26. Figure 4.27 depicts the unwrapped phase responses. The impulse responses of the three forms of the IIR filter are shown as stem plots in Fig. 4.28. All three responses are identical. Similarly the step responses of the three IIR filters are shown in Fig. 4.29, which are again, identical. Because the filter is a bandpass filter, it does not pass DC and is clear from the plots in Fig. 4.29. Note that the unit step is a constant after $n = 0$. The steady state response of the bandpass to a constant input will be zero. Finally, a white Gaussian noise is applied to the IIR filters, and the corresponding responses are calculated using the DTFT. Since the given filter is LTI, the DTFT of the output of the filter to a given input is the product of the DTFTs of the filter and its input. First we compute the DTFT of the input noise sequence and then multiply it point by point with the DTFT of the filter to obtain the DTFT of the output. The white Gaussian sequence in the discrete-time domain is shown in Fig. 4.30. The magnitudes of the DTFTs of the Gaussian noise and the filter output are shown in the top and bottom plots in Fig. 4.31. Because the input noise is white, its DTFT appears flat over the frequency range between zero and half the sampling frequency. In the figure, half the sampling frequency corresponds to π. The DTFT of the filtered noise appears bandpass in shape because the filter is bandpass.

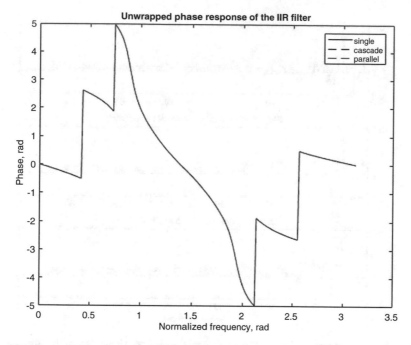

Fig. 4.27 Unwrapped phase of the DTFT in Fig. 4.26

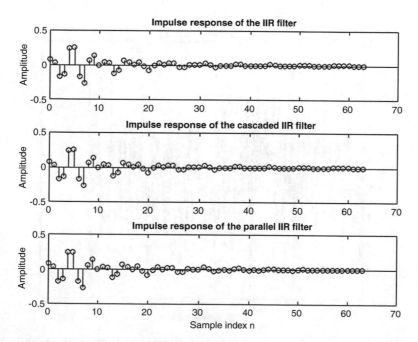

Fig. 4.28 Impulse response of the IIR filter of Example 4.16. Top: single section. Middle: cascade form. Bottom: parallel form

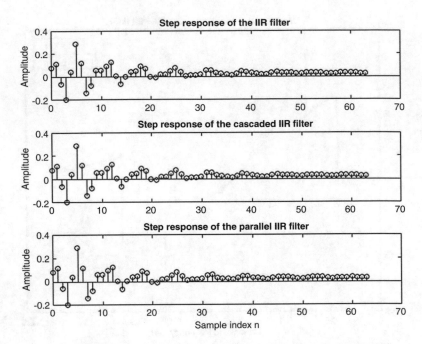

Fig. 4.29 Step response of the IIR filter of Example 4.16. Top: single section. Middle: cascade form. Bottom: parallel form

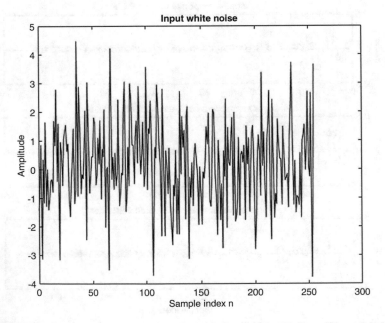

Fig. 4.30 White Gaussian noise sequence applied to the input of the IIR filter of Example 4.16. The standard deviation is 1.5

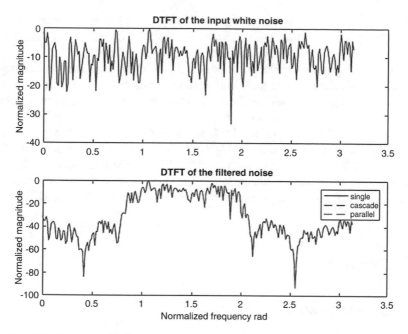

Fig. 4.31 Magnitudes of the DTFT of the input noise and filtered output noise. Top: magnitude of the DTFT of the input noise. Bottom: magnitude of the DTFT of the filtered output noise

4.6 Summary

We have described the mapping of discrete-time signals and systems in the frequency domain using what is known as the discrete-time Fourier transform. The DTFT of discrete-time signals is periodic with period 2π and is a continuous function of the normalized frequency Ω in radians. Because the DTFT is periodic, its inverse is the discrete-time sequence. In other words, the inverse DTFT (IDTFT) of the DTFT of a sequence is the coefficients of the Fourier series expansion of the given DTFT. We described how to determine the DTFT of several commonly encountered sequences. We also learned the properties of DTFT, which are useful in the determination of the DTFT of more complex sequences. Several examples are worked out to illustrate the definition and the use of the properties of the DTFT. As the discrete-time system is concerned, the DTFT of an LTI discrete-time system gives rise to its frequency domain transfer function. We further observed that the DTFT of a sequence or an LTI discrete-time system results in a rational polynomial. Examples are given to illustrate the process of obtaining the IDTFT. In particular, we used the partial fraction expansion to determine the response of an LTI discrete-time system to a specific input sequence. We further described the concept of filtering through the DTFT. In particular, we defined the characteristics of ideal lowpass, highpass, bandpass, and bandstop filters in terms of the DTFT. To understand the significance of filtering, we showed a few examples of lowpass and bandpass filtering using MATLAB. Further, we used an image-filtering example to demonstrate the effect of lowpass filtering of a gray-scale image using MATLAB functions.

The next chapter deals with the discrete version of the frequency domain representation of discrete-time signals and systems, known as the *discrete Fourier transform* (DFT). As we will see in detail, DFT deals with the frequency domain representation of finite-length sequences. It is also a computational tool to calculate the DFT of finite-length sequences. DFT can be used to perform filtering operations. There is a computationally efficient algorithm known as the fast Fourier transform (FFT), which is widely used in practice. We will describe two aspects of the FFT algorithm and show how much saving in the number of arithmetic operations is achieved in comparison with the brute force method of calculating the DFT.

4.7   Problems

1. Find the DTFT of the sequence $x[n] = \begin{cases} 0.5^{|n|}, & |n| \leq 7 \\ 0, & \text{otherwise} \end{cases}$.

2. Find the DTFT of the sequence $x[n] = \begin{cases} \alpha^{|n|}, & |n| \leq M \\ 0, & \text{otherwise} \end{cases}$.

3. Determine the DTFT of the following sequences:

   (a) $x[n] = 2n\alpha^n u[n]$, with $|\alpha| < 1$.
   (b) $g[n] = \alpha^n u[n + 1]$, $|\alpha| < 1$
   (c) $h[n] = \beta^n u[n - 1]$, $|\beta| < 1$

4. Find the DTFT of the sequence $x[n] = B\beta^n \cos(n\Omega_0 + \theta)u[n]$, where B, β, Ω_0, and θ are real constants and $|\beta| < 1$.

5. Find the DTFT of the sequence described by $x[n] = u[n + 1] - u[n - 4]$.

6. Determine the DTFT of the finite-length sequence $x[n] = \cos\left(\frac{n\pi}{2N}\right)$, $-N \leq n \leq N$

7. If $x[n] = \begin{cases} 1 - \dfrac{|n|}{N}, & -N \leq n \leq N \\ 0, & \text{otherwise} \end{cases}$, what is its DTFT?

8. Find the IDTFT of $X(e^{j\Omega}) = \frac{-\alpha e^{-j\Omega}}{(1 - \alpha e^{-j\Omega})^2}$, $|\alpha| < 1$.

9. Determine the sequence whose DTFT is given by $H(e^{j\Omega}) = \cos(4\Omega)$.

10. If $x[n] \leftrightarrow X(e^{j\Omega})$ with $x[n]$ a real sequence, find in terms of the sequence $x[n]$, the sequence $y[n]$ whose DTFT is given by $Y(e^{j\Omega}) = X(e^{j4\Omega})$.

11. If $X(e^{j\Omega})$ is the DTFT of a real sequence $x[n]$, then what is the IDTFT of the function $G(e^{j\Omega}) = \frac{1}{2}\left[X\left(e^{j\Omega/3}\right) + X\left(-e^{j\Omega/3}\right)\right]$.

12. Given an N-point sequence $\{x[n]\}$, $0 \leq n \leq N - 1$, obtain the M-point sequence $\{y[n]\}$ with $M > N$, by appending the sequence $\{x[n]\}$ with M-N zeros. Find the DTFT $Y(e^{j\Omega})$ of $y[n]$ in terms of the DTFT $X(e^{j\Omega})$ of the sequence $x[n]$.

13. Consider the sequences (a) $x[n] = 2n, \ -N \le n \le N$, (b) $g[n] = \begin{cases} 0, for\ n\ even \\ \dfrac{1}{n\pi}, for\ n\ odd \end{cases}$.
   Determine which of the above sequences has real-valued DTFT and which has the imaginary-valued DTFT without computing the DTFT.

14. The transfer function of an LTI discrete-time system is given by $H(e^{j\Omega}) = \frac{1}{1-\alpha e^{-j\Omega}}$, $|\alpha| \prec 1$. Calculate its response to an input $x[n] = \beta^n u[n]$, $|\beta| \prec 1$ using DTFT. Note that α and β are different.

15. The transfer function of an LTI discrete-time system is given by $H(e^{j\Omega}) = \frac{1}{1-0.75e^{-j\Omega}}$. Calculate its response to an input $x[n] = 0.5^n u[n]$ using DTFT.

References

1. Mitra SK (2011) Digital signal processing: a computer-based approach, 4th edn. McGraw Hill
2. Oppenheim AV, Shafer RW (1989) Discrete-time signal processing. Prentice-Hall, Englewood Cliffs
3. Oppenheim AV, Willsky AS, Young IT (1983) Signals and systems. Prentice-Hall, Englewood Cliffs
4. Rabiner LR, Gold B (1975) Theory and application of digital signal processing. Prentice-Hall, Englewood Cliffs, NJ

Chapter 5
Discrete Fourier Transform

5.1 Introduction

Discrete Fourier transform (DFT) is a frequency domain representation of finite-length discrete-time signals. It is also used to represent FIR discrete-time systems in the frequency domain. As the name implies, DFT is a discrete set of frequency samples uniformly distributed around the unit circle in the complex frequency plane that characterizes a discrete-time sequence of finite duration. DFT is also intrinsically related to the DTFT, as we will see in this chapter. Because DFT is a finite set of frequency samples, it is a computational tool to perform filtering and related operations. There is an efficient algorithm known as the fast Fourier transform (FFT) to perform filtering of long sequences, power spectrum estimation, and related tasks. We will learn about the FFT in this chapter as well.

5.2 Definition of DFT

The DFT of an N-point or length-N sequence $x[n]$ is defined as

$$X(k) = \sum_{n=0}^{N-1} x[n] e^{-j\left(\frac{2\pi}{N}\right)nk}, 0 \leq k \leq N - 1 \tag{5.1}$$

It is customary to use the notation $W_N \equiv e^{-j\left(\frac{2\pi}{N}\right)}$. We can, therefore, rewrite (5.1) as

Electronic supplementary material: The online version of this article (https://doi.org/10.1007/978-3-319-76029-2_5) contains supplementary material, which is available to authorized users.

© Springer International Publishing AG, part of Springer Nature 2019   151
K. S. Thyagarajan, *Introduction to Digital Signal Processing Using MATLAB with Application to Digital Communications*,
https://doi.org/10.1007/978-3-319-76029-2_5

$$X(k) = \sum_{n=0}^{N-1} x[n] W_N^{nk}, 0 \le k \le N - 1 \qquad (5.2)$$

From the definition, it is clear that the DFT of an N-point sequence is indeed discrete and has the same length as the sequence. Remember that 2π corresponds to the sampling frequency. So, the DFT is a set of frequency samples spaced uniformly over one sampling frequency. We also observe that the DFT is periodic with period N because of the fact that

$$X(k + N) = \sum_{n=0}^{N-1} x[n] W_N^{n(k+N)} = \sum_{n=0}^{N-1} x[n] W_N^{nk} e^{-j2n\pi} = X(k) \qquad (5.3)$$

5.3   Relationship Between DTFT and DFT

Recall the definition of DTFT of an N-point sequence $x[n]$, which is

$$X\left(e^{j\Omega}\right) = \sum_{n=0}^{N-1} x[n] e^{-jn\Omega} \qquad (5.4)$$

If we sample the DTFT at N points equally spaced around the unit circle $e^{j\Omega}$, we get

$$X(k) = X\left(e^{j\Omega}\right)\Big|_{\Omega=\frac{2\pi}{N}k} = \sum_{n=0}^{N-1} x[n] e^{-j\frac{2\pi}{N}nk}, 0 \le k \le N - 1, \qquad (5.5)$$

which is the DFT of the N-point sequence. Thus, the DFT of an N-point sequence is the DTFT of that sequence sampled at N points equally spaced around the unit circle in the frequency domain. In other words, the DFT is the sampled version of the DTFT with the samples spaced uniformly around the unit circle in the frequency domain.

Example 5.1 Find the DFT of the sequence $x[n] = 0.5^n$, $0 \le n \le N - 1$ and compare it with the DTFT of the same sequence.

Solution Using the definition of the DFT, we can write

$$X[k] = \sum_{n=0}^{N-1} 0.5^n W_N^{nk}, 0 \le k \le N - 1 \qquad (5.6)$$

By collecting the two factors inside the summation in Eq. (5.6), we can express the DFT as

$$X(k) = \sum_{n=0}^{N-1} \left(0.5W_N^k\right)^n, 0 \le k \le N-1 \tag{5.7}$$

The right-hand side of (5.7) is a geometric series and, so, can be written in closed form as

$$X(k) = \frac{1 - \left(0.5W_N^k\right)^N}{1 - 0.5W_N^k}, 0 \le k \le N-1 \tag{5.8}$$

We can express the DFT in (5.8) in magnitude-phase form as given below:

$$|X(k)| = \frac{\left|1 - \left(0.5W_N^k\right)^N\right|}{\left|1 - 0.5W_N^k\right|} = \frac{1 - 0.5^N}{\sqrt{1.25 - \cos\left(\frac{2\pi k}{N}\right)}}, 0 \le k \le N-1 \tag{5.9a}$$

$$\theta(k) = -\tan^{-1} \frac{0.5 \sin\left(\frac{2\pi k}{N}\right)}{1 - 0.5 \cos\left(\frac{2\pi k}{N}\right)}, 0 \le k \le N-1 \tag{5.9b}$$

The magnitude and phase of the DFT of the sequence in Example 5.1 are shown in Figs. 5.1a, b, respectively. The plots also show the DTFT of the same sequence. It is evident that the DFT is the sampled version of the DTFT. The sequence length chosen is 32 samples.

5.4   Inverse DFT

The process of recovering a sequence from its DFT is called the inverse discrete Fourier transform (IDFT). If the IDFT does not exist, then there is no use for the DFT. Fortunately, the IDFT does exist and is defined as

$$x[n] = \frac{1}{N} \sum_{k=0}^{N-1} X(k) W_N^{-nk}, 0 \le n \le N-1 \tag{5.10}$$

Proof To prove that the IDFT is indeed $x[n]$, let us substitute for $X(k)$ in (5.10) to get

$$x[n] = \frac{1}{N} \sum_{k=0}^{N-1} \left\{ \sum_{m=0}^{N-1} x[m] W_N^{mk} \right\} W_N^{-nk}, 0 \le n \le N-1 \tag{5.11}$$

By interchanging the order of summation in (5.11), we have

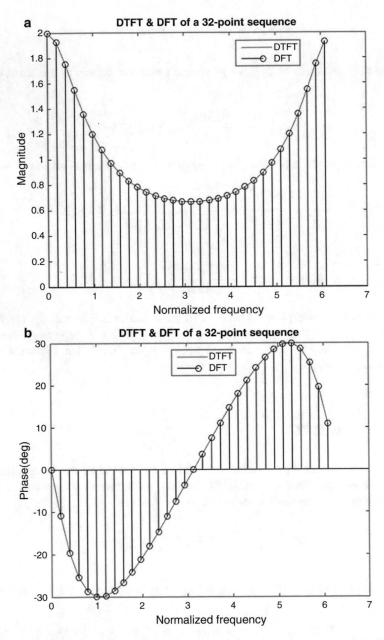

Fig. 5.1 DFT and DTFT of an N-point sequence in Example 5.1: (**a**) magnitude of the DFT and DTFT and (**b**) phase of the DFT and DTFT

$$x[n] = \sum_{m=0}^{N-1} x[m] \left\{ \frac{1}{N} \sum_{k=0}^{N-1} W_N^{k(m-n)} \right\}, 0 \leq n \leq N - 1 \tag{5.12}$$

But the inner summation results in

$$\frac{1}{N} \sum_{k=0}^{N-1} W_N^{k(m-n)} = \left\{ \begin{array}{c} 1, if \ m = n + rN, r \in Z \\ 0, otherwise \end{array} \right. \tag{5.13}$$

Therefore, the right-hand side of (5.12) equals $x[n]$, hence, the result.

5.5   Effect of Sampling the DTFT on the Reconstructed Sequence

We saw in Chap. 2 that the effect of sampling a continuous-time signal is to replicate the frequency spectrum of the continuous-time signal in the frequency domain. So, if the continuous-time signal is strictly band-limited, then the replicated spectra do not overlap in the frequency domain. Therefore, there is no aliasing distortion, and the continuous-time signal can be recovered exactly from the corresponding discrete-time sequence. The duality of this result is as follows. If we sample the frequency spectrum or DTFT of a discrete-time sequence at a finite set of points equally spaced around the unit circle in the complex plane, then the reconstructed sequence using IDFT is a replicated version of the original sequence. If the length of the original sequence is equal or smaller than the number of frequency samples, then the reconstructed sequence from the IDFT will have no distortion. Let us prove this result as follows. Consider a sequence $x[n]$, whose DTFT is given by

$$X(e^{j\Omega}) = \text{DTFT}\{x[n]\} = \sum_{n=-\infty}^{\infty} x[n]e^{-jn\Omega} \tag{5.14}$$

If we sample $X(e^{j\Omega})$ at M points spaced uniformly around the unit circle, we will have an M-point DFT sequence. Denote this sequence by $Y(k)$. Therefore,

$$Y(k) = X(e^{j\Omega})\big|_{\Omega=\frac{2\pi k}{M}} = \sum_{n=-\infty}^{\infty} x[n]e^{-jn\frac{2\pi k}{M}}, 0 \leq k \leq M - 1 \tag{5.15}$$

The sequence $y[n]$ can be found by performing the IDFT of $Y(k)$, which is

$$y[n] = \text{IDFT}\{Y(k)\} = \frac{1}{M} \sum_{k=0}^{M-1} Y(k) W_M^{-nk} \tag{5.16}$$

Using Eq. (5.15) in (5.16), we can rewrite (5.16) as

$$y[n] = \frac{1}{M} \sum_{k=0}^{M-1} \left\{ \sum_{p=-\infty}^{\infty} x[p] W_M^{pk} \right\} W_M^{-nk} \qquad (5.17)$$

By interchanging the order of summation in (5.17), we obtain

$$y[n] = \sum_{p=-\infty}^{\infty} x[p] \left\{ \frac{1}{M} \sum_{k=0}^{M-1} W_M^{-(n-p)k} \right\} \qquad (5.18)$$

However, since

$$\frac{1}{M} \sum_{k=0}^{M-1} W_M^{-k(n-p)} = \begin{cases} 1, & \text{if } p = n + mM \\ 0, & \text{otherwise} \end{cases} \qquad (5.19)$$

we have

$$y[n] = \sum_{m=-\infty}^{\infty} x[n + mM] \qquad (5.20)$$

Thus we find that the reconstructed sequence $y[n]$ is the sum of the replicas of the sequence $x[n]$ shifted by integer multiples of the number of frequency samples. If the sequence $x[n]$ is time-limited to $N \leq M$, then $y[n]$ is exactly equal to $x[n]$ between 0 and M-1. Otherwise, the replicas overlap and cause distortion in the sequence $y[n]$. Let us illustrate the idea of frequency sampling by the following example.

Example 5.2 Find the M-point DFT of an N-point sequence of your choice. Then calculate the M-point IDFT and compare the two sequences to see if there is any distortion.

Solution Let the sequence be described by

$$x[n] = \sin(0.2\pi n) + 2\cos(0.3\pi n), 0 \leq n \leq N - 1 \qquad (5.21)$$

We can use the MATLAB to calculate the DTFT of the sequence in (5.21). The function to use is *freqz*, which accepts the coefficients of the polynomials of the numerator and denominator of a transfer function. However, in this example since the sequence is of finite length, its DTFT is just a polynomial with the coefficients corresponding to the given sequence. To calculate the DTFT, we use the statement $[H, W] = freqz(x, A, N, 'whole')$. The DTFT is returned in H at N points uniformly spaced around the unit circle. Here A is a vector of the same size as the sequence and is all zeros except the first element, which is unity. W is the set of N normalized frequency points between zero and 2π. Next we sample the DTFT at M equally spaced points around the unit circle. This is done by retaining every N/M samples of the DTFT that we just calculated. The M-point sequence is recovered by performing the inverse DFT of the M samples of the DTFT. The IDFT of the M-point DFT X can be computed using the MATLAB function *ifft*. The actual function call is $y = ifft(X)$.

The two sequences can then be compared to see if there is distortion in the sequence y. The thirty-two-point sequence in (5.21) is plotted and shown in the top plot in Fig. 5.2. The sequence obtained from the IDFT of the DFT of the thirty-two-point sequence with M = 32 is shown in the bottom plot of Fig. 5.2. Since N = M, we find no distortion in the recovered sequence. Next, the sequence obtained from eight-point IDFT is shown in the bottom plot of Fig. 5.3. As can be seen from the figure, since M < N, there is a significant distortion in the sequence y[n]. For N = 32 and M = 8, the sequence in the discrete-time domain is obtained by adding the replicas of the original sequence shifted left by 0, M, 2 M, and 3 M as given by

$$y[n] = x[n] + x[n + M] + x[n + 2M] + x[n + 3M], 0 \leq n \leq M - 1 \qquad (5.22)$$

The sequence obtained from (5.22) is plotted and shown in the bottom figure of Fig. 5.4. As a comparison, the top plot in Fig. 5.4 is the same as that shown in the bottom plot in Fig. 5.3. This shows that when $M < N$, distortion occurs in the reconstructed sequence because of overlapping of the replicas. This is further demonstrated by plotting the magnitude of the spectrum of the sequence obtained

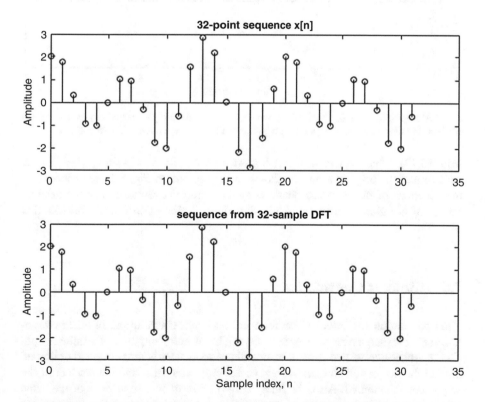

Fig. 5.2 Thirty-two-point IDFT from the thirty-two-point sequence: top plot, actual sequence of length 32; bottom plot, sequence obtained from thirty-two-point DFT of the sequence in Example 5.2

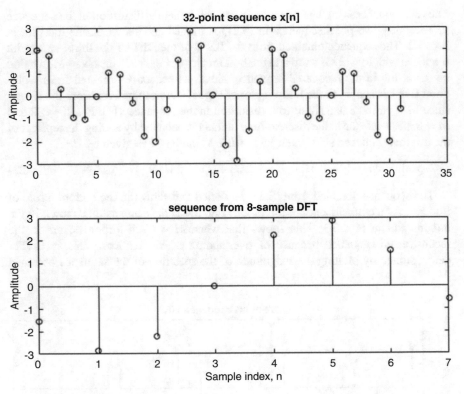

Fig. 5.3 Eight-point IDFT from the thirty-two-point sequence: top plot, actual sequence of length 32; bottom plot, sequence obtained from eight-point DFT of the sequence in Example 5.2

from (5.22), which is shown in the bottom plot in Fig. 5.5. In comparison to the spectrum of the original sequence shown in the top plot in Fig. 5.5, we observe that the spectrum of the sequence obtained by summing the replication of the original sequence is quite different, implying that there is distortion due to the fact that $M < N$.

5.6   Circular Convolution

The convolution we described earlier is called linear convolution. In linear convolution the two sequences to be convolved can be of finite length or of infinite length. To linearly convolve two sequences, one of the sequences is time-reversed or flipped around the origin and slid left or right one sample at a time, and the product of the sequences is summed. Also, the linear convolution of two sequences of the same length-N results in a sequence of length $2N - 1$. Circular convolution is meant for finite-length sequences. In circular convolution, the two sequences to be convolved are of the same length. If not, the sequence of smaller length is zero-padded to be of

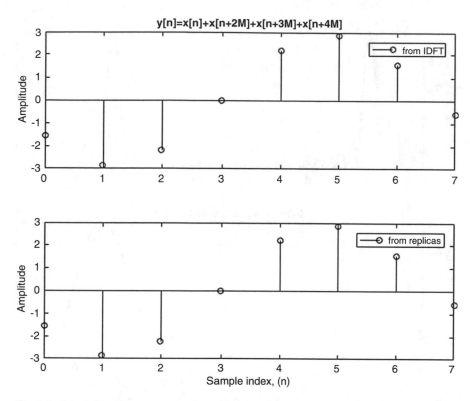

Fig. 5.4 Calculation of the sequence using equation 5.20: top plot, sequence obtained by IDFT of the DFT samples; bottom plot, sequence obtained from Eq. (5.20)

the same length. The circular convolution of the two sequences of the same length-N results in a sequence, which is also of length N. In circular convolution, one sequence is shifted circularly over the other sequence one sample at a time, the two are multiplied, and the product summed to yield the convolved sequence. Circular shift implies that the circular convolution is periodic with period N, which is the length of the sequences. Having described circular convolution qualitatively, let us give the formal definition of circular convolution.

Definition of Circular Convolution The circular convolution of two N-point sequences is another N-point sequence and is defined as

$$y[n] = \sum_{k=0}^{N-1} x[k]h[<n-k>_N], 0 \le n \le N-1 \tag{5.23}$$

In Eq. (5.23), the notation $<n-k>_N$ refers to modulo-N operation. For instance, five modulo four is $<5>_4 = 5 - \lfloor \frac{5}{4} \rfloor 4 = 5 - 4 = 1$. To distinguish from linear

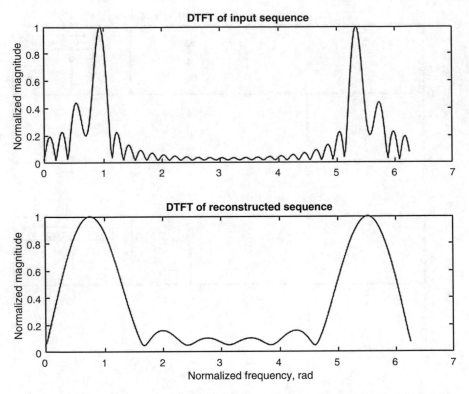

Fig. 5.5 Spectra of the original sequence and that of the sequence obtained from (5.22): top plot, spectrum of the original sequence; bottom plot, spectrum of the sequence obtained from (5.22)

convolution, we will use the symbol $\bigoplus$ to denote circular convolution. We can, therefore, write (5.23) in terms of the symbol for circular convolution as

$$y[n] = x[n] \bigoplus h[n] \tag{5.24}$$

Matrix Representation of Circular Convolution The circular convolution in Eq. (5.23) can also be represented in matrix form as given below:

$$y[n] = \begin{bmatrix} h[0] \ h[N-1] \ h[N-2] \ldots\ldots.h[1] \\ h[1] \ h[0] \ h[N-1] \ldots\ldots\ldots.h[2] \\ h[2] \ h[1] \ h[0] \ldots\ldots\ldots\ldots.h[3] \\ \ldots\ldots\ldots\ldots\ldots\ldots\ldots\ldots\ldots \\ h[N-1] \ h[N-2] \ldots\ldots\ldots\ldots.h[0] \end{bmatrix} \begin{bmatrix} x[0] \\ x[1] \\ \vdots \\ x[N-1] \end{bmatrix} \tag{5.25}$$

We will explain the circular convolution of two finite-length sequences by an example.

Example 5.3 Calculate the circular convolution of the sequences $x[n] = [1\ 2\ 3\ 4]$ and $h[n] = [4\ 3\ 2\ 1]$.

Solution From Eq. (5.23), we can write the circular convolution as

$$y[0] = x[0]h[0] + x[1]h[3] + x[2]h[2] + x[3]h[1] = 1*4 + 2*1 + 3*2 + 4*3 = 24$$

$$y[1] = x[0]h[1] + x[1]h[0] + x[2]h[3] + x[3]h[2] = 22$$

$$y[2] = x[0]h[2] + x[1]h[1] + x[2]h[0] + x[3]h[3] = 24$$

$$y[3] = x[0]h[3] + x[1]h[2] + x[2]h[1] + x[3]h[0] = 30$$

The following figure is a graphical representation of computing the circular convolution of the two sequences in Example 5.3. Here, the sequence h[n] is rotated counterclockwise one sample at a time, multiplied point by point by x[n], and then the product added to obtain the circular convolution at sample index n. The process is continued until the N-point circular convolution is completed (Fig. 5.6).

The same can also be obtained via matrix equation, which is

$$\bar{y} = \begin{bmatrix} 4 & 1 & 2 & 3 \\ 3 & 4 & 1 & 2 \\ 2 & 3 & 4 & 1 \\ 1 & 2 & 3 & 4 \end{bmatrix} \begin{bmatrix} 1 \\ 2 \\ 3 \\ 4 \end{bmatrix} = \begin{bmatrix} 24 \\ 22 \\ 24 \\ 30 \end{bmatrix} \tag{5.26}$$

5.7 Properties of the DFT

As in Z-transform and DTFT, one can exploit the properties of the DFT in solving problems with elegance and efficiency. In this section we will describe some properties of the DFT with proof where necessary.

Linearity The DFT is a linear transform. That means that the following equation holds

$$Y(k) = DFT\{\alpha x_1[n] + \beta x_2[n]\} = \alpha X_1(k) + \beta X_2(k) \tag{5.27}$$

In the above equation, both $\{x_1[n]\}$ and $\{x_2[n]\}$ are N-point sequences, $X_i[k] = DFT\{x_i[n]\}$, $i = 1, 2$, and α and β are constants.

Circular Time-Shifting If an N-point sequence $x[n]$ is circularly shifted in time by n_0, then the DFT of the time-shifted sequence is described by

$$DFT\{x[<n - n_0>_N]\} = W_N^{kn_0} X(k), n_0 \in Z \tag{5.28}$$

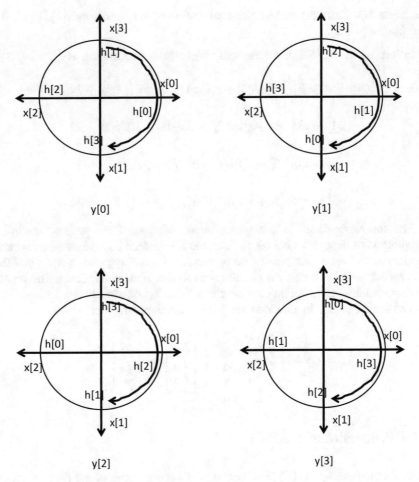

Fig. 5.6 Graphical representation of the computation of circular convolution of the two sequences specified in Example 5.3

Proof From the definition of DFT, we have

$$DFT\{x[\ <n-n_0>_N\} = \sum_{n=0}^{N-1} x[n-n_0]W_N^{nk} \qquad (5.29)$$

By using $m = n - n_0$ in the above equation, we can rewrite (5.29) as

$$DFT\{x[\ <n-n_0>_N\} = \sum_{m=-n_0}^{N-1-n_0} x[m]W_N^{(m+n_0)k} = W_N^{n_0k}X(k) \qquad (5.30)$$

Circular Frequency-Shifting If the DFT of an N-point sequence is circularly shifted by an integer k_0, then the corresponding discrete-time sequence is given by

$$IDFT\{X[<k-k_0>_N]\} = W_N^{-nk_0}x[n], 0 \le n \le N-1 \qquad (5.31)$$

The proof is similar to that used for circular time-shifting property.

Circular Convolution Theorem The DFT of the circular convolution of two N-point sequences is the product of the two DFTs, that is,

$$Y(k) = DFT\left\{x[n] \bigoplus h[n]\right\} = X(k)H(k), 0 \le k \le N-1 \qquad (5.32)$$

The proof is similar to that given for the linear convolution.

Modulation The modulation theorem states that the IDFT of the circular convolution of two N-point DFTs results in the product of the corresponding discrete-time sequences. In other words,

$$IDFT\left\{\frac{1}{N}\sum_{l=0}^{N-1}X[l]H[<k-l>_N]\right\} = x[n]h[n], 0 \le n \le N-1 \qquad (5.33)$$

Proof Using the definition of IDFT, we can write the IDFT of the circular convolution found on the left-hand side of (5.33) as

$$IDFT\left\{\frac{1}{N}\sum_{l=0}^{N-1}X[l]H[<k-l>_N]\right\} = \frac{1}{N}\sum_{k=0}^{N-1}\left\{\frac{1}{N}\sum_{l=0}^{N-1}X[l]H[<k-l>_N]\right\}W_N^{-nk} \qquad (5.34)$$

By interchanging the summation order, we can rewrite (5.34) as

$$IDFT\left\{\frac{1}{N}\sum_{l=0}^{N-1}X[l]H[<k-l>_N]\right\} = \frac{1}{N}\sum_{l=0}^{N-1}X[l]\left\{\frac{1}{N}\sum_{k=0}^{N-1}H[<k-l>_N W_N^{-nk}]\right\} \qquad (5.35)$$

Substitute $m = k - l$ in the above equation, which results in

$$IDFT\left\{\frac{1}{N}\sum_{l=0}^{N-1}X[l]H[<k-l>_N]\right\} = \frac{1}{N}\sum_{l=0}^{N-1}X[l]\left\{\frac{1}{N}\sum_{m=-l}^{N-1-l}H[m]W_N^{-n(m+l)}\right\}$$

$$= \frac{1}{N}\sum_{l=0}^{N-1}X[l] \times \left\{\frac{1}{N}\sum_{m=-l}^{N-1-l}H[m]W_N^{-nm}\right\}W_N^{-nl}$$

$$= \left\{\frac{1}{N}\sum_{l=0}^{N-1}X[l]W_N^{-nl}\right\}\left\{\frac{1}{N}\sum_{m=-l}^{N-1-l}H[m]W_N^{-nm}\right\} = x[n]h[n] \qquad (5.36)$$

Energy Conservation (Parseval's Theorem) This property states that the energy of a sequence in the discrete-time domain is conserved in the frequency domain. The implication is that one can calculate the energy in a sequence either in the time domain or equivalently in the DFT domain. The result is the same. In mathematical terms, Parseval's theorem implies

$$\sum_{n=0}^{N-1} |x[n]|^2 = \frac{1}{N} \sum_{k=0}^{N-1} |X[k]|^2 \tag{5.37}$$

Proof To be sure that the above statement is true, let us write the energy of an N-point sequence as

$$\sum_{n=0}^{N-1} |x[n]|^2 = \sum_{n=0}^{N-1} x[n] x^*[n] \tag{5.38}$$

In (5.38), replace $x[n]$ by its IDFT, so we can rewrite (5.38) as

$$\sum_{n=0}^{N-1} |x[n]|^2 = \sum_{n=0}^{N-1} \left\{ \frac{1}{N} \sum_{k=0}^{N-1} X[k] W_N^{-nk} \right\} x^*[n] \tag{5.39}$$

Interchanging the order of summation in (5.39), we have

$$\sum_{n=0}^{N-1} |x[n]|^2 = \frac{1}{N} \sum_{k=0}^{N-1} X[k] \left\{ \sum_{n=0}^{N-1} x^*[n] W_N^{-nk} \right\} = \frac{1}{N} \sum_{k=0}^{N-1} X[k] \left\{ \sum_{n=0}^{N-1} x[n] W_N^{nk} \right\}^*$$
$$= \frac{1}{N} \sum_{k=0}^{N-1} X[k] \times X^*[k] = \frac{1}{N} \sum_{k=0}^{N-1} |X[k]|^2 \tag{5.40}$$

The following Table lists the properties of DFT for easy reference (Table 5.1).

Table 5.1 Properties of DFT

Property	Length-N sequence	N-point DFT
	$x[n]$	$X[k]$
	$h[n]$	$H[k]$
Linearity	$ax[n] + bh[n]$	$aX[k] + bH[k]$
Circular time-shifting	$x[\langle n - n_0 \rangle_N]$	$W_N^{kn_0} X[k]$
Circular frequency-shifting	$W_N^{-nk_0} x[n]$	$X[\langle k - k_0 \rangle_N]$
Circular convolution	$\sum_{m=0}^{N-1} x[m] h[\langle n - m \rangle_N]$	$X[k]H[k]$
Modulation	$x[n]h[n]$	$\frac{1}{N} \sum_{m=0}^{N-1} X[m] H[\langle k - m \rangle_N]$
Parseval's theorem	$\sum_{n=0}^{N-1} |x[n]|^2 = \frac{1}{N} \sum_{k=0}^{N-1} |X[k]|^2$	

Example 5.4 Compute the circular convolution of the sequences in Example 5.3 using the circular convolution property of DFT.

Solution Let y[n] be the circular convolution of x[n] and h[n], which are of length 4. Therefore, y[n] is also of length 4. The DFT of y[n] is given by

$$Y[k] = X[k]H[k], 0 \leq k \leq 3 \tag{5.41}$$

To compute X[k] and H[k], we can use the matrix equation for the DFTs. The DFT of the four-point sequence x[n] can be expressed in matrix equation as

$$X = \{W_4^{nk}\} \underline{x} \tag{5.42}$$

In (5.42), $\{W_4^{nk}\}$ is a 4 × 4 DFT matrix with the row index corresponding to n and column index corresponding to k. So,

$$\{W_4^{nk}\} = \begin{bmatrix} 1 & 1 & 1 & 1 \\ 1 & -j & -1 & j \\ 1 & -1 & 1 & -1 \\ 1 & j & -1 & -j \end{bmatrix} \tag{5.43}$$

Using (5.43) in (5.42), we get

$$X = \begin{bmatrix} 1 & 1 & 1 & 1 \\ 1 & -j & -1 & j \\ 1 & -1 & 1 & -1 \\ 1 & j & -1 & -j \end{bmatrix} \begin{bmatrix} 1 \\ 2 \\ 3 \\ 4 \end{bmatrix} = \begin{bmatrix} 10 \\ -2+2j \\ -2 \\ -2-2j \end{bmatrix} \tag{5.44}$$

Similarly, the DFT of h[n] using the matrix equation is given by

$$H = \begin{bmatrix} 1 & 1 & 1 & 1 \\ 1 & -j & -1 & j \\ 1 & -1 & 1 & -1 \\ 1 & j & -1 & -j \end{bmatrix} \begin{bmatrix} 4 \\ 3 \\ 2 \\ 1 \end{bmatrix} = \begin{bmatrix} 10 \\ 2-2j \\ 2 \\ 2+2j \end{bmatrix} \tag{5.45}$$

Then, by circular convolution property, the DFT of y[n] is given by

$$Y = XH = \begin{bmatrix} X[0]H[0] \\ X[1]H[1] \\ X[2]H[2] \\ X[3]H[3] \end{bmatrix} = \begin{bmatrix} 100 \\ 8j \\ -4 \\ -8j \end{bmatrix} \tag{5.46}$$

The inverse DFT of Y will give us the sequence y[n], which is the circular convolution of x[n] and h[n]. Therefore, we have

$$y[n] = \{W_4^{nk}\}^{-1} Y[k] \tag{5.47a}$$

The DFT matrix is orthogonal, which means that the inverse of the DFT matrix is its own conjugate transpose, that is,

$$\left\{ W_4^{nk} \right\}^{-1} = \left\{ W_4^{nk} \right\}^{*T} \tag{5.47b}$$

Therefore, we have

$$y[n] = IDFT\{Y\}$$

$$= \frac{1}{4} \left\{ W_4^{nk} \right\}^{*T} Y = \frac{1}{4} \begin{bmatrix} 1 & 1 & 1 & 1 \\ 1 & j & -1 & -j \\ 1 & -1 & 1 & -1 \\ 1 & -j & -1 & j \end{bmatrix} \begin{bmatrix} 100 \\ 8j \\ -4 \\ -8j \end{bmatrix} = \begin{bmatrix} 24 \\ 22 \\ 24 \\ 30 \end{bmatrix} \tag{5.47c}$$

This is what we got by carrying out the circular convolution in the discrete-time domain.

5.8   Linear Convolution Using Circular Convolution

We mentioned that since DFT is discrete, meaning that it is defined over a finite set of points, it is amenable to digital computation. Then, how can we use DFT to perform linear convolution? The answer is as follows. Let us consider two sequences x[n] of length M and h[n] of length N. Then the linear convolution of x[n] and h[n] is a sequence of length L = M + N-1. The circular convolution of two sequences of the same length is a sequence, also, of the same length. Therefore, to perform the linear convolution of x[n] and h[n], we must make the lengths of the two sequences equal to L. This is achieved by appending the sequence x[n] with L-M zeros and h[n] with L-N zeros. Once the lengths of the sequences are made equal, we can compute the DFTs of the two L-point sequences, multiply them point by point, and then perform the inverse DFT to obtain the linear convolution.

Example 5.5 Compute the linear convolution of the two sequences in Example 5.3 using circular convolution.

Solution The two sequences in Example 5.3 are of length 4. Therefore, the linear convolution will result in a sequence of length 7. We must first make the lengths of the sequences equal to 7 by appending three zeros to each sequence. We can use the matrix equation as in (5.25) to obtain the linear convolution. It is given by

$$y[n] = \begin{bmatrix} 4 & 0 & 0 & 0 & 1 & 2 & 3 \\ 3 & 4 & 0 & 0 & 0 & 1 & 2 \\ 2 & 3 & 4 & 0 & 0 & 0 & 1 \\ 1 & 2 & 3 & 4 & 0 & 0 & 0 \\ 0 & 1 & 2 & 3 & 4 & 0 & 0 \\ 0 & 0 & 1 & 2 & 3 & 4 & 0 \\ 0 & 0 & 0 & 1 & 2 & 3 & 4 \end{bmatrix} \begin{bmatrix} 1 \\ 2 \\ 3 \\ 4 \\ 0 \\ 0 \\ 0 \end{bmatrix} = \begin{bmatrix} 4 \\ 11 \\ 20 \\ 30 \\ 20 \\ 11 \\ 4 \end{bmatrix}$$

The DFT is also performed using FFT algorithm. The FFT algorithm is efficient when the sequence length is an integer power of 2. In this example, since the length of the linear convolution is 7, which is not an integer power of 2, we will use 8, instead. So, the two zero-padded sequences are

$$x_e[n] = [1\ 2\ 3\ 4\ 0\ 0\ 0\ 0], h_e[n] = [4\ 3\ 2\ 1\ 0\ 0\ 0\ 0]$$

We, then, compute the DFTs of length-8 $x_e[n]$ and $h_e[n]$ using the FFT algorithm. As mentioned before, the MATLAB function *fft* computes the DFT of a sequence. After computing the DFTs, we have to multiply them point by point to obtain the DFT of the linear convolution of the two zero-appended sequences. The discrete-time sequence is then found by taking the IDFT of the DFT product. Again, this is achieved by the inverse FFT. The corresponding MATLAB function is *ifft*. The eight-point DFTs of the two zero-padded sequences are shown in Fig. 5.7. The DFT of the circular convolution of the two zero-padded sequences are shown in the top plot in Fig. 5.8, and the linear convolution using circular convolution is shown in the bottom plot of Fig. 5.8. The first seven samples are the valid samples of the linear convolution.

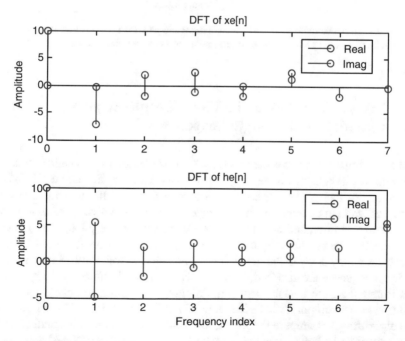

Fig. 5.7 DFTs of the zero-padded sequences in Example 5.5: top plot, DFT of the sequence $x_e[n]$; bottom plot, DFT of the sequence $h_e[n]$

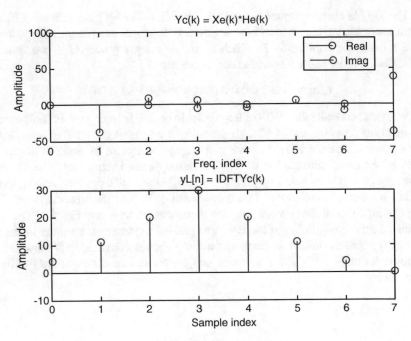

Fig. 5.8 Linear convolution via circular convolution: top plot, DFT of the circular convolution of $x_e[n]$ and $h_e[n]$; bottom plot, sequence corresponding to the linear convolution

5.9   Linear Convolution of a Finite-Length Sequence with an Infinite-Length Sequence

In the previous section, we described how to use circular convolution to achieve linear convolution of two finite-length sequences. This was fine and dandy because the lengths of the two sequences were assumed to be small, and so padding with zeros was okay. In practice, one of the sequences to be convolved is relatively small, such as an FIR filter, and the other sequence is relatively very long. In such cases, zero-padding the smaller-length sequence is not computationally efficient. It also introduces a long delay in the output because one has to wait until all the samples of the input sequence are acquired. Is there a way out of this situation? The answer is yes. Remember, the reason for using circular convolution to enable linear convolution is the computational efficiency of the FFT algorithm. We will describe two algorithms to compute the linear convolution of a relatively small sequence with an infinitely long sequence, which are called *overlap and add* and *overlap and save* methods.

5.9.1 Overlap and Add

A qualitative statement of the overlap and add method is as follows. In this method, the longer sequence is divided into nonoverlapping segments, and each segment is circularly convolved with the smaller-length sequence. Of course, we must zero-pad the smaller-length sequence and the segmented sequence to make the length conforming to the linear convolution. As a result of the linear convolution, the length of the output will be larger than the segment length. Therefore, it will overlap with the next segment. The overlapping samples are added to produce the correct linear convolution, hence, the name overlap and add. Let us now describe the procedure in detail. Let $x[n]$ be the infinitely long input sequence, and let $h[n]$ be the length-N sequence. Divide the input sequence into nonoverlapping blocks of length M samples each, where $M > N$. We can, therefore, express the input sequence in terms of the segmented blocks, as given by

$$x[n] = \sum_{m=0}^{\infty} x_m[n - mM] \tag{5.48}$$

where the segmented sequences are given by

$$x_m[n] = \begin{cases} x[n + mM], 0 \le n \le M - 1 \\ 0, otherwise \end{cases} \tag{5.49}$$

We next perform the linear convolution of each segment $x_m[n]$ with $h[n]$. The overall output can be expressed as

$$y[n] = \sum_{k=0}^{N-1} h[k]x[n - k] \tag{5.50}$$

Using Eq. (5.48) in (5.50), we can rewrite (5.50) as

$$y[n] = \sum_{k=0}^{N-1} h[k] \sum_{m=0}^{\infty} x_m[n - k - mM] \tag{5.51}$$

By interchanging the order of summation in (5.51), we have

$$y[n] = \sum_{m=0}^{\infty} \left\{ \sum_{k=0}^{N-1} h[k]x_m[n - k - mM] \right\} = \sum_{m=0}^{\infty} y_m[n - mM] \tag{5.52}$$

where

$$y_m[n] = h[n] \otimes x_m[n] = \sum_{k=0}^{N-1} h[k]x_m[n - k - mM] \tag{5.53}$$

So, we have expressed the linear convolution of $x[n]$ and $h[n]$ as an infinite sum of the linear convolutions of the nonoverlapping segments $x_m[n]$ and $h[n]$. The length of each linear convolution is $L = M + N$ -1. There are, therefore, N-1 samples that extend beyond each segment. These last N-1 samples fall into the first N-1 samples of the next segment. Therefore, these samples must be added to the first N-1 samples of the linear convolution of the next segment and so on. We must mention the fact that the linear convolution of each segment is computed using circular convolution via FFT for computational efficiency. The lengths of the two sequences in the circular convolution must be the same. Therefore, we must zero-pad the length-N sequence with M-1 samples and the length-M sequence with N-1 samples so that both sequences are of length $L = M + N$-1. Figure 5.9 shows the input and output sequences indicating which samples overlap in the output so that they are added to produce the correct result. We can list the steps involved in the calculation of the linear convolution of an infinitely long input sequence with a short sequence using the overlap-add method as follows:

1. Segment the long input sequence $x[n]$ into nonoverlapping blocks of length M samples long. Call each segment as $x_m[n]$, $m = 0, 1, 2$, *etc.*
2. Append the length-N sequence $h[n]$ with M-1 zeros.
3. Compute the L-point DFT of the zero-padded sequence $h[n]$. This is a one-time computation.
4. For each M, append the input segment $x_m[n]$ with N-1 zeros to make its length $L = M + N$-1, and then compute its L-point DFT.

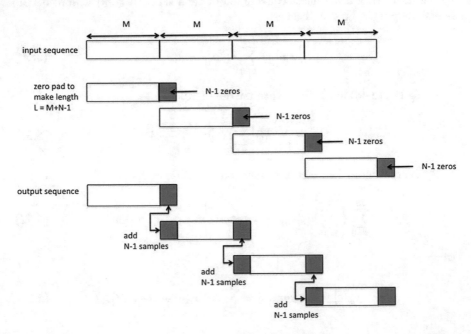

Fig. 5.9 Diagram illustrating the input segmentation and forming the output sequence by adding the overlapping samples

5. Multiply the two L-point DFTs point by point and perform the IDFT of the product to produce the output segment $y_m[n]$.
6. Write the output segments sequentially by adding the last N-1 samples of the previous output segment to the first N-1 samples of the current output segment and so on.

To make the procedure clearer, let us work out an example.

Example 5.6 Consider an input sequence, which is a sum of two sinusoids of frequencies 200 and 1100 Hz and amplitudes 1.5 and 2, respectively. The sampling frequency is 5000 Hz. Let the length of the input sequence be 2048 samples. Filter this sequence through an FIR lowpass filter of order 14 and a cutoff frequency of 625 Hz.

Solution   Using the specifications, the input sequence is expressed as

$$x[n] = 1.5\cos\left(\frac{2\pi*200n}{5000}\right) + 2\cos\left(\frac{2\pi*1100n}{5000}\right), 0 \le n \le 2047 \qquad (5.54)$$

Next we need to design a lowpass FIR filter with a cutoff frequency of 625 Hz. We will defer the discussion on FIR filter design to a later chapter. For now, we will use the MATLAB function *fir1*, which accepts as input the cutoff frequency normalized to half the sampling frequency, filter order, and window function. Since we have not learnt FIR filter design, we will not discuss windowing method or windows here. The MATLAB function call to design a lowpass FIR filter of order 14 and a normalized cutoff frequency of $625/2500 = 0.25$ is $B = fir1(14,0.25,window$ $(@hamming,15))$. The filter impulse response is returned in B, which is of length 15. The filtering of the long input sequence through a short FIR filter using the overlap and add technique can be carried out in MATLAB using the function *fftfilt*. It accepts the FIR filter impulse response B and the input sequence x and returns the filtered sequence using the overlap and add technique. Or we can write a routine based on the above-listed procedure to carry out the overlap-add method of linear convolution. The MATLAB code to perform the overlap and add filtering is in the M-file named *Overlap_add.m*. Figure 5.10 shows the plots of the input and filtered sequences as top and bottom plots, respectively, over the first 256 samples. The corresponding frequency spectra are shown in Fig. 5.11. From the figures, it is clear that the FIR filter has passed the 200 Hz component and rejected the 1100 Hz component.

5.9.2   Overlap and Save

This is an alternative method of linear convolution of a long sequence with a relatively short length FIR filter using circular convolution. Unlike the overlap and add method, the input sequence in the overlap and save method is segmented into

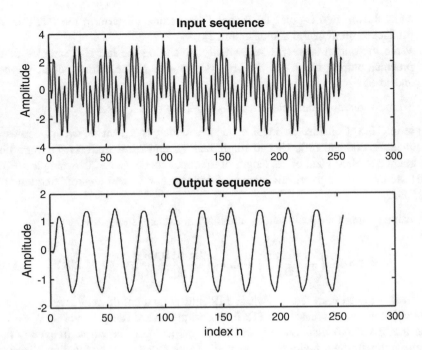

Fig. 5.10 Result of overlap and add filtering of the input sequence in Example 5.6. Top plot, first 256 samples of the 2048-length input sequence; bottom plot, first 256 samples of the filtered sequence

overlapping blocks of length M samples. The last N-1 samples of the previous block overlap the first N-1 samples of the current block and so on. Here, N is the length of the FIR filter impulse response. Since we are performing linear convolution of each length-M block and length-N filter, the length of each output block is L = M + N-1 samples. Because we are using circular convolution to perform linear convolution, both the FIR filter impulse response and the input blocks must be padded with appropriate number of zeros to make the length equal to the length of the linear convolution, which is L. However, the first input block is padded with N-1 zeros at the front, while the rest of the blocks overlap with adjacent blocks over N-1 samples at the beginning, as shown in Fig. 5.12. Because of this overlap, the first N-1 samples in the output blocks are discarded, as shown in Fig. 5.12.

The procedure for the overlap and save method is as follows:

1. Insert N-1 zeros at the beginning of the input sequence $x[n]$.
2. Segment the zero-padded input sequence into overlapping blocks $x_m[n]$ of length L = M + N -1 samples, where the last N-1 samples of the previous block overlap with the first N-1 samples of the current block, and so on.
3. Append the length-N FIR filter $h[n]$ with M-1 zeros to make its length L.
4. Compute the L-point DFT $H[k]$, of $h[n]$.
5. For each M:

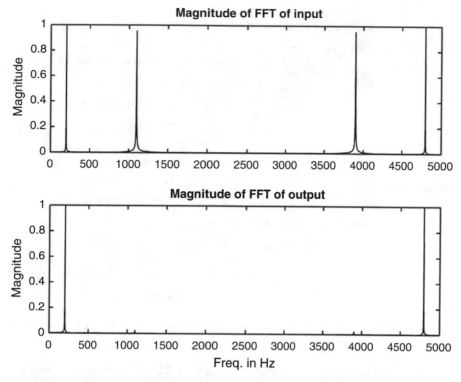

Fig. 5.11 Frequency spectra of the input and filtered sequences of Example 5.6. Top plot, input frequency spectrum showing two frequency components at 200 and 1100 Hz; bottom plot, frequency spectrum of the filtered sequence containing only the 200 Hz component

    (a) Compute the L-point DFT $X_m[k]$.
    (b) Calculate $Y_m[k] = H[k]X_m[k]$, $0 \le k \le L - 1$.
    (c) Compute the IDFT $y_m[n] = IDFT\{Y_m[k]\}$, $0 \le n \le L - 1$.
    (d) Discard the first N-1 samples of each $y_m[n]$.

6. Form the true output sequence $y[n]$ by concatenating the last M samples in each output block $y_m[n]$.

We will make it clear by working out an example.

Example 5.7 The input sequence and the FIR filter for this example are the same as in Example 5.6. We have to filter the long input sequence through a short FIR filter using overlap and save method.

Solution Let us use the input block length to be 64 and N = 15. The first block is zero-padded with 14 zeros followed by 50 samples to make its length 64. The subsequent blocks overlap as shown in Fig. 5.12. The FIR filter impulse response is appended with M − 1 = 49 zeros to make its length equal to 64. The DFT of the FIR filter impulse response is computed and is one time only. The rest of the

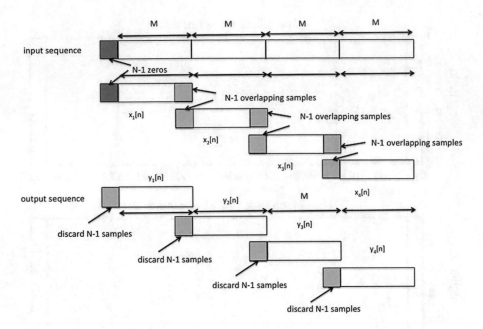

Fig. 5.12 Diagram illustrating the input segmentation and forming the output sequence by discarding the first N-1 samples

calculations follow the procedure listed above. The input sequence and the filtered sequence are plotted over the first 256 samples and shown in the top and bottom plots in Fig. 5.13. The corresponding frequency spectra are shown in Fig. 5.14. The results of overlap and save method are in perfect agreement with those obtained from the overlap and add method. The MATLAB codes for this example are in the M-file named *Overlap_save*.m.

5.9.3   DFT Leakage

The DFT, as we have described above, represents an N-point discrete-time sequence in the frequency domain precisely at the discrete frequencies $\frac{kf_s}{N}$, $0 \leq k \leq N-1$, where f_s is the sampling frequency. In the DSP jargon, $k\frac{f_s}{N}$ are the frequency *bins*. One reason for using DFT is to determine the spectrum of a discrete-time signal. If a frequency in the input signal does not coincide with one of the DFT bins, then the magnitude of the DFT of that particular signal will not be an impulse-like in shape. Instead, it will spread over the entire DFT bins. That is to say that the energy of the input discrete-time sequence in a particular frequency will leak into other neighboring bins. It smears the impulse-like frequency over the entire frequency bins. Therefore, the corresponding spectrum is only approximate. This is what is called the DFT leakage. Since we do not know the exact frequencies contained in a discrete-time sequence, we can only obtain an approximate

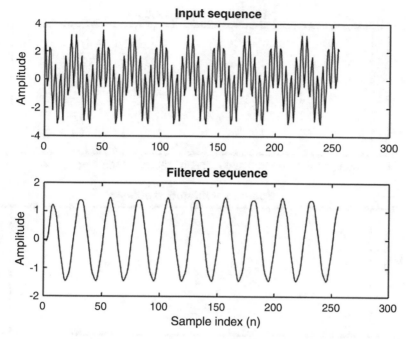

Fig. 5.13 Result of overlap and save filtering of the input sequence in Example 5.7. Top plot, first 256 samples of the 2048-length input sequence; bottom plot, first 256 samples of the filtered sequence

spectrum of an input sequence. However, one can minimize the DFT leakage and obtain a better approximation to the spectrum by windowing the finite-length input sequence using a suitable window function. We will illustrate this phenomenon by an example using MATLAB.

Example 5.8 Illustration of DFT leakage

Consider a sinusoid at a frequency of 10 Hz. First, we sample the sinusoid at a frequency f_s and retain $N = 64$ samples. Choose $f_s = \frac{fN}{4.5}$, compute the DFT of the sampled sinusoid, and plot the magnitude of the computed DFT. Next, choose a different sampling frequency, for instance, $f_{s1} = \frac{fN}{4}$, and compute the DFT over the same 64 points and plot its magnitude. Compare the two DFTs.

Solution The sinusoidal sequence is described by $x[n] = \sin\left(\frac{2\pi f n}{f_s}\right), 0 \leq n \leq 63$. The DFT of the sinusoid is computed using the MATLAB function $X = fft(x)$. The M-file to solve this problem is named *Example5_8.m*. The input sinusoid sampled at f_s and f_{s1} is shown in Fig. 5.15 in the top and bottom plots, respectively. The magnitudes of the corresponding DFTs are shown in Fig. 5.16 in the top and bottom plots, respectively. We see the leakage of energy into neighboring DFT bins in the top plot of Fig. 5.16, which is due to the fact that the input frequency does not

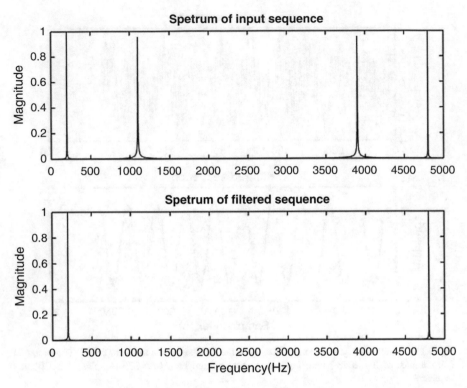

Fig. 5.14 Frequency spectra of the input and filtered sequences of Example 5.7. Top plot, input frequency spectrum showing two frequency components at 200 and 1100 Hz; bottom plot, frequency spectrum of the filtered sequence containing only the 200 Hz component

correspond to a DFT bin. We see no leakage in the bottom plot of Fig. 5.16 because the input frequency of 10 Hz corresponds to the fifth DFT bin, which is $4*\frac{f_{s1}}{N} = 10\,Hz$. By windowing the input sinusoid with a Kaiser window with $\beta = 3.75$, we are able to minimize the DFT leakage. This is clear from the plot in Fig. 5.17.

5.10   Discrete Transforms

Discrete Fourier transform is a method of representing finite-length sequences in the frequency domain. It is also used to perform filtering of both finite and long duration sequences. There are efficient methods to compute the DFT, which we will consider in another chapter. In general, frequency domain transforms used in digital signal and image processing fall under the category of *unitary* transforms. Let a finite-length sequence be denoted by $\{x[n]\}, 0 \leq n \leq N - 1$. In vector form, the sequence is described by

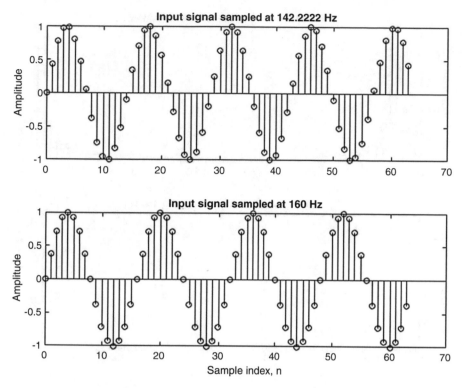

Fig. 5.15 Input sinusoidal sequence of Example 5.8: top, sampling frequency is 142.2 Hz; bottom, sampling frequency is 160 Hz

$$\mathbf{x} = [x[n], 0 \leq n \leq N - 1]^T \tag{5.55}$$

In (5.55), the superscript T denotes the transpose operation. The linear transformation of $x[n]$ can then be represented in vector form by

$$\mathbf{y} = \mathbf{A}\mathbf{x} \tag{5.56a}$$

Or in expanded form, the linear transformation in (5.56a) can be written as

$$y[k] = \sum_{n=0}^{N-1} a[k,n]x[n], 0 \leq k \leq N - 1 \tag{5.56b}$$

5.10.1 Unitary Transform

The linear transformation described in (5.56a) or (5.56b) is termed *unitary* if the following condition is satisfied:

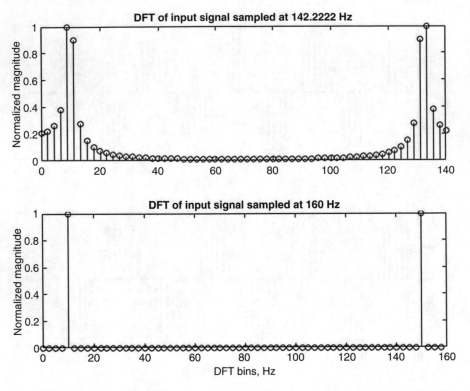

Fig. 5.16 Magnitude of the DFT of the input sinusoid: top, sampling frequency is 142.2 Hz; bottom, sampling frequency of 160 Hz

$$A^{-1} = A^{*T} \tag{5.57}$$

That is, if the inverse of the matrix A is its own conjugate transpose, then the linear transformation in (5.56a) is unitary. Otherwise it is non-unitary. The unitary matrix A is called the *kernel* matrix, and its elements may be real or complex. The vector y is known as the transformed vector, and its elements are referred to as the transform coefficients.

Example 5.9   Consider a four-point DFT. Is it a unitary transform?

Solution From our earlier definition of the DFT of an N-point sequence, it is given by

$$X[k] = \sum_{n=0}^{N-1} x[n] W_N^{nk}, 0 \leq k \leq N-1 \tag{5.58}$$

For N = 4, the DFT kernel matrix is found to be

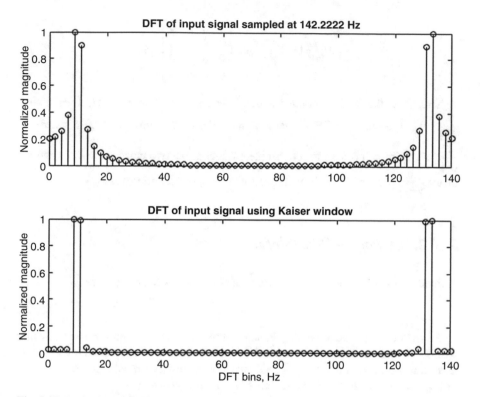

Fig. 5.17 Reduction of DFT leakage using Kaiser window

$$\{a[k,n]\} = \left\{W_4^{nk}\right\} = \left\{e^{-j\frac{2\pi}{4}nk}\right\} = \begin{bmatrix} 1 & 1 & 1 & 1 \\ 1 & -j & -1 & j \\ 1 & -1 & 1 & -1 \\ 1 & j & -1 & -j \end{bmatrix} \qquad (5.59)$$

The rows of the kernel matrix in (5.59) correspond to the frequency points $\{k, 0 \leq k \leq N-1\}$ and the columns to the time index n. The conjugate transpose of the matrix in (5.59) is given by

$$\left\{W_4^{nk}\right\}^{*T} = \begin{bmatrix} 1 & 1 & 1 & 1 \\ 1 & j & -1 & -j \\ 1 & -1 & 1 & -1 \\ 1 & -j & -1 & j \end{bmatrix} \qquad (5.60)$$

The product of the matrices in (5.59) and (5.60) results in

$$\{W_4^{nk}\}\{W_4^{nk}\}^{*T} = \begin{bmatrix} 4 & 0 & 0 & 0 \\ 0 & 4 & 0 & 0 \\ 0 & 0 & 4 & 0 \\ 0 & 0 & 0 & 4 \end{bmatrix} \neq I \tag{5.61}$$

Therefore, the DFT as defined above is not unitary. However, the DFT matrix $\frac{1}{\sqrt{N}}\{W_N^{nk}\}$ is unitary. For instance, if we redefine the four-point DFT matrix as $\frac{1}{\sqrt{4}}\{W_4^{nk}\}$, then it is easy to see that the product of the two matrices in (5.61) with a factor $\frac{1}{\sqrt{4}}$ introduced in each matrix results in a 4×4 identity matrix.

5.10.2   Orthogonal Transform

A linear transform is said to be *orthogonal* if it satisfies the condition

$$A^{-1} = A^T \tag{5.62}$$

From (5.62), it is clear that the elements of the kernel matrix must be real. In other words, a linear transform is orthogonal if its coefficients are real and its inverse is its own transpose. It must be pointed out that a unitary transform need not be orthogonal. For instance, consider the 2×2 matrix given by

$$A = \frac{4}{5}\begin{bmatrix} {}^{3}\!/_{4} & -j \\ j & -{}^{3}\!/_{4} \end{bmatrix} \tag{5.63}$$

Then, its conjugate transpose takes the form

$$A^{*T} = \frac{4}{5}\begin{bmatrix} {}^{3}\!/_{4} & -j \\ j & -{}^{3}\!/_{4} \end{bmatrix} \tag{5.64}$$

From (5.63) and (5.64), it is found that $AA^{*T} = I$. Therefore it is unitary. However, we find that

$$AA^T = \frac{16}{25}\begin{bmatrix} -{}^{7}\!/_{16} & {}^{j3}\!/_{2} \\ {}^{j3}\!/_{2} & -{}^{7}\!/_{16} \end{bmatrix} \neq I \tag{5.65}$$

Therefore, the unitary transform in (5.63) is not orthogonal, which proves the statement.

5.10.3   Discrete Cosine Transform

Given a finite-length sequence $\{x[n], 0 \le n \le N-1\}$, its discrete cosine transform (DCT) is defined as

$$X[k] = \alpha[k] \sum_{n=0}^{N-1} x[n] \cos \left[\frac{(2n+1)\pi k}{2N} \right], 0 \le k \le N-1 \qquad (5.66)$$

In (5.66), $\alpha[k]$ is defined as

$$\alpha[k] = \begin{cases} \frac{1}{\sqrt{N}}, for\ k = 0 \\ \sqrt{\frac{2}{N}}, for\ 1 \le k \le N-1 \end{cases} \qquad (5.67)$$

The elements of the N-point DCT kernel matrix are given by

$$a_{DCT}[k,n] = \begin{cases} \frac{1}{\sqrt{N}}, for\ k = 0\ and\ 0 \le n \le N-1 \\ \sqrt{\frac{2}{N}} \cos \left[\frac{(2n+1)\pi k}{2N} \right], 0 \le k, n \le N-1 \end{cases} \qquad (5.68)$$

For example, the 4×4 DCT kernel matrix is determined from (5.68) and is given by

$$A_{DCT} = \frac{1}{2} \begin{bmatrix} 1 & 1 & 1 & 1 \\ \sqrt{2}\cos\left(\frac{\pi}{8}\right) & \sqrt{2}\cos\left(\frac{3\pi}{8}\right) & \sqrt{2}\cos\left(\frac{5\pi}{8}\right) & 0\sqrt{2}\cos\left(\frac{7\pi}{8}\right) \\ \sqrt{2}\cos\left(\frac{\pi}{4}\right) & \sqrt{2}\cos\left(\frac{3\pi}{4}\right) & \sqrt{2}\cos\left(\frac{5\pi}{4}\right) & \sqrt{2}\cos\left(\frac{7\pi}{4}\right) \\ \sqrt{2}\cos\left(\frac{3\pi}{8}\right) & \sqrt{2}\cos\left(\frac{9\pi}{8}\right) & \sqrt{2}\cos\left(\frac{15\pi}{8}\right) & \sqrt{2}\cos\left(\frac{21\pi}{8}\right) \end{bmatrix}$$

$$(5.69)$$

The sequence $\{x[n], 0 \le n \le N-1\}$ can be obtained from the unitary transform coefficients as follows:

$$x = A^{*T} y \qquad (5.70)$$

In the case of DCT, the sequence is obtained by simply pre-multiplying the transformed vector by the transpose of the NxN DCT matrix.

Energy Conservation and Compaction Properties of Unitary Transforms Consider an N-point sequence as a vector x, its transformed vector y using the unitary transform A. Let E_x and E_y denote the energies in the two vectors. Then,

$$E_y = \|y\|^2 = y^{*T}y = x^{*T}A^{*T}Ax = x^{*T}Ix = x^{*T}x = \|x\|^2 = E_x \qquad (5.71)$$

The above equation implies that the energy in the finite-length discrete-time sequence is preserved in the transform domain. This is the energy conservation property of a unitary transform.

Even though the energy is preserved in the transform domain, the amount of energy contained in each transform coefficient is different. That is to say that the energy is unevenly distributed in the transform coefficients. Some coefficients may carry more energy than other coefficients. To be more specific, the variance in the transform coefficients can be defined as

$$E\left\{\|\mathbf{y} - \boldsymbol{\mu}_x\|^2\right\} = E\left\{(\mathbf{y} - \mathbf{y}_x)^{*T}(\mathbf{y} - \boldsymbol{\mu}_x)\right\} = \sum_{k=0}^{N-1} \sigma_y^2 \qquad (5.72)$$

where, $\boldsymbol{\mu}_y$ is the vector of mean values of the vector $\mathbf{y}$. Using (5.56a) in (5.72), we obtain

$$\sum_{k=0}^{N-1} \sigma_y^2 = E\left\{[A(\mathbf{x} - \boldsymbol{\mu}_x)]^{*T}[A(\mathbf{x} - \boldsymbol{\mu}_x)]\right\} \qquad (5.73)$$

But,

$$[A(\mathbf{x} - \boldsymbol{\mu}_x)]^{*T} = (\mathbf{x} - \boldsymbol{\mu}_x)^{*T}A^{*T} \qquad (5.74)$$

where $\boldsymbol{\mu}_x$ is the vector of mean values of the vector $\mathbf{x}$. Therefore, (5.73) reduces to

$$\sum_{k=0}^{N-1} \sigma_y^2 = E\left\{(\mathbf{x} - \boldsymbol{\mu}_x)^{*T}A^{*T}A(\mathbf{x} - \boldsymbol{\mu}_x)\right\} = \sum_{n=0}^{N-1} \sigma_x^2 \qquad (5.75)$$

Thus, the sum of the variances of the transform coefficients equals the sum of the variances of the elements of the input vector.

5.10.4   Hadamard Transform

The DFT and DCT transforms use sinusoids as basis functions. The Hadamard transform uses rectangular basis functions. The kernel matrix of Hadamard transform for a two-point sequence is defined as

$$A_H = \frac{1}{\sqrt{2}}\begin{bmatrix} 1 & 1 \\ 1 & -1 \end{bmatrix} \qquad (5.76)$$

Therefore, the transformed vector can be determined to be

$$\mathbf{y} = A_H\mathbf{x} = \frac{1}{\sqrt{2}}\begin{bmatrix} 1 & 1 \\ 1 & -1 \end{bmatrix}\begin{bmatrix} x[0] \\ x[1] \end{bmatrix} = \frac{1}{\sqrt{2}}\begin{bmatrix} x[0] + x[1] \\ x[0] - x[1] \end{bmatrix} \qquad (5.77)$$

It is easy to verify that Hadamard transform is orthogonal. An interesting property of Hadamard transform is that higher-order transforms can be generated recursively as follows: Given the NxN Hadamard kernel matrix, the 2Nx2N Hadamard kernel matrix is obtained by replacing the elements of the NxN matrix by the NxN kernel matrix, that is,

$$A_{H2N} = \begin{bmatrix} A_{HN} & A_{HN} \\ A_{HN} & -A_{HN} \end{bmatrix} \tag{5.78}$$

A 4×4 Hadamard matrix is obtained using (5.76) in (5.78) and is described by

$$A_{H4} = \frac{1}{2} \begin{bmatrix} 1 & 1 & 1 & 1 \\ 1 & -1 & 1 & -1 \\ 1 & 1 & 1 & 1 \\ 1 & -1 & 1 & -1 \end{bmatrix} \tag{5.79}$$

Example 5.10 Given an N-point sequence, compute its DCT and Hadamard transform. Also compute the inverse DCT and inverse Hadamard transform and compare the sequence obtained from the inverse transform with the input sequence. Also plot the percent of total energy contained in each transform coefficient against the sample index. Use MATLAB to solve the problem.

Solution The input sequence used in this example is described by

$$x[n] = 1.5 \sin\left(\frac{2\pi f_1 n}{f_s}\right) - 0.5 \cos\left(\frac{2\pi f_2 n}{f_s}\right) + \sin\left(\frac{2\pi f_3 n}{f_s}\right), 0 \le n \le 63, \text{where}$$

$f_1 = 10\,Hz, f_2 = 23\,Hz, f_3 = 33\,Hz$, and $f_s = 100\,Hz$. The DCT of the input sequence is obtained in MATLAB by calling the function *dct* with input argument x and output argument X_dct. The function call to compute the Hadamard transform is X_h = *fwht(x)*. The corresponding inverse transforms are obtained using the functions *idct* and *ifwht*, respectively. The M-file to solve this problem is named *Example5_10.m*. The energy of the input signal contained in each transform coefficient is the absolute square of the respective transform coefficients. The input sequence and its DCT and IDCT are shown as stem plots in Fig. 5.18 in the top, middle, and bottom plots, respectively. Similarly, Fig. 5.19 shows the input sequence, Hadamard transform, and the inverse Hadamard transform as stem plots in the top, middle, and bottom plots, respectively. Finally, the percentage of total energy contained in each transform coefficient is plotted against the frequency index for the two transforms and is shown as bar chart in Fig. 5.20. From the figure, we observe that the DCT contains most of the energy in a fewer transform coefficients than what the Hadamard transform does. That means that the DCT has a better energy compaction property than the Hadamard transform. The total energy of the input sequence is expressed as

$$E = \sum_{n=0}^{N-1} |x[n]|^2$$

It is also equal to the sum of the absolute square of the DCT as obtained from MATLAB and is given below.

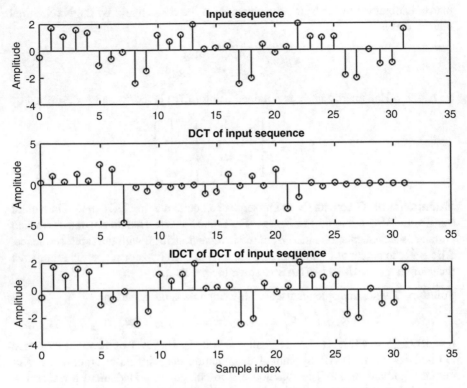

Fig. 5.18 Discrete cosine transform: top, input sequence; middle, the DCT of the input sequence; bottom, IDCT of the DCT sequence in the middle plot

$$E = \sum_{n=0}^{N-1} |x[n]|^2 = \sum_{k=0}^{N-1} |X_{\text{dct}}[k]|^2$$

For the Hadamard transform, it is found that

$$E = \sum_{n=0}^{N-1} |x[n]|^2 = N \sum_{k=0}^{N-1} |X_{Hadamard}[k]|^2$$

5.11 Summary

The DFT plays an important role in digital signal processing. We defined the N-point DFT of a sequence and established its relationship to the discrete-time Fourier transform. Some useful properties of DFT were described, and some examples

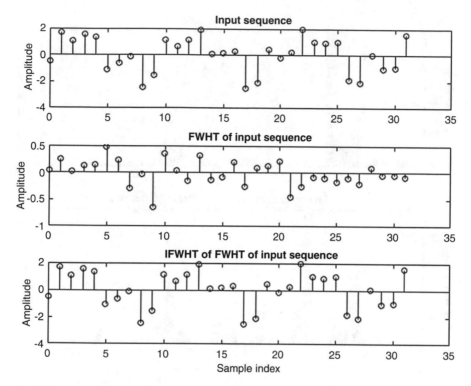

Fig. 5.19 Discrete Hadamard transform: top, input sequence; middle, Hadamard transform of the input sequence; bottom, inverse Hadamard transform of the Hadamard transform sequence in the middle plot

worked out to elucidate the usefulness of these properties. Circular convolution was defined, and we showed how it can be computed using the DFT. Since LTI discrete-time systems employ linear convolution, we showed how linear convolution could be accomplished via circular convolution. In processing real-time signals, one first acquires the signal in digital form and then processes it. In such cases, it is more meaningful to process the input signal as it evolves. To this end, we showed how linear convolution of such a real-time signal could be computed by segmenting the input sequence into smaller-length blocks and then circularly convolving the blocks. Two methods – overlap and add and overlap and save – were explained with examples. We introduced the concept of unitary and orthogonal transforms. In fact DFT is a unitary transform. Two orthogonal transforms, namely, DCT and Hadamard transforms, were introduced. These transforms play an important role in speech and image data compression as they compact the total energy in the input sequence or image into the transform coefficients in an unevenly manner. We will illustrate the energy compaction property of unitary and orthogonal transforms in a later chapter. In the next chapter, we will deal with the design of IIR digital filters.

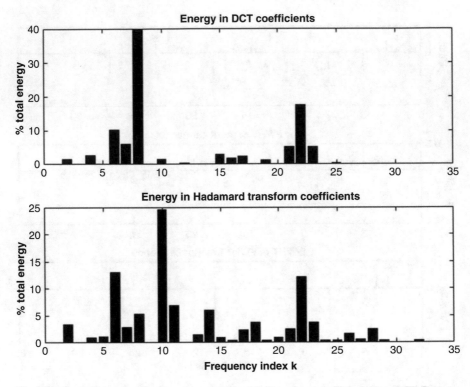

Fig. 5.20 Percentage of total energy compacted in each Transform coefficient; Top: DCT, Bottom: Hadamard Transform

5.12   Problems

1. If the convolution $\tilde{y}[n] = \sum_{k=0}^{N-1} \tilde{x}[k]\tilde{h}[n-k]$, where $\tilde{x}[n]$ and $\tilde{h}[n]$ are both periodic with period N, show that $\tilde{y}[n]$ is also periodic with period N.

2. Determine the N-point DFT of $x[n] = \alpha^n$, $0 \le n \le$ N-1.

3. Determine the five-point DFT of $x[n] = 0.25^n$, $0 \le n \le 4$.

4. Show that the circular convolution is (a) commutative and (b) associative.

5. Let $\{x[n]\} = \{-3, 2, -1, 4\}$ and $\{h[n]\} = \{1, 3, 2, -2\}$ be two length-4 sequences for $0 \le n \le 3$. Compute the circular convolution of x[n] and h[n] using the matrix equation.

6. Let x[n] and X[k] be the N-point DFT pairs. Find the sequence whose DFT is given by $Y[k] = \alpha X\left[\langle k - m_1 \rangle_N\right] + \beta X\left[\langle k - m_2 \rangle_N\right]$, where m_1 and m_2 are positive integers less than N.

7. Find the N-point DFT of the sequence $x[n] = \sin^2\left(\frac{2n\pi}{N}\right), 0 \le n \le N - 1$.

8. If $X[k]$, $0 \leq k \leq N - 1$ is the DFT of the sequence $x[n]$, $0 \leq n \leq N - 1$, then express the 2 N-point DFT of the sequence $h[n] = \begin{cases} x[n], 0 \leq n \leq N - 1 \\ 0, N \leq n \leq 2N - 1 \end{cases}$.

9. Let $X[k]$, $0 \leq k \leq N - 1$ be the DFT of the sequence $x[n]$, $0 \leq n \leq N - 1$. Determine the MN-point DFT of the sequence $y[n] = \begin{cases} x\left[\dfrac{n}{M}\right], n = 0, M, 2N, \cdots (N - 1)M \\ 0, otherwise \end{cases}$ in terms of $X[k]$.

10. Let $x[n]$ be an N-point sequence whose DFT is $X[k]$, $0 \leq k \leq N - 1$ with N even. If $X[2j] = 0$ for $0 \leq j \leq \frac{N}{2} - 1$, show that $x[n] = -x\left[\langle n + \frac{N}{2}\rangle_N\right]$.

11. What is the DFT of the sequence $g[n] = (-1)^n x[n]$ if $x[n]$, $0 \leq n \leq N - 1$, N and $X[k]$, $0 \leq k \leq N - 1$ form a DFT pair with N even.

12. The circular cross-correlation of the two N-point sequences $x[n]$, $0 \leq n \leq N - 1$ and $y[n]$, $0 \leq n \leq N - 1$ is defined as $C_{xy}[k] = \sum_{n=0}^{N-1} x[n] y\left[\langle n - k\rangle_N\right]$, $0 \leq k \leq N - 1$. Determine $C_{xy}[k]$ in terms of the DFTs of $x[n]$ and $y[n]$.

13. Consider the sequence $x[n]$, $0 \leq n \leq N - 1$ with its N-point DFT denoted by $X[k]$, $0 \leq k \leq N - 1$. Let $g[n] = x[3n]$, $0 \leq n \leq \frac{N}{3} - 1$, where it is assumed that N is divisible by 3. Find the $\frac{N}{3}$ − point DFT of $g[n]$ in terms of $X[k]$.

14. Given an N-point sequence $x[n]$, $0 \leq n \leq N - 1$, define a zero-padded sequence $y[n] = \begin{cases} x[n], & 0 \leq n \leq N - 1 \\ 0, & N \leq n \leq 2N - 1 \end{cases}$. Express the 2 N–point DFT $Y[k]$ in terms of $X[k]$.

15. We want to compute the linear convolution of a length-35 filter sequence with an input sequence of length 2000. (a) Determine the smallest number of DFTs and IDFTs needed to compute the linear convolution using the overlap-add method. (b) Repeat (a) for overlap-save method.

References

1. Ahmed N, Natarajan T, Rao KR (1974) Discrete cosine transform. IEEE Trans on Computers C-23:90–93
2. Ansari R (1985) An extension of the discrete Fourier transform. IEEE Trans Circuits Sys CAS32:618–619
3. Bagchi S, Mitra SK (1998) Nonuniform discrete Fourier transform and its signal processing applications. Kluwer Academic Publishers, Norwell
4. Bellanger M (2000) Digital processing of signals: theory and practice, 3rd edn. Wiley, New York
5. Blahut RE (1985) Fast algorithms for digital signal processing. Addison-Wesley, Reading
6. Cochran WT et al (1967) What is the fast Fourier transform. Proc IEEE 55(10):164–174

7. Cooley JW, Tukey JW (1965) An algorithm for the machine calculation of complex Fourier series. Math Comput 19:297–301
8. Hadamard J (1893) Resolution d'une question relative aux determinants. Bull Sci Math Ser 2 (17) Part I:240–246
9. Narasimha MJ, Peterson AM (1978) On the computation of the discrete cosine transform. IEEE Trans Comm COM-26(6):934–936
10. Oraintara S, Chen Y-J, Nguyen TQ (2002) Integer fast Fourier transform. IEEE Trans Signal Process 50:607–618

Chapter 6
IIR Digital Filters

6.1 Introduction

In this chapter we will describe the design of *infinite impulse response* (IIR) digital filters. The impulse response of an IIR digital filter has an infinite extent or length or duration, hence the name IIR filters. Design of an IIR filter amounts to the determination of its impulse response sequence $\{h[n]\}$ in the discrete-time domain or to the determination of its transfer function $H(e^{j\Omega})$ in the frequency domain. The design can also be accomplished in the Z-domain. In fact, this is the most commonly used domain. The theory of analog filters preceded that of digital filters. Elegant design techniques for analog filters in the frequency domain were developed much earlier than the development of digital filters. As a result, we will adopt some of the techniques used to design analog filters in designing an IIR digital filter. In order to facilitate the design of an IIR digital filter, one must specify certain parameters of the desired filter. These parameters can be in the discrete-time domain or in the frequency domain. Once the parameters or specifications are known, the task is to come up with either the impulse response sequence or the transfer function that approximates the specifications of the desired filter as closely as possible. In the discrete-time domain, one of the design techniques is known as the *impulse invariance* method. In the frequency domain, the design will yield a *Butterworth* or *Chebyshev* or *elliptic* filter. These three design procedures will result in a closed-form solution. Similarly, the impulse invariance technique will also result in a closed-form solution to the design of IIR digital filters. In addition to these analytical methods, an IIR digital filter can also be designed using iterative techniques. These are called the *computer-aided* design. Let us first describe the impulse invariance

Electronic supplementary material: The online version of this article (https://doi.org/10.1007/978-3-319-76029-2_6) contains supplementary material, which is available to authorized users.

© Springer International Publishing AG, part of Springer Nature 2019
K. S. Thyagarajan, *Introduction to Digital Signal Processing Using MATLAB with Application to Digital Communications*,
https://doi.org/10.1007/978-3-319-76029-2_6

method of designing an IIR digital filter. We will then deal with the design in the frequency domain and the computer-aided design.

6.2   Impulse Invariance Technique

This technique implies that the impulse response of the desired IIR digital filter is the sampled version of the impulse response of an appropriate analog filter. Therefore, one has to specify either the transfer function of an appropriate analog filter or its impulse response. It is customary to specify the transfer function of an analog filter in terms of the Laplace variable s. The analog transfer function $H_A(s) = \frac{N(s)}{D(s)}$ is a rational polynomial in the Laplace variable s with the degree of the numerator polynomial less than or equal to that of the denominator. In order to determine the impulse response $h_A(t)$ of the analog filter, we must find the inverse Laplace transform of the corresponding transfer function. In practice though, the inverse Laplace transform is found by first expressing the given analog transfer function in partial fraction form, then finding the time-domain function corresponding to each pole factor, and then adding the terms. We can enumerate the steps involved in the impulse invariance method as follows:

Impulse Invariance Design Procedure   Given the analog filter transfer function $H_A(s)$, do the following.

1. Express the transfer function in partial fraction form as given by

$$H_A(s) = \sum_{i=1}^{N} \frac{A_i}{s + p_i},\qquad(6.1)$$

where $-p_i$ are the poles, A_i are the corresponding residues, and N is the filter order. If a pole is complex, it will occur with its conjugate. Combine the two complex conjugate poles with the corresponding complex conjugate residues and express as a second-order function.

2. Find the corresponding time-domain analog function. Each term in Eq. (6.1) will correspond to the time-domain function

$$\frac{A_i}{s + p_i} \leftrightarrow A_i e^{-p_i t} u(t)\qquad(6.2)$$

3. Add all the N time-domain functions in the partial fraction expansion to obtain $h_A(t)$.
4. Convert the analog impulse response to the impulse response $h[n]$ of the IIR digital filter by sampling $h_A(t)$ with appropriate sampling interval T. Thus,

$$h[n] = h_A(t)|_{t=nT} = \sum_{i=1}^{N} A_i e^{-p_i nT} \quad n \geq 0 \tag{6.3}$$

To make it clearer, consider the following example.

Example 6.1 Design an IIR digital filter using the impulse invariance method. The analog filter transfer function is given as $H_A(s) = \frac{s+2}{s^2+2s+10}$. Use a sampling frequency of 4 Hz.

Solution Since the given analog transfer function is of order 2, there are two poles. The poles correspond to the roots of the denominator of the analog transfer function. That is,

$$s^2 + 2s + 10 = 0 \Longrightarrow (s+1)^2 + 9 = 0 \Longrightarrow s_{1,2} = -1 \pm j3$$

In terms of the pole factors, we have

$$H_A(s) = \frac{A}{s+1-j3} + \frac{A^*}{s+1+j3}$$

The residues are found from

$$A = (s+1-j3)H_A(s)|_{s=-1+j3} = \left.\frac{s+2}{s+1+j3}\right|_{s=-1+j3} = \frac{1+j3}{j6} = \frac{3-j}{6}$$

and $A^* = \frac{3+j}{6}$. The impulse response of the analog filter is the inverse Laplace transform of its transfer function. From the partial fractions, we have

$$\frac{(3-j)/6}{s+1-j3} \Longrightarrow \frac{3-j}{6}e^{-(1-j3)t}u(t)$$

Similarly, the inverse Laplace transform of the second factor in the partial fraction expansion is found to be

$$\frac{(3+j)/6}{s+1+j3} \Longrightarrow \frac{3+j}{6}e^{-(1+j3)t}u(t)$$

By combining the two complex conjugate terms in the partial fraction expansion, we can express the impulse response of the analog filter as

$$h_A(t) = e^{-t}\left\{ \cos 3t + \frac{1}{3}\sin 3t \right\}u(t) \tag{6.4}$$

Alternatively, we can use the fact that

$$\mathcal{L}\{e^{-t}\cos(3t)u(t)\} = \frac{s+1}{s^2+2s+10}$$

$$\mathcal{L}\{e^{-t}\sin{(3t)}u(t)\} = \frac{3}{s^2 + 2s + 10}$$

and rewrite H(s) as

$$\frac{s+2}{s^2+2s+10} = \frac{s+1}{s^2+2s+10} + \frac{\left(\frac{1}{3}\right)3}{s^2+2s+10}$$

and then identify each term with the respective time-domain function as shown above. Finally, the impulse response of the desired digital filter is obtained by sampling the impulse response in Eq. (6.4) at intervals of $\frac{1}{4} = 0.25$ s and is given by

$$h[n] = e^{-0.25n}\left\{\cos{(0.75n)} + \frac{1}{3}\sin{(0.75n)}\right\}u[n] \qquad (6.5)$$

In order to compare the frequency responses of the analog and digital filters, we need to express the impulse response h[n] in the Z-domain. By taking the Z-transform of Eq. (6.5) and with some algebraic manipulation, we get

$$H(z) = \frac{1 - 0.3929z^{-1}}{1 - 1.1397z^{-1} + 0.6065z^{-2}} \qquad (6.6)$$

In Fig. 6.1, the impulse responses of the digital and analog filters are shown in the top and bottom plots, respectively. The impulse response of the digital filter is seen

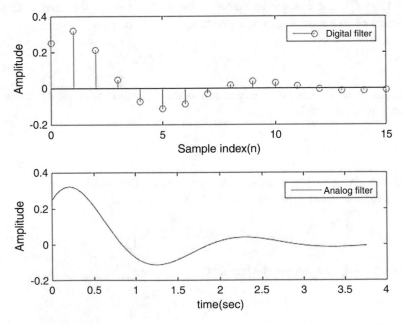

Fig. 6.1 Impulse response of the digital filter in Example 6.1 obtained using the impulse invariance method. Top plot: impulse response of the digital filter. Bottom plot: impulse response of the analog filter

to be the sampled version of that of the analog filter. The frequency response of both filters is shown in Fig. 6.2. The two are not identical. Guess why? Because the continuous-time impulse response of the analog filter is sampled to obtain the discrete-time impulse response, we will expect aliasing distortion if the frequency response of the analog time-domain function is not band limited. In fact, the frequency response of the analog filter is of infinite extent, though it decays as the frequency increases. Therefore, no matter how large the sampling frequency is, there will be some aliasing distortion. We can also use the MATLAB function *impinvar* to design the digital filter using the impulse invariance method. The actual function call is *[Bz,Az]* = *impinvar(B,A,Fs)*, where B and A are the coefficient vectors corresponding to the numerator and denominator polynomials of the analog filter transfer function, both in ascending order of the Laplace variable s, Fs is the sampling frequency, and Bz and Az are the coefficient vectors of the numerator and denominator polynomials of the transfer function of the digital filter, both in ascending powers of the variable z^{-1}. The frequency response of the digital filter obtained using the MATLAB function *impinvar* is also shown in Fig. 6.2. As expected, the frequency responses of the digital filters obtained using MATLAB and analysis are identical (Fig. 6.2).

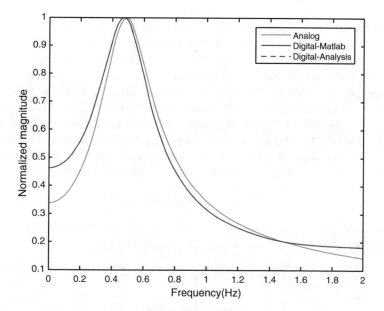

Fig. 6.2 Frequency responses of the analog and digital filters in Example 6.1

6.3   Design of IIR Digital Filters in the Frequency Domain

Instead of designing an IIR digital filter in the time domain, one can design in the frequency domain. In fact, the design in the frequency domain makes more sense than designing an IIR digital filter in the discrete-time domain because (a) we are more accustomed to changes in frequencies than to changes in time functions and (b) design techniques are more elegant and powerful in the frequency domain. When designing an IIR digital filter in the frequency domain, it is customary to design a lowpass filter. Other types of filters, namely, highpass, bandpass, and bandstop filters, can be designed by suitable frequency transformations. As we pointed out earlier, the theory of analog filters preceded that of digital filters. As a consequence, various types of analog filter transfer functions are used in the design of IIR digital filters. For instance, by converting an analog Butterworth filter into an IIR digital filter, we have an IIR Butterworth digital filter. However, in order to preserve the shape of the frequency response of an analog Butterworth filter in the digital filter domain, one has to start with an analog Butterworth filter transfer function and use an appropriate transformation to convert it to the digital domain. This technique is known as the *bilinear transformation*, which we will describe next.

6.3.1   Digital Filter Frequency Specifications

It is not possible to meet the characteristics of an ideal lowpass filter, be it analog or digital. By ideal we mean a brick wall type. Therefore, one has to specify tolerance limits to the frequency specifications. Then one can come up with an approximation to the frequency specifications. The approximation is achieved by a proper transfer function. So, let us first look at the frequency specifications of an IIR digital filter. Figure 6.3 shows a typical magnitude response of a digital lowpass filter. The parameter δ_p is called the peak passband ripple and δ_s the peak stopband ripple. Instead of the actual magnitude of the frequency response, it is also common to express the magnitude in decibels (dB). More specifically, the attenuation or loss function in dB is related to the magnitude of the frequency response by

$$A(\Omega) = -20log_{10}\left|H\left(e^{j\Omega}\right)\right| \ dB \tag{6.7}$$

The peak passband and stopband ripples in dB are denoted by α_p and α_s, respectively, and are related to the actual ripples δ_p and δ_s by

$$\alpha_p = -20log_{10}\left(1 - \delta_p\right) \ dB \tag{6.8a}$$

$$\alpha_s = -20log_{10}\delta_s \ dB \tag{6.8b}$$

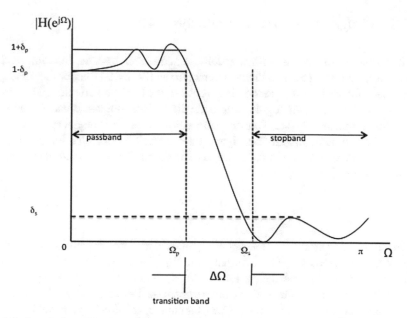

Fig. 6.3 A typical specification of a lowpass digital filter in the frequency domain. The ordinate is the magnitude of the desired frequency response, and the abscissa is the frequency $\frac{2\pi f}{F_s}$; π corresponds to half the sampling frequency

As shown in the figure, Ω_p and Ω_s are called the passband and stopband edges, respectively. The transition band or width is then expressed as $\Delta\Omega = \Omega_s - \Omega_p$. In terms of the actual frequencies, the passband and stopband edges are written as $\Omega_p = \frac{2\pi f_p}{F_s}$ and $\Omega_s = \frac{2\pi f_s}{F_s}$, respectively, where f_p in Hz is the passband edge, f_s in Hz is the stopband edge, and F_s in Hz is the sampling frequency. The following example will clarify the definition of the ripples in actual values and in dB.

Example 6.2   Find the peak ripples δ_p and δ_s corresponding to the peak passband ripple $\alpha_p = 0.24$ *dB* and minimum stopband attenuation $\alpha_s = 49$ *dB*.

Solution

$$\alpha_p = -20log_{10}\left(1 - \delta_p\right) \Rightarrow log_{10}\left(1 - \delta_p\right) = -\frac{\alpha_p}{20} = -0.012$$

Therefore,

$$1 - \delta_p = 10^{-0.012} \Rightarrow \delta_p = 1 - 10^{-0.012} = 0.0272528$$

Corresponding to the minimum attenuation, we have

$$\alpha_s = -20log_{10}\delta_s \Rightarrow \delta_s = 10^{-0.05\alpha_s} = 10^{-0.05*49} = 0.00354813$$

6.3.2   Design Using Bilinear Transformation

In this method, one starts with a specific analog filter transfer function such as Butterworth, elliptic, etc., and then converts it to an IIR digital filter of the same type. The analog filter's transfer function is in terms of the Laplace variable s. The process of conversion from the analog domain to the digital filter domain is carried out using the bilinear transformation. This transformation maps the frequency axes in the analog domain into that in the digital filter domain. Equivalently, the Laplace variable s is transformed into the z variable. More specifically, the bilinear transformation is defined as

$$s = \frac{1}{T}\left(\frac{1 - z^{-1}}{1 + z^{-1}}\right) \tag{6.9}$$

In the above equation, T is the sampling interval. Before we embark on the design procedure, let us look at the transformation in Eq. (6.9) more closely. Let us see how the complex s-plane in the analog domain is mapped into the complex z-plane in the digital filter domain. To this end, we have to express the complex variable z in terms of the complex variable s. From Eq. (6.9), the inverse relationship can be found to be

$$z = \frac{\left(1 + \frac{T}{2}s\right)}{\left(1 - \frac{T}{2}s\right)} \tag{6.10}$$

The complex variable in the analog domain is described as

$$s = \sigma + j\omega \tag{6.11}$$

where the variable of the real axis (abscissa) is σ and the variable of the imaginary axis (ordinate) is the sinusoidal frequency ω. Substituting for s from (6.11) in (6.10), we can write (6.10) as

$$z = \frac{1 + \frac{T}{2}(\sigma + j\omega)}{1 - \frac{T}{2}(\sigma + j\omega)} = \frac{\left(1 + \frac{T}{2}\sigma\right) + j\frac{T}{2}\omega}{\left(1 - \frac{T}{2}\sigma\right) - j\frac{T}{2}\omega} \tag{6.12}$$

The square of the magnitude of z from (6.12) is found to be

$$|z|^2 = \frac{\left(1 + \frac{T}{2}\sigma\right)^2 + \left(\frac{T}{2}\omega\right)^2}{\left(1 - \frac{T}{2}\sigma\right)^2 + \left(\frac{T}{2}\omega\right)^2} \tag{6.13}$$

From Eq. (6.13) we observe that

$$|z|^2 = \begin{cases} < 1, for\ \sigma < 0 \\ \quad 1, for\ \sigma = 0 \\ > 1, for\ \sigma > 0 \end{cases} \tag{6.14}$$

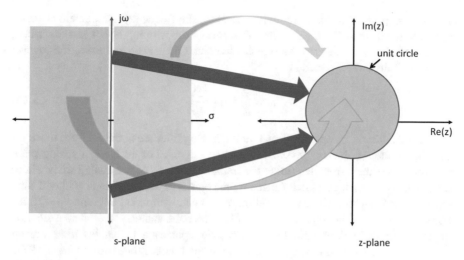

Fig. 6.4 Bilinear mapping of the s-plane onto the z-plane

The interpretation of (6.14) is as follows: When σ is negative, we are in the left half of the complex s-plane, which is mapped on to the interior of the unit circle in the complex z-plane. When σ is positive, we are in the right half of the complex s-plane, which is mapped on to the exterior of the unit circle in the complex z-plane. The entire imaginary axis in the complex s-plane corresponds to $\sigma = 0$, which is mapped on to the circumference of the unit circle in the complex z-plane as shown in Fig. 6.4. In fact, the positive frequencies are mapped on to the upper half of the circumference of the unit circle in counterclockwise manner, and the negative frequencies in the s-plane are mapped on to the lower half of the circumference of the unit circle in clockwise direction in the z-plane (refer to Eq. 6.15). We have to make sure that a stable analog filter will result in a stable digital filter when the bilinear transformation is used. How do we verify this? An analog filter is stable in the BIBO sense if all its poles are on the left half of the s-plane. We just observed that the entire left half of the s-plane maps into the interior of the z-plane. It is well known that an IIR digital filter is stable in the BIBO sense if all its poles are inside the unit circle in the z-plane. Thus, the bilinear transform maintains the stability in the Z-domain. The only remaining thing is to see how the imaginary axis in the s-plane is related to the points on the unit circle in the z-plane. Our intuition indicates that the relationship between the frequencies in the analog and digital filter domains is not linear because of the presence of the denominator factor in Eq. (6.10). When $s = j\omega$, the variable in the complex z-plane takes the form $z = e^{j\Omega}$, and so

$$e^{j\Omega} = \frac{1 + j\frac{T}{2}\Omega}{1 - j\frac{T}{2}\Omega} = \frac{e^{j\tan^{-1}\left(\frac{\omega T}{2}\right)}}{e^{-j\tan^{-1}\left(\frac{\omega T}{2}\right)}} = e^{j2\tan^{-1}\left(\frac{\omega T}{2}\right)} \Rightarrow \omega = 2\tan^{-1}\left(\frac{\omega T}{2}\right) \quad (6.15)$$

As can be seen from (6.15), the relationship between the frequency variables Ω and ω is nonlinear. Therefore, the frequencies in the analog domain are *warped* in the

z-plane. As pointed out in the previous section, the frequency specifications correspond to the desired IIR digital filter. Because of the warping effect just described,
the specified frequencies for the digital filter must be *prewarped* using the inverse
relationship in (6.15), which is

$$\omega = \frac{2}{T} \tan\left(\frac{\Omega}{2}\right) \tag{6.16}$$

Once the frequency specifications for the analog filter are obtained from those for
the digital filter via (6.16), we must determine the order of the chosen analog filter.
Of course, we must then determine the transfer function of the analog filter whose
order has been just determined. As mentioned earlier, a wealth of analog filter design
exists in the literature. We can, therefore, use such information to come up with the
transfer function of the required analog filter. Once the analog filter transfer function
expressed in the Laplace variable s is found, we replace s in the analog filter transfer
function by the term on the right-hand side of the bilinear transform of Eq. (6.9) to
obtain the transfer function of the IIR digital filter. Thus,

$$H(z) = H_A(s)\Big|_{s=\frac{1}{T}\frac{(1-z^{-1})}{(1+z^{-1})}} \tag{6.17}$$

We enumerate the design procedure used in converting an analog filter into an IIR
digital filter using the bilinear transform method as follows: Given the frequency
domain specifications of a lowpass IIR digital filter, do the following.

1. Choose an appropriate analog filter type such as, Butterworth, Chebyshev, etc.
2. Prewarp the critical frequencies of the digital filter using Eq. (6.16).
3. Determine the required order of the analog filter.
4. Determine the corresponding analog filter transfer function.
5. Convert the analog filter transfer function into the digital filter transfer function
   via Eq. (6.17).

6.3.3 Butterworth Lowpass IIR Digital Filter

A Butterworth lowpass filter is also known as the maximally flat filter because its
magnitude of the frequency response has the largest number of derivatives with
respect to ω equal to zero at zero frequency. In other words, the magnitude of its
frequency response is as flat as possible at DC. An Nth-order lowpass analog
Butterworth filter with a *cutoff* frequency ω_c has the magnitude squared of frequency
response expressed as

$$|H_A(\omega)|^2 = \frac{1}{1 + \left(\frac{\omega}{\omega_c}\right)^{2N}} \tag{6.18}$$

Note that at $\omega = \omega_c$, the magnitude squared of the frequency response of the Butterworth lowpass analog filter is half or is -3 dB. Therefore, the cutoff frequency here is also called the *half power* frequency or 3 dB frequency. Typical specifications in the frequency domain for a lowpass filter are the passband edge ω_p, stopband edge ω_s, passband ripple, and minimum stopband attenuation, both in dB. The magnitude squared of the Butterworth lowpass filter function in (6.18) drops to

$$|H_A(\omega_p)|^2 = \frac{1}{1 + \left(\frac{\omega_p}{\omega_c}\right)^{2N}} = \frac{1}{1 + \varepsilon^2} \tag{6.19}$$

at the passband edge frequency. The passband ripple α in dB is related to ε by

$$-20log_{10}\left(\frac{1}{\sqrt{1 + \varepsilon^2}}\right) = \alpha \ dB \Longrightarrow \varepsilon = \sqrt{10^{0.1\alpha} - 1} \tag{6.20}$$

The actual minimum stopband attenuation is related to the attenuation in dB by

$$A = 10^{0.05A_{dB}} \tag{6.21}$$

The filter order N is related to ε, A, ω_p, and ω_s through the following:

$$\frac{1}{k} \equiv \frac{\omega_s}{\omega_p} \tag{6.22a}$$

$$\frac{1}{k_1} \equiv \frac{\sqrt{A^2 - 1}}{\varepsilon} \tag{6.22b}$$

$$N = \left\lceil \frac{log_{10}\left(\frac{1}{k_1}\right)}{log_{10}\left(\frac{1}{k}\right)} \right\rceil \tag{6.23}$$

Finally, in order to find the Nth-order transfer function of the analog lowpass Butterworth filter, we need to know the cutoff frequency. This can be found from Eq. (6.19) as

$$\omega_c = \varepsilon^{-\frac{1}{N}}\omega_p \tag{6.24}$$

Alternatively, the cutoff frequency can also be found from

$$|H_A(\omega_s)|^2 = \frac{1}{1 + \left(\frac{\omega_s}{\omega_c}\right)^{2N}} = \frac{1}{A^2} \tag{6.25}$$

With the filter order and the cutoff frequency known, the corresponding lowpass Butterworth filter transfer function is determined from

$$H_A(s) = \frac{1}{D_N\left(\frac{s}{\omega_c}\right)} \tag{6.26}$$

where $D_N(s)$ is the Nth-order Butterworth polynomial. Butterworth polynomials can be found in standard textbooks on analog filter design. We can also express the analog filter transfer function in (6.26) in terms of its pole factors as given by

$$H_A(s) = \frac{\omega_c^N}{\prod\limits_{m=1}^{N}(s - p_m)} \tag{6.27}$$

where the poles are described by

$$p_m = \omega_c e^{j\left(\frac{(N+2m-1)\pi}{2N}\right)}, 1 \leq m \leq N \tag{6.28}$$

It should be mentioned that the coefficients of the polynomials of the transfer function are real. Therefore, if a pole is complex, it should occur with its complex conjugate. Finally, the desired IIR digital filter transfer function is found from Eq. (6.17). Let us look at an example to clarify the design procedure described thus far.

Example 6.3 Design an IIR lowpass Butterworth digital filter with the following frequency specifications: passband edge $\Omega_p = \frac{\pi}{4}$, stopband edge $\Omega_s = \frac{\pi}{2}$, passband ripple $\alpha_p = 0.5 \ dB$, and a minimum stopband attenuation of 20 dB.

Solution Because the given specifications are for the digital filter, we have to first prewarp the critical frequencies of the digital filter using Eq. (6.16). Since the sampling frequency is not specified, we can omit the scaling factor $\frac{2}{T}$ in Eq. (6.16). This is the same as normalizing the frequencies in the analog domain by twice the sampling frequency. This will not affect the critical frequencies of the digital filter because the frequencies in the analog domain will be re-warped after applying the bilinear transformation to the analog filter transfer function. Therefore, the passband edge and the stopband edge frequencies of the analog filter without the 2/T factor are

$$\omega_p = \tan\left(\frac{\Omega_p}{2}\right) = \tan\left(\frac{\pi}{8}\right) \approx 0.414214 \tag{6.29a}$$

$$\omega_s = \tan\left(\frac{\Omega_s}{2}\right) = \tan\left(\frac{\pi}{4}\right) = 1 \tag{6.29b}$$

Corresponding to the passband ripple and minimum stopband attenuation, we have

$$\varepsilon^2 = 10^{0.1\alpha_p} - 1 = 10^{0.05} - 1 = 0.122018 \tag{6.30a}$$

$$20log_{10}A = 20 \Rightarrow A = 10 \tag{6.30b}$$

Next we find the filter order as follows:

$$\frac{1}{k} = \frac{\omega_s}{\omega_p} = \frac{1}{0.414214} = 2.41421 \tag{6.31a}$$

$$\frac{1}{k_1} = \frac{\sqrt{A^2 - 1}}{\varepsilon} = 28.4843 \tag{6.31b}$$

$$N = \left\lceil \frac{log_{10}(28.4843)}{log_{10}(2.41421)} \right\rceil = \lceil 3.8002 \rceil = 4 \tag{6.31c}$$

Note that in Eq. (6.31c), we need to round the number to the integer that is greater than or equal to the number within the ceiling operator. Having determined the required order of the Butterworth lowpass filter, we need to determine the cutoff frequency of the lowpass filter, which is obtained from (6.24):

$$\omega_c = \varepsilon^{-\frac{1}{N}}\omega_p = 0.538793 \tag{6.32}$$

The poles of the fourth-order Butterworth lowpass analog filter are found from Eq. (6.28):

$$p_m = \omega_c e^{j\pi\frac{(4+2m-1)}{8}}, 1 \leq m \leq 4 \tag{6.33}$$

We notice from (6.33) that p_1 and p_4 are complex conjugates of each other and, similarly, p_2 and p_3 are complex conjugates of each other. Using (6.27) with algebraic simplification, we obtain the analog transfer function of the lowpass Butterworth filter as

$$H_A(s) = \frac{H_0}{(s^2 + a_1 s + a_2)(s^2 + b_1 s + b_2)} \tag{6.34}$$

where the constants are $H_0 = 0.0842722$, $a_1 = 0.412374$, $a_2 = 0.290297$, $b_1 = 0.995558$, and $b_2 = 0.290297$. We finally obtain the transfer function of the desired IIR digital Butterworth lowpass filter using Eq. (6.17), which is given in (6.35) after algebraic manipulation:

$$H(z) = \frac{0.0217(1 + 4z^{-1} + 6z^{-2} + 4z^{-3} + z^{-4})}{(1 - 0.8336z^{-1} + 0.5156z^{-2})(1 - 0.6209z^{-1} + 0.12189z^{-2})} \tag{6.35}$$

We can also use MATLAB to design the lowpass Butterworth digital filter. The MATLAB function *butter* designs either an analog or a digital filter of the type Butterworth. The actual function call is *[z,p,k] = butter(N,wc,'s')*, where N is the lowpass filter order, wc is the cutoff frequency, z is a vector of zeros, p is a vector of poles, k is a gain constant, and the letter s within single quotes implies that the filter is

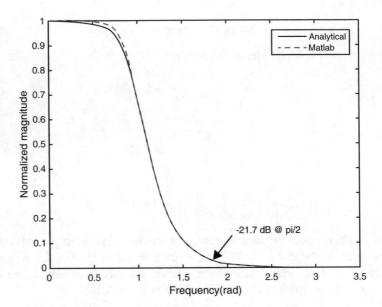

Fig. 6.5 Frequency response of a fourth-order lowpass Butterworth IIR digital filter obtained using the bilinear transformation

analog. If this argument is missing, then it is a digital filter. Next, we apply the bilinear transform to the designed filter using the MATLAB function *bilinear*. This function converts the analog filter to a digital filter. The actual function call is *[zd,pd,kd] = bilinear(z,p,Fs)*, where z and p are the zeros and poles of the analog filter, respectively, Fs is the sampling frequency, and zd, pd, and kd are, respectively, the zeros, poles, and gain of the digital filter. Since the sampling frequency is not specified in the problem, we will use a value ½ for the sampling frequency. This makes the normalization factor $2/T = 1$, as we wanted. Finally, we will convert the zeros and poles of the digital filter into the transfer function using the MATLAB function *zp2tf*. The actual function call is *[Nd, Dd] = zp2tf(zd,pd,kd)*. The magnitude of the frequency response of the desired IIR lowpass Butterworth digital filter is plotted for frequencies between 0 and half the sampling frequency and is shown in Fig. 6.5. The figure also shows the frequency response of the Butterworth digital filter that was designed using MATLAB. Due to the differences in the accuracy of arithmetic calculations, we notice some discrepancy between the two frequency responses.

6.3.4   Chebyshev Type I Lowpass IIR Digital Filter

A Chebyshev type I filter has an equiripple characteristic in the passband and falls off monotonically in the stopband. More specifically, an Nth-order Chebyshev type I analog lowpass filter with a passband edge ω_p has the transfer function whose magnitude square is described by

$$|H_A(\omega)|^2 = \frac{1}{1 + \varepsilon^2 T_N^2\left(\frac{\omega}{\omega_p}\right)} \tag{6.36}$$

where the Nth-order Chebyshev polynomial is defined as

$$T_N(x) = \begin{cases} \cos\left(N\cos^{-1}(x)\right), |x| \leq 1 \\ \cosh\left(N\cosh^{-1}(x)\right), |x| > 1 \end{cases} \tag{6.37}$$

From Eq. (6.37), we observe that the Chebyshev polynomial alternates between $+1$ and -1 in the interval between -1 and $+1$ N times, and for $|x| > 1$, it increases monotonically. Therefore, the lowpass Chebyshev filter's frequency response has ripples in the passband. The filter order can be determined from the following equations.

$$|H_A(\omega_s)|^2 = \frac{1}{1 + \varepsilon^2 T_N^2\left(\frac{\omega_s}{\omega_p}\right)} = \frac{1}{A^2} \Rightarrow T_N^2\left(\frac{\omega_s}{\omega_p}\right) = \frac{A^2 - 1}{\varepsilon^2} \tag{6.38}$$

where ω_s is the stopband edge and A is the minimum stopband attenuation. Since $\omega_s > \omega_p$, we have from (6.38)

$$T_N\left(\frac{1}{k}\right) = \cosh\left(N\cosh^{-1}\left(\frac{1}{k}\right)\right) = \frac{\sqrt{A^2 - 1}}{\varepsilon} \equiv \frac{1}{k_1} \tag{6.39}$$

where we have used the fact that

$$\frac{1}{k} = \frac{\omega_s}{\omega_p} \tag{6.40}$$

From Eqs. (6.39) and (6.40), we determine the value of N as

$$N = \frac{\cosh^{-1}\left(\frac{1}{k_1}\right)}{\cosh^{-1}\left(\frac{1}{k}\right)} = \frac{\ln\left(\frac{1}{k_1} + \sqrt{\left(\frac{1}{k_1}\right)^2 - 1}\right)}{\ln\left(\frac{1}{k} + \sqrt{\left(\frac{1}{k}\right)^2 - 1}\right)} \tag{6.41}$$

Once the filter order is found, the transfer function of the lowpass Chebyshev type I analog filter in terms of its poles are obtained from

$$H_A(s) = \frac{H_0}{\prod\limits_{n=1}^{N} (s - p_n)} \tag{6.42}$$

where H_0 is a normalization factor and the poles are

$$p_n = \sigma_n + j\omega_n, 1 \leq n \leq N \tag{6.43a}$$

$$\sigma_n = -\omega_p a_1 \sin\left(\frac{(2n-1)\pi}{2N}\right), 1 \leq n \leq N \tag{6.43b}$$

$$\omega_n = \omega_p a_2 \cos\left(\frac{(2n-1)\pi}{2N}\right), 1 \leq n \leq N \tag{6.43c}$$

$$a_1 = \frac{\gamma^2 - 1}{2\gamma} \tag{6.43d}$$

$$a_2 = \frac{\gamma^2 + 1}{2\gamma} \tag{6.43e}$$

$$\gamma = \left(\frac{1 + \sqrt{1 + \varepsilon^2}}{\varepsilon}\right)^{\frac{1}{N}} \tag{6.43f}$$

Similar to the lowpass Butterworth filter, the lowpass Chebyshev type I analog filter has no zeros in the finite s-plane. Hence the numerator is just a constant. Finally, we apply the bilinear transformation to the analog filter to obtain the transfer function of the lowpass Chebyshev type I IIR digital filter. The procedural steps in the design of a lowpass Chebyshev type I filter are similar to those used for the Butterworth filter except to use the appropriate equations.

Example 6.4 Design a lowpass IIR Chebyshev type I digital filter using bilinear transformation. Use the same frequency specifications as in Example 6.3.

Solution First, we have to prewarp the critical frequencies of the digital filter. Therefore, we have

$$\omega_p = \tan\left(\frac{\Omega_p}{2}\right) = \tan\frac{\pi}{8} = 0.414214$$

$$\omega_s = \tan\left(\frac{\Omega_s}{2}\right) = \tan\frac{\pi}{4} = 1$$

Corresponding to 0.5 dB passband ripple, we have

$$\varepsilon^2 = 10^{0.1*0.5} - 1 = 0.122018$$

We already know that the minimum stopband attenuation A is 20 dB or 10 in actual value. From (6.39) and (6.40), we determine $\frac{1}{k_1} = 28.4843$ and $\frac{1}{k} = 2.41421$. Using these values in (6.41), we find the filter order $N = \lceil 2.6444 \rceil = 3$. Note that the order of the lowpass Chebyshev type I analog filter is less than that of the Butterworth filter for the same frequency specifications. We now determine the poles of the Chebyshev type I analog filter using Eq. (6.43) and are given by

$$p_1 = -0.1297 + j0.4233 \tag{6.44a}$$

$$p_2 = -0.2595 \tag{6.44b}$$

$$p_3 = p_1^* = -0.1297 - j0.4233 \tag{6.44c}$$

From these poles we obtain the analog transfer function of the lowpass Chebyshev type I filter as

$$H_A(s) = \frac{H_0}{(s + 0.2595)(s^2 + 0.2594s + 0.196)} \tag{6.45}$$

After applying the bilinear transformation with $\frac{2}{T} = 1$ to the analog filter function in (6.45) and with some algebraic manipulation, we get

$$H(z) = H_A(s)\big|_{s=\frac{1-z^{-1}}{1+z^{-1}}} = \frac{0.5455(1 + 3z^{-1} + 3z^2 + z^{-3})}{(1 - 0.5879z^{-1})(1 - 1.1049z^{-1} + 0.6435z^{-2})} \tag{6.46}$$

The magnitude of the frequency response of the lowpass Chebyshev type I IIR digital filter is shown in Fig. 6.6. As mentioned above, the magnitude has ripples in the passband and decreases monotonically in the stopband. It is found that the minimum stopband attenuation for the third-order lowpass Chebyshev type I filter is 24.7 dB. Compare this with that for the Butterworth filter, which is 3 dB less! We

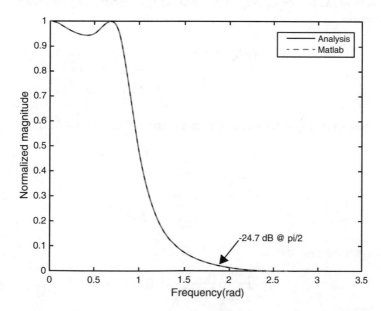

Fig. 6.6 Frequency response of a third-order lowpass Chebyshev type I IIR digital filter obtained using the bilinear transformation. The frequency specifications are the same as in Example 6.3

can also design the Chebyshev type I digital filter using MATLAB. To design an Nth-order lowpass analog Chebyshev type I filter, we use the function *cheby1*. The actual function call is *[z,p,k] = cheby1(N,ap,wp, 's')*, where the arguments are the filter order, passband ripple in dB, passband edge, and the letter s in single quote to imply analog filter. The function returns the zeros, poles, and gain of the analog filter, in that order. As in Example 6.3, we pass the zeros, poles, and the gain of the analog filter to the function *bilinear* to obtain the zeros, poles, and the gain of the digital filter. Finally, the transfer function of the Chebyshev type I lowpass IIR digital filter is obtained using the function *zp2tf*. As a comparison to the results obtained from the analysis, the magnitude of the frequency response of the Chebyshev type I digital filter designed using MATLAB is also shown in Fig. 6.6. It is identical to the analytical result. The MATLAB code for this example is in the M-file named *Example6_4.m*.

6.3.5   *Chebyshev Type II Lowpass IIR Digital Filter*

The Chebyshev type II lowpass filter is similar to that of the type I filter except that the magnitude square of the frequency response falls off monotonically in the passband and alternates in the stopband. Another difference is that the type II Chebyshev lowpass analog filter has zeros in the finite s-plane as well. The magnitude square of the frequency response of type II Chebyshev lowpass analog filter is expressed as

$$|H_A(\omega)|^2 = \cfrac{1}{1 + \varepsilon^2 \left[\cfrac{T_N\left(\frac{\omega_s}{\omega_p}\right)}{T_N\left(\frac{\omega_s}{\omega}\right)} \right]^2} \tag{6.47}$$

The corresponding transfer function can be expressed in terms of zeros and poles as

$$H_A(s) = H_0 \frac{\displaystyle\prod_{m=1}^{N} (s - z_m)}{\displaystyle\prod_{m=1}^{N} (s - p_m)} \tag{6.48}$$

The zeros are described as

$$z_m = j\frac{\omega_s}{\cos\left(\frac{(2m-1)\pi}{2N}\right)}, 1 \le m \le N \tag{6.49}$$

As can be seen from (6.49), the zeros lie on the imaginary axis. Further, if the filter order N is odd, then the zero is at infinity for $m = \frac{(N+1)}{2}$. It implies that that zero factor is absent. The poles are described as follows:

$$p_m = \sigma_m + j\omega_m, 1 \leq m \leq N \qquad (6.50)$$

where

$$\sigma_m = \frac{(\omega_s \alpha_m)}{(\alpha_m^2 + \beta_m^2)}, \omega_m = \frac{(\omega_s \beta_m)}{(\alpha_m^2 + \beta_m^2)}, \qquad (6.51a)$$

$$\alpha_m = -\omega_p \epsilon_1 \sin\left(\frac{(2m-1)\pi}{2N}\right), \beta_m = -\omega_p \epsilon_2 \cos\left(\frac{(2m-1)\pi}{2N}\right) \qquad (6.51b)$$

$$\epsilon_1 = \frac{(\gamma^2 - 1)}{2\gamma}, \epsilon_2 = \frac{(\gamma^2 + 1)}{2\gamma}, \gamma = \left(A + \sqrt{A^2 - 1}\right)^{\frac{1}{N}} \qquad (6.51c)$$

The filter order N is determined from Eq. (6.41), as in type I case. The following example will illustrate the design of type II Chebyshev filter.

Example 6.5 Design a type II Chebyshev lowpass IIR digital filter using the same specifications as in Example 6.3.

Solution Since the frequency specifications are the same as for the Chebyshev type I filter, we obtain the filter order for type II Chebyshev filter N = 3. The transfer function of the lowpass analog filter is found to be

$$H_A(s) = K_0 \frac{s^2 + 1.3333}{s^3 + 1.4054s^2 + 0.9421s + 0.4020} \qquad (6.52)$$

where K_0 is a normalization constant. Note that since the filter order is an odd number, the numerator has only two finite zeros, as expected. Finally, the transfer function of the type II Chebyshev lowpass IIR digital filter is derived from that of the analog filter using the normalized bilinear transform and is given by

$$H(z) = \frac{1 + 1.2857z^{-1} + 1.2857z^{-2} + z^{-3}}{1 - 0.6020z^{-1} + 0.4957z^{-2} - 0.0359z^{-3}} \qquad (6.53)$$

The magnitude of the frequency response of the lowpass digital filter in (6.53) is plotted against normalized frequency and is shown in Fig. 6.7. As a comparison, the frequency response of the filter designed using MATLAB is also plotted and is shown in the same figure in dotted red line. They are identical. It is also found that the attenuation is only 20 dB at the stopband edge. The MATLAB function used in the design is *cheby2*. The steps involved here are similar to the ones in Example 6.4. See the MATLAB M-file named *Example6_5.m*.

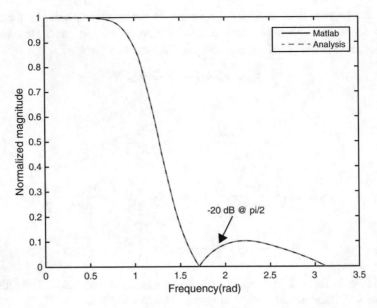

Fig. 6.7 Frequency response of a third-order lowpass Chebyshev type II IIR digital filter obtained using the bilinear transformation. The frequency specifications are the same as in Example 6.3

6.3.6 Elliptic Lowpass IIR Digital Filter

An elliptic filter, also known as a *Cauer filter*, exhibits ripples in both the passband and stopband. For the same given frequency specifications of a lowpass filter, an elliptic filter has the smallest filter order. Assuming the frequency specifications of a lowpass elliptic analog filter to be passband edge ω_p, passband ripple ε, stopband edge ω_s, and minimum stopband attenuation A, the magnitude square of the frequency response can be described by

$$|H_A(\omega)|^2 = \frac{1}{1 + \varepsilon^2 E_N^2\left(\frac{\omega}{\omega_p}\right)},\qquad(6.54)$$

where $E_N(\omega)$ is a rational function of order N. The theory of elliptic filter approximation is quite involved, and so we will not deal with it here. Instead, we will resort to the MATLAB. However, it is necessary to determine the filter order satisfying the given frequency specifications in order to come up with the transfer function using the MATLAB function *ellip*. For the specifications mentioned above, the order of a lowpass elliptic analog filter is given approximately by

$$N \cong \frac{2log_{10}\left(\frac{4}{k_1}\right)}{log_{10}\left(\frac{1}{\rho}\right)},\qquad(6.55)$$

where

$$k_1 = \frac{\varepsilon}{\sqrt{A^2 - 1}}, \tag{6.56a}$$

$$k' = \sqrt{1 - k^2}, \tag{6.56b}$$

$$k = \frac{\omega_p}{\omega_s}, \tag{6.56c}$$

$$\rho_0 = \frac{1 - \sqrt{k'}}{2\left(1 + \sqrt{k'}\right)}, \tag{6.56d}$$

$$\rho = \rho_0 + 2(\rho_0)^5 + 15(\rho_0)^9 + 150(\rho_0)^{13} \tag{6.56e}$$

Example 6.6 Design an elliptic lowpass IIR digital filter using the same specifications as in Example 6.3.

Solution   Corresponding to the passband ripple of 0.5 dB, we find

$$-20log_{10}\left(\frac{1}{\sqrt{1 + \varepsilon^2}}\right) = -0.5 \Longrightarrow \varepsilon \cong 0.34931$$

Therefore, $k_1 \cong 0.035107$. Using the passband and stopband edges, we have $k = \tan\frac{\pi}{8}$. Next, we find $k' \cong 0.91018$, $\rho_0 = 0.011762$, and $\rho \cong \rho_0 = 0.011762$. Using the values of k_1 and ρ in (6.55), we get $N = \lceil 2.1318 \rceil = 3$. So, a third-order elliptic filter will satisfy the given specifications. We use the MATLAB function *ellip*, which accepts the filter order, passband ripple in dB, minimum stopband attenuation in dB, passband edge, and the letter s in single quotes. We can use either the statement $[z,p,k] = ellip(N,rp,rs,wp,\text{'}s\text{'})$, which calculates the zeros, poles, and the gain, or the statement $[Na,Da] = ellip(N,rp,rs,wp,\text{'}s\text{'})$, which calculates the transfer function. With either the analog transfer function or the zeros-poles-gain, we call the function *bilinear* to calculate the transfer function of the corresponding digital filter, as we did in the previous examples. Figure 6.8 shows the magnitude of the frequency response of a third-order lowpass IIR elliptic digital filter. In Fig. 6.9, the magnitude of the frequency responses of the Butterworth, Chebyshev types I and II, and the elliptic filters for the same specifications as in Example 6.3 are shown. As expected, the elliptic filter has the smallest transition width, meaning that it has the steepest response between the passband and stopband. The M-file to solve this problem is named *Example6_6.m*.

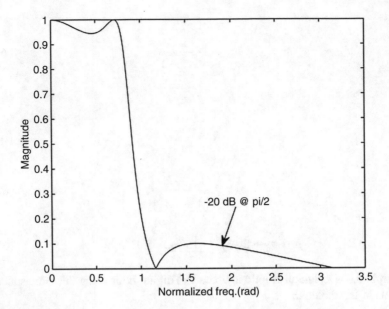

Fig. 6.8 Frequency response of a third-order lowpass elliptic IIR digital filter obtained using the bilinear transformation. The frequency specifications are the same as in Example 6.3

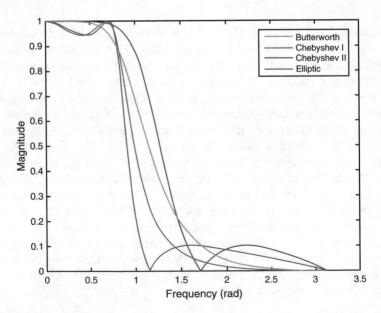

Fig. 6.9 Frequency responses of a third-order lowpass Butterworth, Chebyshev type I and II, and elliptic IIR digital filters obtained using the bilinear transformation. The frequency specifications are the same as in Example 6.3

6.4 Design of IIR Digital Filters Using Frequency Transformation

So far we have discussed the design of a lowpass IIR digital filter. Lowpass filter is not the only type used in practice. Filters such as highpass, bandpass, and bandstop filters are also often used. If we have a prototype lowpass IIR digital filter, we can then convert it to another lowpass, highpass, bandpass, or bandstop filter by using a suitable *spectral transformation* directly in the digital domain. Thus, the method is to replace the complex variable z in the prototype lowpass IIR digital filter $H_P(z)$ by the spectral transformation function $f(\widehat{z})$ as described by

$$H(z) = H_P(z)|_{z=f(\widehat{z})}, \tag{6.57}$$

where the spectral transformation function is defined as

$$f(\widehat{z}) = \pm \prod_{m=1}^{M} \left(\frac{\widehat{z} - \lambda_m}{1 - \lambda_m^* \widehat{z}} \right), |\lambda_m| < 1 \tag{6.58}$$

In (6.58), λ_m may be real or complex, and if complex, it has to occur with its conjugate. For lowpass and highpass conversion, M equals 1, and for bandpass and bandstop, it is 2.

6.4.1 Lowpass-to-Lowpass Conversion

Let us look at converting a lowpass IIR digital filter into another lowpass IIR digital filter. Let the cutoff frequency of the prototype lowpass digital filter $H_{LP}(z)$ be Ω_c. We want to transform this prototype filter to another lowpass filter $H'(\widehat{z})$ with a cutoff frequency $\widehat{\Omega}_c$. Then the spectral transformation will be given by

$$z = f(\widehat{z}) = \frac{\widehat{z} - \lambda}{1 - \lambda\widehat{z}} \tag{6.59}$$

where λ is real. We can find the correspondence between Ω and $\widehat{\Omega}$ as follows. We can rewrite (6.59) as

$$z = \frac{1 - \lambda\widehat{z}^{-1}}{\widehat{z}^{-1} - \lambda} \Rightarrow z^{-1} = \frac{\widehat{z}^{-1} - \lambda}{1 - \lambda\widehat{z}^{-1}} \tag{6.60}$$

Then, when $z = e^{j\Omega}, \widehat{z} = e^{j\widehat{\Omega}}$ and we have, from (6.60)

$$e^{-j\Omega} = \cfrac{\frac{e^{-j\widehat{\Omega}} - \lambda}{1 - \lambda e^{-j\widehat{\Omega}}} = e^{-j\frac{\widehat{\Omega}}{2}} - \lambda e^{j\frac{\widehat{\Omega}}{2}}}{e^{j\frac{\widehat{\Omega}}{2}} - \lambda e^{-j\frac{\widehat{\Omega}}{2}} = e^{-j2\tan^{-1}\left(\left[\frac{1+\lambda}{1-\lambda}\right]\tan\left(\frac{\widehat{\Omega}}{2}\right)\right)}} \tag{6.61}$$

Therefore,

$$\Omega = 2\tan^{-1}\left(\left[\frac{1+\lambda}{1-\lambda}\right]\tan\left(\frac{\widehat{\Omega}}{2}\right)\right) \Rightarrow \tan\frac{\Omega}{2} = \left(\frac{1+\lambda}{1-\lambda}\right)\tan\frac{\widehat{\Omega}}{2} \tag{6.62}$$

The constant λ is found from (6.62) by using the respective cutoff frequencies and is given by

$$\lambda = \frac{\tan\left(\Omega_c/_2\right) - \tan\left(\widehat{\Omega}_c/_2\right)}{\tan\left(\Omega_c/_2\right) + \tan\left(\widehat{\Omega}_c/_2\right)} = \frac{\sin\left(\frac{\Omega_c - \widehat{\Omega}_c}{2}\right)}{\sin\left(\frac{\Omega_c + \widehat{\Omega}_c}{2}\right)} \tag{6.63}$$

Having found the constant in (6.63), we next replace z^{-1} in the prototype lowpass filter by $\left(z^{-1} - \lambda\right)/\left(1 - \widehat{\lambda z}^{-1}\right)$ to obtain the transfer function of the new lowpass IIR digital filter. Let us demonstrate the process of spectral transformation by the following example.

Example 6.7 Design a lowpass Chebyshev type I IIR digital filter with a passband edge of 0.35π, passband ripple of 0.5 dB, stopband edge of $\frac{\pi}{2}$, and a minimum stopband attenuation of 20 dB using the lowpass-to-lowpass spectral transformation of the digital filter designed in Example 6.4.

Solution From Eq. (6.63), the constant is found to be

$$\lambda = \frac{\sin\left(\frac{0.25\pi - 0.35\pi}{2}\right)}{\sin\left(\frac{0.25\pi + 0.35\pi}{2}\right)} = -0.193364 \tag{6.64a}$$

Therefore, the lowpass-to-lowpass transformation takes the form

$$f\left(z^{-1}\right) = \frac{0.19336 + z^{-1}}{1 + 0.19336z^{-1}} \tag{6.64b}$$

The transfer function of the prototype lowpass digital filter is given in Eq. (6.46), which is

$$H_P(z) = \frac{0.5455\left(1 + z^{-1}\right)^3}{\left(1 - 0.5879z^{-1}\right)\left(1 - 1.1049z^{-1} + 0.6435z^{-2}\right)} \tag{6.65}$$

By applying the lowpass-to-lowpass spectral transformation to (6.65), we get the transfer function of the new lowpass IIR digital filter of the same Chebyshev type I filter as given below:

$$H(z) = H_P(z)\big|_{z^{-1} = \frac{(z^{-1} - \lambda)}{(1 - \lambda z^{-1})}}$$

$$= K \frac{(1 + z^{-1})^3}{(1 - 0.445139 z^{-1})(1 - 0.63008 z^{-1} + 0.576549 z^{-2})} \qquad (6.66)$$

The magnitude of the frequency response of the filter in (6.66) is shown in Fig. 6.10a, which is type I Chebyshev characteristic. The figure also shows the magnitude of the frequency response of the prototype filter for comparison. As can be seen from the figure, the new filter has the same characteristic of the prototype filter except that the passband edge is different, as desired. One can also use MATLAB to solve this problem. The MATLAB function *iirlp2lp* applies the lowpass-to-lowpass spectral transformation of Eq. (6.60). We must pass the coefficients of the numerator and denominator polynomials of the prototype lowpass IIR digital filter along with the passband edge of the prototype filter and the passband edge of the desired lowpass filter as arguments to the function *iirlp2lp*. The function returns the coefficients of the numerator and denominator polynomials of the desired lowpass IIR digital filter along with those of the spectral transformation function. Figure 6.10a also shows the magnitude of the frequency response of the desired

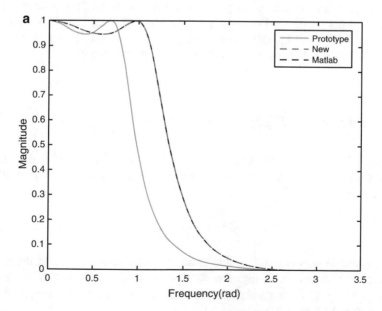

Fig. 6.10a Magnitude of the frequency response of the lowpass type I Chebyshev IIR digital filter of Example 6.7. It also shows the frequency responses of the prototype filter and the filter obtained using MATLAB for comparison

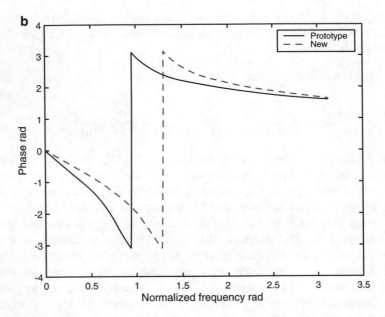

Fig. 6.10b Phase responses of the prototype lowpass and desired lowpass IIR digital filters of Example 6.7

lowpass IIR digital filter. It agrees perfectly with our analytical result. In Fig. 6.10b, the phase responses of the prototype and the desired lowpass filters are shown. The phase response of an IIR digital filter is, in general, nonlinear. Figure 6.10b proves it. The M-file named *Example6_7.m* is used to solve this problem.

6.4.2   Lowpass-to-Highpass Conversion

The spectral transformation used to convert a prototype lowpass IIR digital filter into a highpass digital filter of the same characteristics as the prototype filter is expressed by

$$z^{-1} = -\frac{\widehat{z}^{-1} + \lambda}{1 + \lambda\widehat{z}^{-1}} \tag{6.67}$$

The negative sign in (6.67) is due to the fact that what happens in the lowpass filter at $\Omega = 0$ happens in the highpass filter at $\Omega = \pi$. If Ω_c is the cutoff frequency of the prototype lowpass filter and $\widehat{\Omega}_c$ is that of the highpass filter, then from (6.67), we get the value for λ as

$$\lambda = -\frac{\cos\left(\frac{\Omega_c + \widehat{\Omega}_c}{2}\right)}{\cos\left(\frac{\Omega_c - \widehat{\Omega}_c}{2}\right)} \tag{6.68}$$

We will consider the following example to illustrate the procedure.

Example 6.8 Design a highpass Chebyshev type I IIR digital filter with a passband edge of 0.35π using the lowpass-to-highpass spectral transformation of the digital filter designed in Example 6.4.

Solution Using the given specifications in (6.68), we calculate

$$\lambda = \frac{-\cos\left(\frac{0.25\pi + 0.35\pi}{2}\right)}{\cos\left(\frac{0.25\pi - 0.35\pi}{2}\right)} = -0.59511 \tag{6.69a}$$

Having found λ, the lowpass-to-highpass frequency transformation is given by

$$f\left(z^{-1}\right) = \frac{0.59511 - z^{-1}}{1 - 0.59511 z^{-1}} \tag{6.69b}$$

Next we obtain the transfer function of the highpass digital filter as

$$H_{hp}(z) = H_P(z)\big|_{z^{-1} = -\frac{z^{-1} + \lambda}{1 + \lambda z^{-1}}} \tag{6.70}$$

The magnitude of the frequency response of the highpass Chebyshev type I digital filter is shown in Fig. 6.11a. For the sake of reference, the frequency response of the prototype lowpass digital filter is also plotted in the same figure. Similar to what we did in the previous example, we use the MATLAB function *iirlp2hp* to design the same highpass filter. The arguments to this function are the numerator and denominator polynomial coefficients, the passband edges of the prototype lowpass filter, and the desired highpass filter. It returns the numerator and denominator coefficients of the transfer function of the highpass filter along with those of the lowpass-to-highpass spectral transformation function. The magnitude of the frequency response of the highpass digital filter designed using MATLAB is also plotted for comparison and is shown in Fig. 6.11a. The phase responses of the prototype lowpass and the desired highpass IIR digital filters are shown in Fig. 6.11b. As mentioned before, the phase response of an IIR digital filter is nonlinear. This fact is again ascertained by Fig. 6.11b. The M-file named *Example6_8.m* is used to solve this problem.

6.4.3 Lowpass-to-Bandpass and Bandstop Conversion

The spectral transformation in the digital domain from a lowpass prototype IIR digital filter with a passband edge Ω_p to a bandpass digital filter whose lower and upper passband edges are Ω_{p1} and Ω_{p2}, respectively, is described by

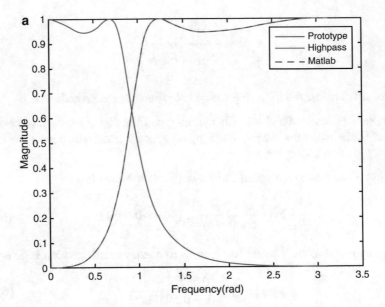

Fig. 6.11a Magnitude of the frequency response of the highpass type I Chebyshev IIR digital filter of Example 6.8. It also shows the frequency responses of the prototype lowpass filter and highpass filter designed using MATLAB for comparison

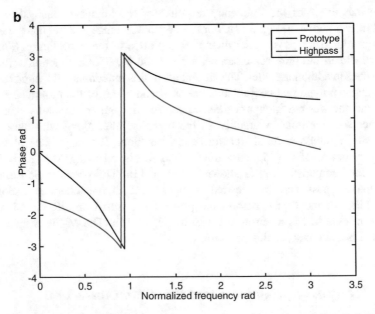

Fig. 6.11b Phase responses of the prototype lowpass and the desired highpass IIR digital filters of Example 6.8

$$f(z) = -\left(\frac{z^{-2} - \frac{2\lambda\rho}{1+\rho}z^{-1} + \frac{\rho-1}{1+\rho}}{1 - \frac{2\lambda\rho}{1+\rho}z^{-1} + \frac{\rho-1}{1+\rho}z^{-2}} \right),$$ (6.71)

where the constants are defined as

$$\lambda = \frac{\cos\left(\frac{\Omega_{p1}+\Omega_{p2}}{2}\right)}{\cos\left(\frac{\Omega_{p2}-\Omega_{p1}}{2}\right)},$$ (6.72a)

$$\rho = \cot\left(\frac{\Omega_{p2} - \Omega_{p1}}{2}\right) \tan\left(\frac{\Omega_p}{2}\right)$$ (6.72b)

The conversion from lowpass-to-bandpass filter is accomplished by the following equation:

$$H_{BP}(z) = H_{LP}(z)|_{z^{-1}=f(z)}$$ (6.73)

The lowpass-to-bandstop spectral transformation is given by

$$f(z) = \left(\frac{z^{-2} - \frac{2\lambda}{1+\rho}z^{-1} + \frac{1-\rho}{1+\rho}}{1 - \frac{2\lambda}{1+\rho}z^{-1} + \frac{1-\rho}{1+\rho}z^{-2}} \right),$$ (6.74)

where

$$\lambda = \frac{\cos\left(\frac{\Omega_{p1}+\Omega_{p2}}{2}\right)}{\cos\left(\frac{\Omega_{p2}-\Omega_{p1}}{2}\right)},$$ (6.75a)

$$\rho = \tan\left(\frac{\Omega_{p2} - \Omega_{p1}}{2}\right) \tan\left(\frac{\Omega_p}{2}\right)$$ (6.75b)

The desired bandstop IIR digital filter is then found by replacing each z^{-1} in the prototype lowpass IIR digital filter by the function specified in (6.74). We will now illustrate the design of bandpass and bandstop IIR digital filters using the spectral transformations described in Eqs. (6.71) and (6.74), respectively, by the following two examples.

Example 6.9 Design a bandpass Chebyshev type I IIR digital filter with lower and upper passband edges of 0.35π and 0.5π, respectively, using the lowpass-to-bandpass spectral transformation of the digital filter designed in Example 6.4.

Solution Using the passband edge frequencies of the lowpass prototype filter and the bandpass filter, we find $\lambda = 0.24008$ and $\rho = 1.72532$. The lowpass-to-bandpass frequency transformation then takes the form

$$f\left(z^{-1}\right) = \frac{-0.26614 + 0.30397z^{-1} - z^{-2}}{1 - 0.30397z^{-1} + 0.26614z^{-2}} \tag{6.76a}$$

Then, the transfer function of the desired bandpass filter is obtained as described by

$$H_{BP}(z)$$
$$= \frac{0.2551 - 0.7653z^{-2} + 0.7653z^{-4} - 0.2551z^{-6}}{1 - 1.2663z^{-1} + 2.8186z^{-2} - 2.0775z^{-3} + 2.332z^{-4} - 0.8535z^{-5} + 0.5559z^{-6}}, \tag{6.76b}$$

where $f(z)$ is defined in Eq. (6.71). The magnitude of the frequency response of the bandpass filter is shown in Fig. 6.12a. We can also use MATLAB to convert a prototype lowpass IIR digital filter to a desired bandpass filter. In particular, we use the function *iirlp2bp*, which accepts the coefficients of the numerator and denominator polynomials of the transfer function of the prototype lowpass IIR digital filter as well as the passband edge of the prototype filter and the lower and upper passband edges of the bandpass filter. It returns the coefficients of the transfer function of the bandpass filter along with those of the lowpass-to-bandpass spectral transformation function. The details are found in the MATLAB code in the M-file named *Example6_9.m*. For the sake of comparison, Fig. 6.12a also shows the magnitude of the frequency response of the bandpass filter designed using MATLAB. The phase response of the designed bandpass IIR digital filter is shown in Fig. 6.12b. As the

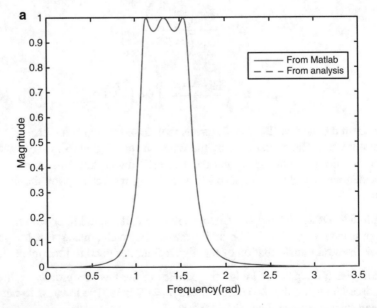

Fig. 6.12a Magnitude of the frequency response of the bandpass type I Chebyshev IIR digital filter of Example 6.9. It also shows the frequency responses of the bandpass filter designed using MATLAB for comparison

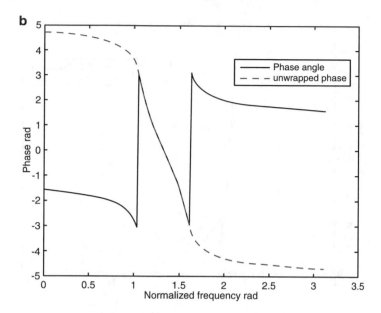

Fig. 6.12b Phase response of the designed bandpass IIR digital filters of Example 6.9

bandpass filter is an IIR digital filter, its phase response is nonlinear and is verified by the plot in Fig. 6.12b. The figure also shows the unwrapped phase response of the bandpass filter. Note that the unwrapped phase is a continuous function of the frequency.

Example 6.10   Design a bandstop Chebyshev type I IIR digital filter with lower and upper passband edges of 0.35π and 0.6π, respectively, using the lowpass-to-bandstop spectral transformation of the digital filter designed in Example 6.4.

Solution   From Eqs. (6.75a) and (6.75b), we obtain $\lambda = 0.08492$ and $\rho = 0.17157$. With these values, the lowpass-to-bandstop frequency transformation is found to be

$$f\left(z^{-1}\right) = \frac{0.70711 - 0.14497z^{-1} + z^{-2}}{1 - 0.14497z^{-1} + 0.70711z^{-2}}$$

The transfer function of the desired bandstop filter is then obtained by replacing z^{-1} in the prototype filter by the above function $f(z^{-1})$ and is found to be

$H_{BS}(z)$
$$= \frac{15.7527 - 8.0267z^{-1} + 48.6215z^{-2} - 16.1305z^{-3} + 48.6215z^{-4} - 8.0267z^{-5} + 15.7527z^{-6}}{1 - 0.3769z^{-1} + 1.4871z^{-2} - 0.4033z^{-3} + 0.9488z^{-4} - 0.1132z^{-5} + 0.1368z^{-6}}$$

The magnitude of the frequency response of the desired bandstop IIR digital filter is shown in Fig. 6.13a. As in the previous example, we can also design a bandstop IIR digital filter using MATLAB. It is similar to the design of a bandpass filter. The MATLAB function to use is *iirlp2bs*. The arguments to the function are the same as

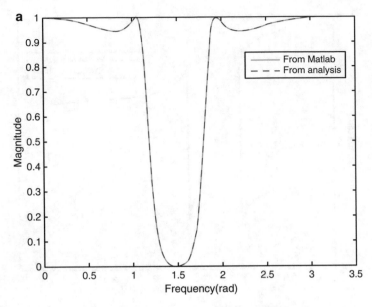

Fig. 6.13a Magnitude of the frequency response of the bandstop type I Chebyshev IIR digital filter of Example 6.10. It also shows the frequency responses of the bandstop filter designed using MATLAB for comparison

those for the bandpass case. The MATLAB code is in the M-file named *Example6_10.m*. Figure 6.13a also shows the magnitude of the frequency response of the bandstop IIR digital filter designed using MATLAB. The phase response of the designed bandpass IIR digital filter is shown in Fig. 6.13b. As can be seen from the figure, the phase response of the bandstop IIR digital filter is also nonlinear.

The following Table 6.1 lists the transformation type and the corresponding frequency transformation function along with the respective parameters for easy reference.

6.5   Computer-Aided Design of IIR Digital Filters

The design techniques we have seen so far are analytical. That is, these techniques give analytical or closed-form expressions for the poles or transfer functions in the analog domain to approximate ideal lowpass filter characteristics. Each type of filter tries to approximate the ideal characteristics in a unique way. For instance, Butterworth filter approximates the ideal lowpass characteristics in as flat as possible in the passband and falls off monotonically in the stopband. Chebyshev filter on the other hand, approximates the ideal lowpass characteristics in an equiripple manner.

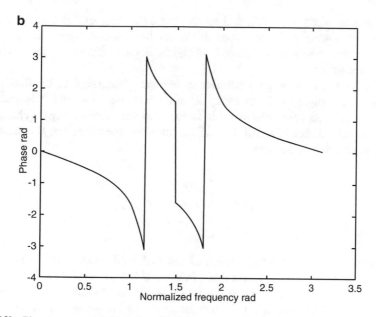

Fig. 6.13b Phase response of the designed bandstop IIR digital filter of Example 6.10

Table 6.1 Spectral transformations to convert a lowpass IIR digital filter into another LP, HP, BP, and BS IIR digital filters

Conversion type	Spectral transformation $f(z^{-1})$	Parameters
LP to LP	$\dfrac{z^{-1} - \lambda}{1 - \lambda z^{-1}}$	$\lambda = \dfrac{\sin\left(\frac{\Omega_c - \widehat{\Omega}_c}{2}\right)}{\sin\left(\frac{\Omega_c + \widehat{\Omega}_c}{2}\right)}$
LP to HP	$-\dfrac{\lambda + z^{-1}}{1 + \lambda z^{-1}}$	$\lambda = -\dfrac{\cos\left(\frac{\Omega_c + \widehat{\Omega}_c}{2}\right)}{\cos\left(\frac{\Omega_c - \widehat{\Omega}_c}{2}\right)}$
LP to BP	$-\left(\dfrac{z^{-2} - \frac{2\lambda\rho}{1+\rho}z^{-1} + \frac{\rho-1}{1+\rho}}{1 - \frac{2\lambda\rho}{1+\rho}z^{-1} + \frac{\rho-1}{1+\rho}z^{-2}}\right)$	$\lambda = \dfrac{\cos\left(\frac{\Omega_{p2} + \Omega_{p1}}{2}\right)}{\cos\left(\frac{\Omega_{p2} - \Omega_{p1}}{2}\right)}$
		$\rho = \cot\left(\dfrac{\Omega_{p2} - \Omega_{p1}}{2}\right)\tan\left(\dfrac{\Omega_p}{2}\right)$
LP to BS	$\dfrac{z^{-2} - \frac{2\lambda}{1+\rho}z^{-1} + \frac{1-\rho}{1+\rho}}{1 - \frac{2\lambda}{1+\rho}z^{-1} + \frac{1-\rho}{1+\rho}z^{-2}}$	$\lambda = \dfrac{\cos\left(\frac{\Omega_{p2} + \Omega_{p1}}{2}\right)}{\cos\left(\frac{\Omega_{p2} - \Omega_{p1}}{2}\right)}$
		$\rho = \tan\left(\dfrac{\Omega_{p2} - \Omega_{p1}}{2}\right)\tan\left(\dfrac{\Omega_p}{2}\right)$

These analog filters are converted to the corresponding digital filters using the bilinear transformation. Other types of digital filters are designed using spectral transformations either in the analog or digital domain. These are also based on analytical solutions.

In addition to these elegant techniques, one can also design an IIR digital filter directly in the digital domain using iterative solutions. Because this technique involves numerical computations, it is called *computer-aided design* technique. In computer-aided design of an IIR digital filter, one fixes the filter order N and assumes a transfer function of the type

$$H(z) = \frac{\sum\limits_{j=0}^{M} b_j z^{-j}}{\sum\limits_{k=0}^{N} a_k z^{-k}}, M \leq N \tag{6.77}$$

The end result is the determination of the real coefficients in (6.77) so that the frequency response $H(e^{j\Omega})$ meets that of the specified response $G(e^{j\Omega})$ in a certain manner. Typically, a *cost function* involving the above two functions is minimized. The general cost function to be minimized over a set of frequency points is expressed as the sum of the pth power of the weighted absolute difference between the desired and realized frequency responses, as

$$\varepsilon = \sum_{l=1}^{L} \left| W\left(e^{j\Omega_l}\right)\left(G\left(e^{j\Omega_l}\right) - H\left(e^{j\Omega_l}\right)\right)\right|^p \tag{6.78}$$

The most commonly used value for p is 2, and the resulting solution is called the *least squares* solution. Another cost function is the maximum of the weighted absolute difference between the desired and realized frequency responses. Let us not go into the details of the iterative procedure to determine the filter coefficients. Instead, we will use MATLAB to solve the problem. MATLAB has the function named *yulewalk* that designs an IIR digital filter of a given order and a set of frequency points in the interval between 0 and 1 along with the corresponding magnitudes of the frequency response. Depending on the frequency specifications, it can design a lowpass, highpass, bandpass, or bandstop IIR digital filter. Let us illustrate the computer-aided design using MATLAB by an example.

Example 6.11 Design a fifth-order IIR lowpass digital filter with a passband ripple of 0.1 dB, minimum stopband attenuation of 40 dB, passband edge of $\frac{\pi}{4}$, and stopband edge of $\frac{\pi}{2}$ using MATLAB.

Solution The MATLAB function *yulewalk* accepts the filter order N, a vector of frequencies in the interval [0,1] and the corresponding magnitudes of the desired frequency response, in that sequence. It returns the vectors B and A of the coefficients of the numerator and denominator polynomials of the digital filter transfer function. The approximation will be better if we supply enough frequency points and corresponding magnitudes. First we have to convert the passband ripple and the stopband attenuation in dB to actual values. Therefore, the passband ripple and stopband attenuation in actual values are

$$\delta_p = 1 - 10^{-\frac{\alpha_p}{20}} = 0.011447 \quad \text{and} \quad \delta_s = 10^{-\frac{\alpha_s}{20}} = 0.01.$$

Let us choose the frequency points between 0 and 1 as given by

$$F = \left[0, \frac{\Omega_p}{8}, \frac{\Omega_p}{4}, \frac{\Omega_p}{2}, \frac{3\Omega_p}{4}, \Omega_p, 0.4, \Omega_s, 0.6, 0.7, 0.8, 1\right],$$ where Ω_p and Ω_s are normalized by π.

Let the corresponding magnitudes be specified as

$$M = \left[1 + \frac{\delta_p}{2}, 1 + \frac{3\delta_p}{4}, 1, 1, 1 - \frac{\delta_p}{2}, 1 - \frac{3\delta_p}{4}, 2\delta_s, \delta_s, 0.1\delta_s, 0.1\delta_s, 0.05\delta_s, 0.00001\right]$$

Next, we call the function *[B,A] = yulewalk(5,F,M)*, which returns the coefficients of the transfer function of the IIR digital filter. The transfer function is found to be

$$H(z) = 0.0290 \frac{1 + 2.089z^{-1} + 3.0407z^{-2} + 3.0389z^{-3} + 2.0844z^{-4} + 0.9971z^{-5}}{1 - 1.8307z^{-1} + 2.1003z^{-2} - 1.3981z^{-3} + 0.6005z^{-4} - 0.1192z^{-5}}$$

The magnitude of the frequency response in dB of the designed filter is plotted against the normalized frequency and is shown in Fig. 6.14. Figure 6.15 shows the magnitude of the frequency response of Chebyshev I, elliptic, and Butterworth filters

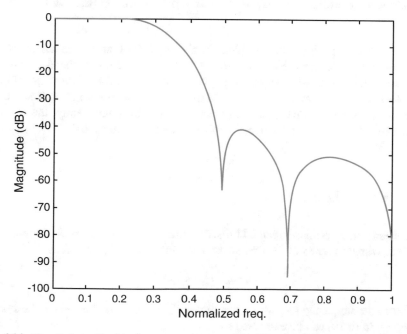

Fig. 6.14 Magnitude in dB of the frequency response of the fifth-order IIR digital filter of Example 6.11 designed using computer-aided technique

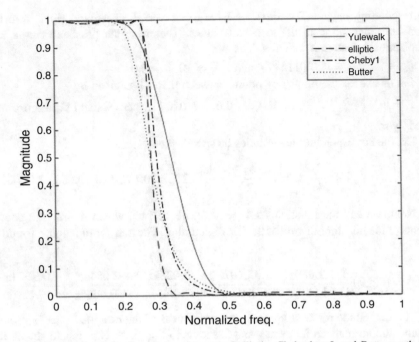

Fig. 6.15 Magnitude of the frequency responses of elliptic, Chebyshev I, and Butterworth IIR digital filters, satisfying the specifications in Example 6.11. For the sake of comparison, the magnitude of the frequency response of the filter designed using computer-aided technique is also shown

with the same specifications as given in Example 6.11 as a comparison to the computer-aided designed filter. It is found that the minimum stopband attenuation for the computer-aided designed filter is 52.08 dB, 48.34 dB for the elliptic, 44.04 dB for the Chebyshev I, and 38.28 dB for the Butterworth filter. The elliptic filter is found to have the smallest transition width or, equivalently, the sharpest response between the passband and stopband edges, as anticipated.

6.6   Group Delay

The steady-state response of an LTI discrete-time system to a sinusoidal input $e^{jn\Omega_0}$ can be written in terms of the frequency response of the LTI system as

$$y[n] = H\left(e^{j\Omega_0}\right)e^{jn\Omega_0} \tag{6.79}$$

Since the frequency response is a complex function of the frequency, the system response in (6.79) can be expressed as

$$y[n] = \left| H\left(e^{j\Omega_0}\right) \right| e^{j\Omega_0(n-n_0)}$$

(6.80)

Thus, the LTI discrete-time system introduces a phase delay or phase lag $n_0\Omega_0$ *rad*, which amounts to a time delay

$$\tau = \frac{phase\ delay}{\Omega_0} = \frac{n_0\Omega_0}{\Omega_0} = n_0\ samples$$

(6.81)

If the phase response of the LTI discrete-time system is not linear, then different frequency components will produce different time delays. If an input sequence consists of several different frequency components, then the LTI system will introduce different amount of delay to the components and the overall effect is a distortion called *phase distortion*. In general, an IIR filter has a nonlinear phase response due to the fact that it has both zeros and poles. Because of the nonlinearity in phase response of an IIR digital filter, we can define the *group delay* as

$$\tau_g = -\frac{d\theta(\Omega)}{d\Omega},$$

(6.82)

where $\theta(\Omega)$ is the phase response of the IIR filter. This group delay refers to the delay of a group of frequency components present in an input signal. Consider a stable second-order transfer function given by

$$H(z) = \frac{b_0 + b_1 z^{-1} + b_2 z^{-2}}{a_0 + a_1 z^{-1} + a_2 z^{-2}},$$

(6.83)

where the coefficients of the polynomials are real. The frequency response corresponding to (6.83) is expressed as

$$H\left(e^{j\Omega}\right) = \frac{b_0 + b_1 e^{-j\Omega} + b_2 e^{-j2\Omega}}{a_0 + a_1 e^{-j\Omega} + a_2 e^{-j2\Omega}}$$

(6.84)

Therefore, the phase response can be written as

$$\theta(\Omega) = -\tan^{-1}\left(\frac{b_1 \sin \Omega + b_2 \sin 2\Omega}{b_0 + b_1 \cos \Omega + b_2 \cos 2\Omega}\right)$$
$$+ \tan^{-1}\left(\frac{a_1 \sin \Omega + a_2 \sin 2\Omega}{a_0 + a_1 \cos \Omega + a_2 \cos 2\Omega}\right)$$

(6.85)

From Eqs. (6.82) and (6.85), we obtain the group delay of a second-order IIR filter transfer function as

$$\tau_g = \left\{ \frac{\left(b_1^2 + 2b_2^2\right) + b_1(b_0 + 3b_2)\cos(\Omega) + 2b_0 b_2 \cos(2\Omega)}{\left(b_0^2 + b_1^2 + b_2^2\right) + 2b_1(b_0 + b_2)\cos(\Omega) + 2b_0 b_2 \cos(2\Omega)} \right\}$$
$$- \left\{ \frac{\left(a_1^2 + 2a_2^2\right) + a_1(a_0 + 3a_2)\cos(\Omega) + 2a_0 a_2 \cos(2\Omega)}{\left(a_0^2 + a_1^2 + a_2^2\right) + 2a_1(a_0 + a_2)\cos(\Omega) + 2a_0 a_2 \cos(2\Omega)} \right\}$$

(6.86)

Let us consider the following example to illustrate how the group delay of a second-order IIR filter looks like.

Example 6.12 Calculate the group delay over the interval $[0, \pi]$ of the Butterworth IIR digital filter of Example 6.3. Also compute its response to an input sequence

$$x[n] = \left\{ \sin\left(\frac{26\pi n}{1000}\right) + \sin\left(\frac{266\pi n}{1000}\right) + \cos\left(\frac{666\pi n}{1000}\right) \right\} u[n] \qquad (6.88)$$

Solution The transfer function of the fourth-order Butterworth IIR digital filter was found to be

$$H(z) = \left(\frac{1 + 2z^{-1} + z^{-2}}{1 - 0.8336z^{-1} + 0.5156z^{-2}}\right)\left(\frac{1 + 2z^{-1} + z^{-2}}{1 - 0.6209z^{-1} + 0.12189z^{-2}}\right) \qquad (6.89)$$

By substituting the values of the coefficients of the numerator and denominator polynomials of each of the second-order functions in (6.89) in (6.86), we determine the group delay of the IIR filter over the frequency interval $[0, \pi]$. We can also use the MATLAB function *grpdelay* to calculate the group delay. The actual function call is $Gd = grpdelay(B,A,W)$, where B and A are the vectors of the coefficients of the numerator and denominator polynomials of the transfer function, W is the set of frequency points in the interval $[0, \pi]$, and Gd is the corresponding group delay in samples. But, before we use the *grpdelay* MATLAB function, we need to convert the two second-order sections to a single transfer function. This is accomplished by the function *sos2tf*. So, we use the statement $[B,A] = sos2tf([B1\ A1;\ B2\ A2],G)$, where B1 and A1 are the coefficients of the numerator and denominator polynomials of the first section, B2 and A2 correspond to the second section of the transfer function, and G is the gain. The function returns the coefficients of the numerator and denominator polynomials of the overall transfer function. Once we obtain the overall transfer function, we can then calculate the group delay. Figure 6.16 shows the phase response of the filter described by (6.89). The group delay is shown in Fig. 6.17. The figure shows the plots obtained using both analytical equation and MATLAB function for the sake of comparison. As expected, both the phase and group delay of the IIR digital filter are nonlinear.

To demonstrate the effect of nonlinear phase of the IIR filter on its input, we calculate the response of the IIR filter to the input sequence specified in (6.88). The response can be calculated either using the difference equation or using the MATLAB functions *fft* and *ifft*. The difference equation corresponding to the overall transfer function can be described by

$$\begin{aligned} y[n] = {}& x[n] + 4x[n-1] + 6x[n-2] + 4x[n-3] + x[n-4] \\ & + 1.4545y[n-1] - 1.1551y[n-2] + 0.4217y[n-3] \\ & - 0.0628y[n-4]. \end{aligned} \qquad (6.90)$$

The initial conditions are assumed to be zero, and the input sequence corresponds to (6.88). We compute the response for a set of consecutive values for n. The other

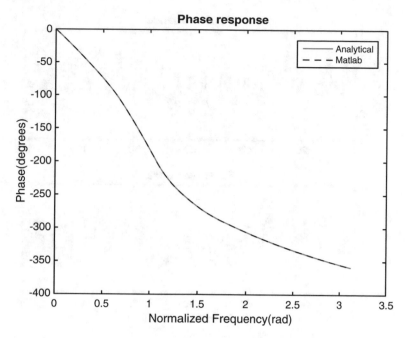

Fig. 6.16 Phase response of the Butterworth IIR digital filter of Example 6.12

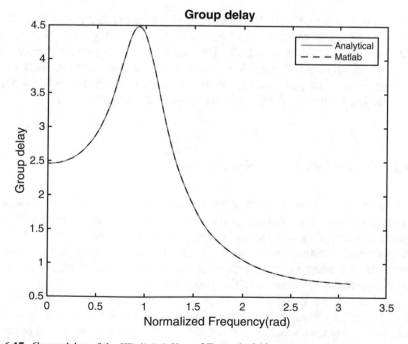

Fig. 6.17 Group delay of the IIR digital filter of Example 6.12

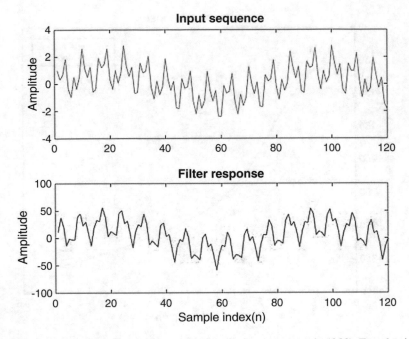

Fig. 6.18 Response of the filter in Example 6.12 to the input sequence in (6.88). Top plot: input sequence. Bottom plot: output sequence

method is to find the DFTs of the filter frequency response and the input sequence over the same set of values of the frequency variable, multiply the two point by point, and then find the inverse DFT. The input and the response are plotted and shown in the upper and lower plots, respectively, in Fig. 6.18. Because of the nonlinear nature of the group delay, we observe some phase distortion in the output sequence. The M-file to solve this problem is named *Example6_12.m*.

6.6.1   Group Delay Equalization

We learnt from the previous section that an IIR filter, in general, has a nonlinear phase response and, as a consequence, has a varying group delay. Some applications require a constant group delay when filtering a signal with an IIR digital filter. In this section we will describe a procedure to obtain a constant group delay out of an IIR filter without altering its magnitude of frequency response. Such a process or procedure is called *equalization* of the group delay.

An *allpass* filter is an IIR filter with its zeros being the inverse of its poles. The poles of an allpass filter must lie inside the unit circle in the Z-domain for it to be stable. Because the poles and zeros of an allpass filter are inverses of each other, the magnitude of its frequency response is constant over the entire frequency interval.

However, the locations of the poles can be adjusted so that the group delay is constant or approximately a constant. The design of such an allpass filter will be carried out using MATLAB. More specifically, the designed IIR digital filter is cascaded with an allpass filter, and then the poles and zeros of the allpass filter are adjusted so that the overall group delay approximates a constant value. The group delay of the IIR filter along with the order of the allpass filter and a set of frequency points over which the overall group delay is a constant are the arguments to the MATLAB function *iirgrpdelay*. It returns the coefficients of the numerator and denominator polynomials of the allpass filter along with a group delay offset. We will make the procedure clear by working out an example.

Example 6.13   Design a fourth-order elliptic IIR digital filter with a passband ripple of 1 dB, passband edge of 0.3π, and a minimum stopband attenuation of 35 dB, and then equalize its group delay in the passband using an eighth-order allpass digital filter.

Solution   First we design the elliptic IIR digital filter by calling the MATLAB function *ellip*, whose input arguments are filter order, passband ripple in dB, minimum stopband attenuation in dB, and the passband edge, in that order. The specified passband edge must be normalized by π, which corresponds to half the sampling frequency. The *ellip* function returns the zeros, poles, and the quotients of the elliptic filter. Before we design the allpass filter, we must convert the zeros and poles of the IIR elliptic filter to second-order sections using the MATLAB function *zp2sos*. This function returns the coefficients of the numerator and denominator polynomials of the second-order sections along with the gain factor. Next we call the function *dfilt.df2sos*, which accepts the second-order sections *SOS* and returns a direct form II filter object, *H*. For the time being, let us not worry about the terms *direct form* I or *direct form* II. We will discuss these in a later chapter. The group delay of this IIR digital filter is next calculated using the MATLAB function *grpdelay*. This group delay of the elliptic IIR filter along with the order of the allpass filter and frequency points is input to the function *iirgrpdelay*, which returns the coefficients of the numerator and denominator polynomials of the allpass filter. The allpass filter is then converted to direct form II and then cascaded with the IIR filter using the MATLAB function *cascade*. The details of the MATLAB code can be found in the M-file named *Example6_13.m*. We can invoke the filter visualization tool *fvtool* to obtain the various plots and filter characteristics. The information obtained using the *fvtool* is given in the form of figures. Figure 6.19 shows the plot of the equalized group delay, which is seen to be a constant in the passband with a small ripple. As a comparison, the group delay of the fourth-order elliptic IIR digital filter is also shown in the figure. The overall frequency response and that of the allpass filter are shown in Fig. 6.20. As stated earlier, the magnitude of the frequency response of the allpass section remains fairly constant, and so the response of the IIR filter is unaltered. We can also use the fvtool to obtain the pole-zero plot of the overall filter and is shown in Fig. 6.21. The impulse response of the cascaded filter and its step response are shown in Figs. 6.22 and 6.23, respectively. A further

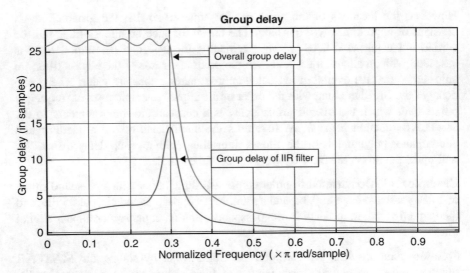

Fig. 6.19 Overall group delay of the cascaded IIR and allpass filters of Example 6.13

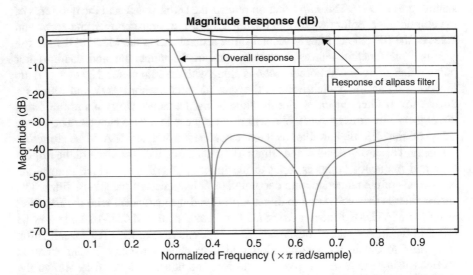

Fig. 6.20 Magnitude of the frequency responses of the fourth-order IIR and eighth-order allpass digital filters of Example 6.13

advantage of using the *fvtool* function is that we can obtain filter coefficients as well as the cost of implementing the overall filter in terms of additions and multiplications operations, which are shown in Figs. 6.24 and 6.25, respectively.

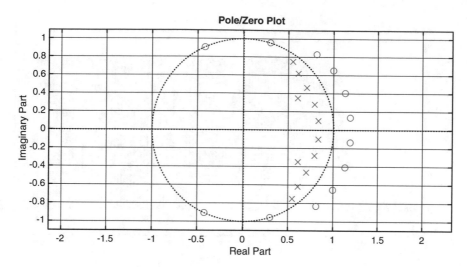

Fig. 6.21   Overall pole-zero plot of the filter in Example 6.13

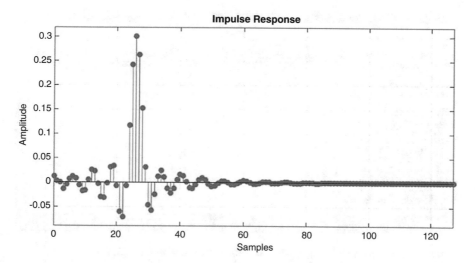

Fig. 6.22   Overall impulse response of the filter in Example 6.13

6.7   Simulation Using Simulink

Thus far we have learnt the design of an IIR digital filter using both analytical and computer-aided techniques. We have shown several examples using MATLAB to illustrate these design methods. In this section we will apply an IIR digital filtering

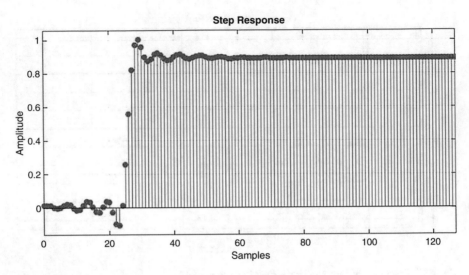

Fig. 6.23 Overall step response of the filter in Example 6.13

```
cascade

Stage #1: df2sos
SOS Matrix:
1    0.83927670235811447   1.0000000000000004  1  -1.2127630561552545   0.4877757810830
1   -0.59320890570160123   1                   1  -1.0905920828063635   0.8569431498369

Scale Values:
0.047026561652913849

Stage #2: df2
Numerator:
   0.26018271528694231
  -2.1367521691422078
   7.9948960965512077
 -17.794253966364032
  25.766380417034878
 -24.862858555665998
  15.621581759818232
  -5.848410988094412
   1
Denominator:
```

Fig. 6.24 Coefficients of the digital filter second-order sections of Example 6.13

process to modulate and demodulate an *amplitude-modulated* (AM) signal, which is used in AM radio broadcast. We will use MATLAB's Simulink to simulate the generation of an AM signal. An AM signal is an analog signal wherein a voice or music signal (modulating waveform) modulates the amplitude of a high-frequency signal (a carrier wave). The modulating signal is also called a message waveform. The carrier wave is of much higher frequency than the maximum frequency of the message signal. The frequency range of an AM broadcast is from 550 kHz to

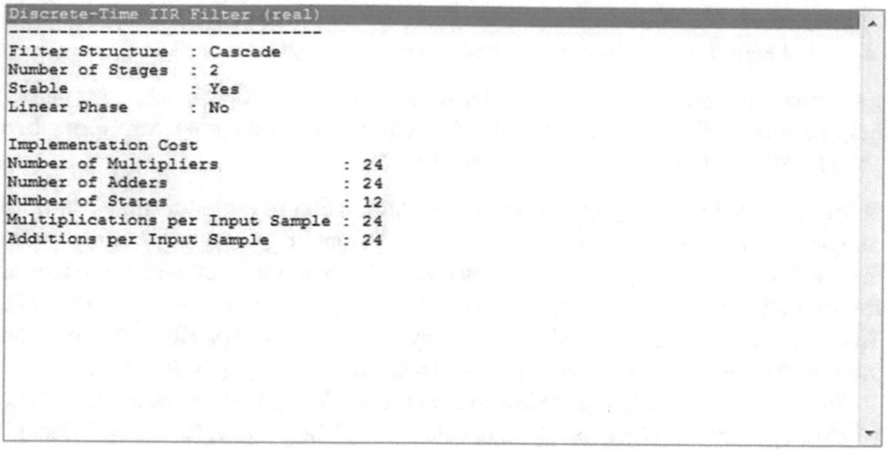

```
Discrete-Time IIR Filter (real)
--------------------------------
Filter Structure    : Cascade
Number of Stages    : 2
Stable              : Yes
Linear Phase        : No

Implementation Cost
Number of Multipliers              : 24
Number of Adders                   : 24
Number of States                   : 12
Multiplications per Input Sample   : 24
Additions per Input Sample         : 24
```

Fig. 6.25 Details of filter implementation cost in terms of arithmetic operations

1600 kHz. Each broadcast program has a specified carrier frequency with a bandwidth of 10 kHz. An AM waveform can be described by

$$x_{AM}(t) = A_c(1 + m(t)) \cos(2\pi f_c t), \tag{6.91}$$

where A_c is the amplitude of the carrier waveform, m(t) is the message waveform, and f_c is the carrier frequency. The magnitude of the message waveform m(t) must be less than or equal to 1. If the message waveform is a pure sinusoid

$$m(t) = A_m \sin(2\pi f_m t), \tag{6.92}$$

then $\frac{A_m}{A_c} \leq 1$ is called the modulation index. If the message waveform is a baseband signal with a maximum frequency $f_m \ll f_c$, then the modulated waveform $x_{AM}(t)$ is a bandpass signal with center frequency f_c and bandwidth $2f_m$. From (6.91) and (6.92), we can deduce that the modulated waveform can be described by

$$x_{AM}(t) = A_c \cos(\omega_c t) + \frac{m}{2} \sin((\omega_c + \omega_m)t) + \frac{m}{2} \sin((\omega_c - \omega_m)t), \tag{6.93}$$

where m is the modulation index. As can be seen from the above equation, the AM signal has a component at the carrier frequency and two *sidebands*, one above and one below the carrier frequency. Thus, the modulated AM signal is a bandpass signal with a bandwidth equal to twice the message bandwidth.

From Eq. (6.91), we find that the AM signal is generated by multiplying a carrier waveform by a message waveform and then adding a carrier signal to the product. In order to recover the message waveform from the AM signal, we can multiply the AM signal by a locally generated carrier signal of the same frequency and phase and then filter it with a lowpass filter of bandwidth equal to that of the message signal. The process of recovering a message waveform from the AM signal is known as *demodulation.* We will demonstrate the processes of modulation and demodulation of an AM signal using Simulink in the following example.

Example 6.14 Show modulation and demodulation of an AM signal with a carrier frequency of 100 kHz. Use a 10 kHz sinusoid as the modulating waveform. Use MATLAB's Simulink as the simulation tool.

Solution Since we are going to perform the simulation in digital form, we have to sample the modulated waveform. The minimum sampling frequency must be twice the maximum frequency of the modulated signal. As pointed out earlier, the largest frequency in the modulated AM signal is $f_{max} = f_c + f_m = 100 + 10 = 110 \ kHz$. Therefore, the minimum sampling frequency must be 220 kHz. However, to be on the safe side, we will use a sampling frequency of $f_s = 2.2f_{max} = 242 \ kHz$.

To start the Simulink, type "Simulink" in the workspace and press return. A new window appears, as shown in Fig. 6.26. Click "File," then "New," and then "Model." A new blank window will appear in which we will create the block diagram of the AM modulator. Next, click "Sources" from the "Libraries" menu. A set of sources

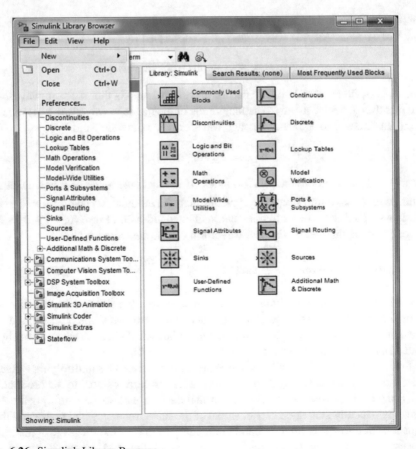

Fig. 6.26 Simulink Library Browser

Fig. 6.27 Simulink Library Browser-Source-Sine Wave

blocks appears on the right side, as shown in Fig. 6.27. Drag the box named "Sine Wave" to the Model window. Double-click the block, and a new window named "Source block parameters: sine wave" appears as shown in Fig. 6.28. Fill in the parameters as shown in the figure. Use "Time based" in "Sine type" and "Use simulation time" in "Time(t)" parameters. Choose an amplitude of 0.75, bias of 1, a frequency of 10,000, 0 rad for phase, and a "Sample Time" of 4.1322e-06. This sample time corresponds to the inverse of 2.2 times 110 kHz. As shown in Fig. 6.29, under Communication System Toolbox – Modulation, click "Analog Passband Modulation," and drag the box named "DSB Modulator Passband" to the Model window. Then double-click the new block. A new window (Fig. 6.30) named "Function Block Parameters: DSB AM Modulator" appears. Enter 0 for the "Input signal offset" parameter, 100,000 for the "Carrier frequency (Hz)," and 0 for "Initial Phase (rad)," and click "Apply" and then "OK." Next, click "Sinks" under "Simulink." A set of sinks appears as shown in Fig. 6.31. Drag the box named "Scope" to the Model window. Drag the same box again to the Model window or you can copy the box already dragged and paste it. So, we have two "Scopes." So far,

Fig. 6.28 Source block
parameters: sine wave

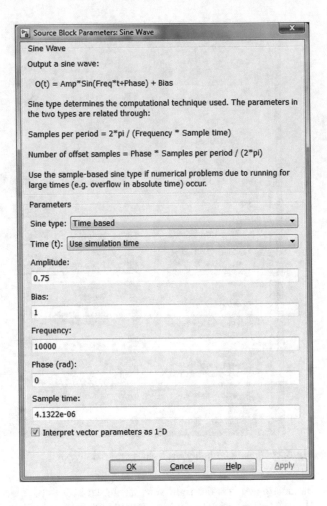

we have dragged the blocks necessary to perform the modulation of a carrier signal
by a low frequency sinusoidal waveform. Next, we have to include the blocks
necessary to perform demodulation to recover the message waveform. As shown
in Fig. 6.32, drag the box named "DSB AM Demodulator Passband" to the Model
window. Double-click the Demodulator Box, and a window named "Function Block
Parameters: DSB AM Demodulator Passband" will appear. Fill in the parameters as
shown in Fig. 6.33. Note that the lowpass filter passes the message waveform and
rejects the components at twice the carrier frequency. Here we have chosen an
eighth-order lowpass Butterworth filter with a cutoff frequency of 12 kHz because
the message waveform is of frequency 10 kHz. Finally, we have to connect all the
blocks as shown in Fig. 6.34. From the figure we see that we have an input sinusoid
fed to the DSB modulator as well as to a scope for viewing, the output of the
modulator is fed to the input of the demodulator and a scope, and finally the output of
the demodulator is fed to a scope. Once the diagram is saved, click the green circle

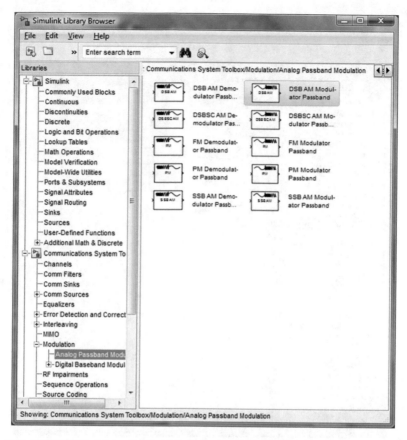

Fig. 6.29   DSB AM Modulator Passband block from Simulink Library Browser

Fig. 6.30   Function Block
Parameters: DSB AM
Modulator

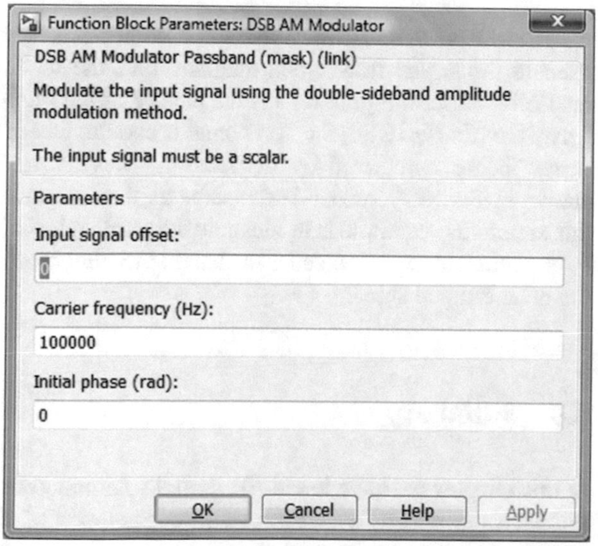

Fig. 6.31 Scope from Simulink Library Browser: Sinks

with a right-pointing arrow to start the simulation. If there is no error, MATLAB will execute the signal flow block diagram, and the results will be displayed on the respective scopes. Figure 6.35 is the picture of the scope displaying the modulating waveform. In Fig. 6.36, the AM signal is shown, where we see the amplitude of the carrier being modulated by the message waveform. The demodulated signal is shown in Fig. 6.37, which is the same as the message waveform. The purpose of this simulation example is to illustrate the application of an IIR lowpass digital filter in communications. You can also change the filter type and order to see its effect on the demodulated signal.

6.8 Summary

In this chapter we have learnt IIR digital filters in general and how to design such filters in particular. We discussed three methods of designing IIR digital filters, namely, the impulse invariance technique, bilinear transformation, and computer-

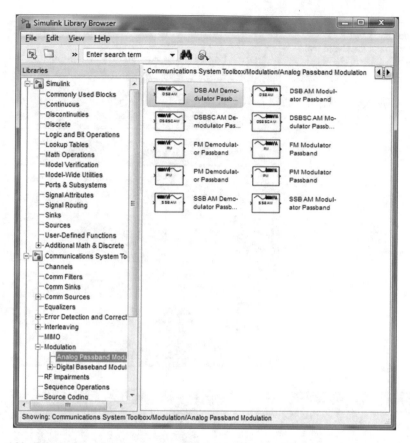

Fig. 6.32 DSB AM Demodulator Passband block from Simulink Library Browser

aided technique. The goal of the impulse invariance technique is to preserve the shape of the impulse response of a specified analog filter in the discrete-time domain. As we are more familiar with signals in the frequency domain than in the time domain, analog filters are designed to satisfy specifications in the frequency domain. In other words, analog filters are designed to meet frequency responses with some tolerances in the passband and stopband. Since elegant techniques exist in the literature for the design of analog filters in the frequency domain, it is redundant to invent such techniques for the design of IIR digital filters. Therefore, simple translation of the existing frequency domain design methods for the analog filters is used for the design of IIR digital filters. One such method is the bilinear transform, which preserves the characteristics of the analog filters in the digital domain. However, due to the nonlinear nature of the bilinear transform, critical frequencies in the digital domain have to be prewarped so that the final IIR digital filter will have the correct critical frequencies. We also described spectral transformations in the digital domain used to convert a prototype lowpass IIR digital filter to another

Fig. 6.33 Function Block
Parameters: DSB AM
Demodulator Passband from
Simulink Library Browser

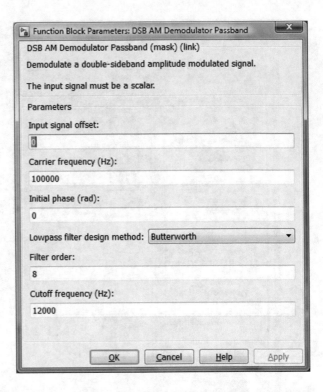

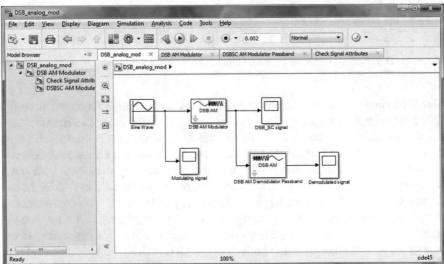

Fig. 6.34 Overall DSB AM Modulator-Demodulator block diagram

Fig. 6.35 Scope displaying
the modulating input signal

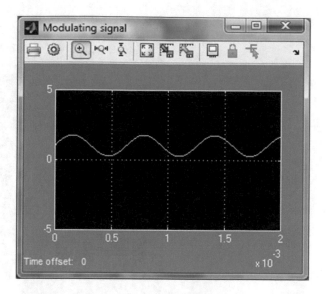

Fig. 6.36 Scope displaying
the DSB AM signal

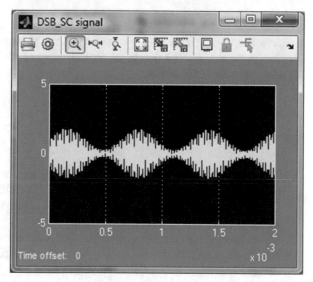

lowpass or highpass or bandpass or bandstop filter. The third technique is called
computer-aided design technique because this technique uses iterative computations.
The basic idea behind computer-aided design is to assume the transfer function of a
given order IIR digital filter and then iteratively alter the filter coefficients so that the
cost of an objective function is minimized. We also discussed the equalization of the
group delay of an IIR digital filter under computer-aided design techniques. Several
examples, both analytical and MATLAB-based were shown to solidify the understanding of the design of IIR digital filters. Finally, the use of IIR digital filter in
communications was illustrated using MATLAB's Simulink tool. More specifically,

Fig. 6.37 Scope displaying the demodulated signal

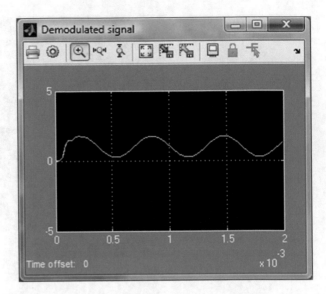

we simulated amplitude modulation and demodulation and exemplified the use of lowpass IIR digital filter in the recovery of the message waveform. In the next chapter, we will study the design of another class of digital filters known as *finite impulse response* (FIR) filters.

6.9   Problems

1. Determine the peak ripple values δ_p and δ_s for the peak passband ripple of 0.14 dB and minimum stopband attenuation of 68 dB.
2. Corresponding to $\delta_p = 0.015$ and $\delta_s = 0.04$, find the peak passband ripple α_p and minimum stopband attenuation α_s in dB.
3. Determine the transfer function of the digital filter using the impulse invariance method corresponding to the analog filter whose transfer function in the Laplace domain is described by $H(s) = \frac{17}{s^2+2s+17}$. Use a sampling interval $T = 0.1$ sec.
4. The transfer function of the digital filter obtained using the impulse invariance method is found to be $H(z) = \frac{0.2z^{-1}}{1-0.8z^{-1}+0.15z^{-2}}$. If the sampling interval is 0.25 sec, find the transfer function of the corresponding analog filter.
5. Using the bilinear transformation, find the digital filter transfer function corresponding to the transfer function of the analog filter $H_a(s) = \frac{2S^2+s-1}{(s+4)(s^2+2s+10)}$. Use T $= 0.25$ sec.
6. Convert the analog transfer $H(s) = \frac{1}{1+3.2361s+5.2361s^2+5.2361s^3+3.2361s^4+s^5}$ into an IIR digital filter using the bilinear transformation. Use a sampling frequency of 5 kHz. Plot the magnitude of the frequency responses.

7. The transfer function $H(z) = \frac{0.1966\left(1+3z^{-1}+3z^{-2}+z^{-3}\right)}{1+0.1541z^{-1}+0.4691z^{-2}-0.0506z^{-3}}$ was obtained using the bilinear transformation of a corresponding analog filter. Determine the analog filter transfer function using a sampling interval of 0.25 ms.

8. Design a Butterworth lowpass IIR digital filter of order $N = 3$ to have a cutoff frequency of 0.3π. Use MATLAB to solve the problem.

9. Starting with the lowpass digital filter obtained in the previous problem as the prototype, design a bandstop digital filter with band edge frequencies 0.4π and 0.7π using the spectral transformation in the Z-domain.

10. Design a lowpass elliptic IIR digital filter with a peak-to-peak passband ripple of 0.1 dB, minimum stopband attenuation of 45 dB, and a passband edge frequency of 0.4π rad using MATLAB. Plot the magnitude of the frequency response in dB against normalized frequency in rad.

11. Using the transfer function obtained in the previous problem as the prototype filter, design a bandstop IIR digital filter with stopband edges at 0.3π rad and 0.5π rad, respectively. Plot the magnitude of the frequency response of the bandstop digital filter.

12. Design a Butterworth digital lowpass filter with a sampling frequency of 10 kHz, 3 dB frequency of 1 kHz, and a minimum stopband attenuation of 40 dB at the frequency 3.5 kHz. First determine the filter order using (6.23), and then design the filter using bilinear transform. You can use MATLAB to solve the problem.

13. For the same filter specifications as in the previous problem, design a type 1 Chebyshev digital filter using the bilinear transform. To determine the filter order, use Eq. (6.42). Plot the magnitude of the frequency response in dB, and compare it with that of the Butterworth filter in Problem 12.

14. Using the same filter order and passband and stopband frequencies as in Problem 12, design a lowpass digital filter using the MATLAB function *yulewalk*. Compare its frequency response with those in Problems 12 and 13.

15. Equalize the group delay of the filter in Problem 13 using MATLAB.

References

1. Antoniou A (1993) Digital filters: analysis, design, and applications, 2nd edn. McGraw-Hill, New York
2. Charalambous C, Antoniou A (1980) Equalization of recursive digital filters. IEE Proc 127(Part G):219–225
3. Constantinides AC (1970) Spectral transformations for digital filters. Proc IEE 117:1585–1590
4. Deczky AG (1972) Synthesis of digital recursive filters using the minimum P error criterion. IEEE Trans Audio Electroacoust AU-20(2):257–263
5. Johnson M (1983) Implement stable IIR filters using minimal hardware, EDN
6. Kaiser JF (1963) Design methods for sampled data filters. Proceedings first annual Allerton conference on circuit and system theory, Chapter 7, pp 221–236
7. Kaiser JF (1965) Some practical considerations in the realization of linear digital filters. Proceedings third annual Allerton conference on circuit and system theory, pp 621–633
8. Kaiser JF (1966) Digital filters. In: Kuo FF, Kaiser JF (eds) System analysis by digital computer. Wiley, New York, pp 218–227

9. Mitra SK (2011) Digital signal processing: a computer-based approach, 4th edn. McGraw-Hill, New York
10. Mitra SK, Hirano K, Nishimura S, Sugahara K (1990) Design of digital bandpass/bandstop digital filters with tunable characteristics, Frequenz, 44, pp 117–121
11. Mitra SK, Neuvo Y, Roivainen H (1990) Design and implementation of recursive digital filters with variable characteristics. Int J Circuit Theory Appl 18:107–119
12. Oppenheim A, Schafer R, Buck J (1999) Discrete-time signal processing, 2nd edn. Prentice Hall, Upper Saddle River
13. Parks TW, Burrus CS (1987) Digital filter design. New York, Wiley
14. Rabiner LR, Gold B (1975) Theory and application of digital signal processing. Prentice Hall, Englewood Cliffs
15. Remez EY (1934) General computational methods of Chebyshev approximations, Atomic Energy Translation 4491, Vol. 198, pp 2063
16. Steiglitz K (1970) Computer-aided design of recursive digital filters. IEEE Trans on Audio Electroacoust 18(2):123
17. Vaidyanathan PP, Regalia PA, Mitra SK (1987) Design of doubly-complementary IIR digital filters using a single complex allpass filter, with multirate applications. IEEE Trans Circuits Syst CAS-34:378–389
18. Williams CS (1986) Designing digital filters. Prentice Hall, Englewood Cliffs

Chapter 7
FIR Digital Filters

7.1 Types of Linear-Phase FIR Filters

A *finite impulse response* (FIR) digital filter, as the name implies, has an impulse response sequence that is of finite duration as opposed to an IIR digital filter, which has an impulse response that is of infinite duration. Therefore, the Z-transform of the impulse response of an FIR digital filter in general can be written as

$$H(z) = \sum_{n=0}^{N-1} h[n]z^{-n} \tag{7.1}$$

If the impulse response sequence $\{h[n]\}$, $0 \leq n \leq N - 1$ satisfies certain requirements, then the corresponding FIR filter will have linear-phase response. In fact, an important aspect of an FIR digital filter is its linear-phase characteristics. Depending on the value of the filter length N and the nature of the impulse response whether symmetric or asymmetric, we can define four types of linear-phase FIR digital filters.

Type I FIR Filter If the filter order M is even (or the filter length $N = M + 1$ is odd) and the impulse response is symmetric, then the FIR filter is termed type I linear-phase FIR filter. In this case, we can express the impulse response sequence as

$$h[n] = h[M - n], 0 \leq n \leq M, \tag{7.2}$$

and its frequency response is found to be

Electronic supplementary material: The online version of this article (https://doi.org/10.1007/978-3-319-76029-2_7) contains supplementary material, which is available to authorized users.

© Springer International Publishing AG, part of Springer Nature 2019
K. S. Thyagarajan, *Introduction to Digital Signal Processing Using MATLAB with Application to Digital Communications*,
https://doi.org/10.1007/978-3-319-76029-2_7

$$H\left(e^{j\Omega}\right) = \sum_{n=0}^{M+1} h[n]e^{-jn\Omega} = \sum_{n=0}^{\frac{M}{2}-1} h[n]\left\{e^{-jn\Omega} + e^{-j(M-n)\Omega}\right\} + h\left[\frac{M}{2}\right]e^{-j\frac{M}{2}\Omega}. \quad (7.3)$$

After simplification of (7.3), we have the frequency response of type I FIR filter as

$$H\left(e^{j\Omega}\right) = e^{-j\frac{M}{2}\Omega}\left\{h\left[\frac{M}{2}\right] + 2\sum_{n=1}^{\frac{M}{2}} h\left[\frac{M}{2} - n\right]\cos\left(n\Omega\right)\right\} = e^{-j\frac{M}{2}\Omega}\widehat{H}\left(e^{j\Omega}\right), \quad (7.4)$$

where $\widehat{H}\left(e^{j\Omega}\right)$ is a real function of Ω. From Eq. (7.4), we find that the phase response of type I FIR filter is linear with a delay of $M/2$ samples.

Type II FIR Filter An FIR filter is said to be type II if its degree or order M is odd and its impulse response sequence is symmetric, as defined in (7.2). Using Eq. (7.2) in (7.3), we can express the frequency response of type II FIR filter after simplification as

$$H\left(e^{j\Omega}\right) = e^{-j\frac{M}{2}\Omega}\left\{2\sum_{n=1}^{\frac{M+1}{2}} h\left[\frac{M+1}{2} - n\right]\cos\left(\left(n - \frac{1}{2}\right)\Omega\right)\right\} \quad (7.5)$$

Again, the quantity within the braces is real, and the phase response is linear with a delay of $\frac{M}{2}$ samples.

Type III FIR Filter In this case the filter order or degree is even, and the impulse response is asymmetric as defined by

$$h[n] = -h[M - n], 0 \le n \le M \quad (7.6)$$

The frequency response of type III FIR filter can be shown to be

$$H\left(e^{j\Omega}\right) = je^{-j\frac{M}{2}\Omega}\left\{2\sum_{n=1}^{\frac{M}{2}} h\left[\frac{M}{2} - n\right]\sin\left(n\Omega\right)\right\} \quad (7.7)$$

The quantity within the braces in (7.7) is real, and the phase response is linear with a delay of $\frac{M}{2}$ samples.

Type IV FIR Filter The type IV FIR filter has an odd degree, and its impulse response is asymmetric as defined in (7.6). Using these facts, we can express its frequency response as

$$H\left(e^{j\Omega}\right) = je^{-j\frac{M}{2}\Omega}\left\{2\sum_{n=1}^{\frac{M+1}{2}} h\left[\frac{M+1}{2} - n\right]\sin\left(\left(n - \frac{1}{2}\right)\Omega\right)\right\} \quad (7.8)$$

As can be seen from (7.8), the quantity within the braces is real, and so the phase response is linear with a delay of $\frac{M}{2}$ samples.

7.2   Linear-Phase FIR Filter Design

The design of an FIR digital filter amounts to the determination of its impulse response sequence. In order for the FIR filter to have linear-phase response, the corresponding impulse response must be one of the four types defined earlier. We will stick to type I FIR filter design in the following discussion. There are two types of design techniques, namely, window-based design and computer-aided design. We will consider both techniques in this chapter.

7.2.1   Lowpass FIR Filter Design

The frequency response of an ideal lowpass filter with a cutoff frequency Ω_c can be described by

$$H_I\left(e^{j\Omega}\right) = \begin{cases} 1, |\Omega| \leq \Omega_c \\ 0, \ -\pi - \Omega_c \leq \Omega \leq \Omega_c + \pi \end{cases} \tag{7.9}$$

The sequence which gives rise to (7.9) corresponds to the inverse DTFT (IDTFT) of (7.9), which is obtained from the definition of IDTFT

$$h_I[n] = \frac{1}{2\pi} \int_{-\pi}^{\pi} H_I\left(e^{j\Omega}\right) e^{jn\Omega} d\Omega = \frac{\sin\left(n\Omega_c\right)}{n\pi} = \frac{\Omega_c}{\pi} sinc\left(\frac{n\Omega_c}{\pi}\right), \ -\infty \leq n \leq \infty,$$

$$\tag{7.10}$$

where the sinc function is defined as

$$sinc(x) = \frac{\sin\left(\pi x\right)}{\pi x} \tag{7.11}$$

As can be seen from Eq. (7.10), the duration of the impulse response sequence of an ideal lowpass digital filter is infinite. But we are looking for an FIR filter, which has a finite duration impulse response. So, the only way to limit the duration of the impulse response in (7.10) is to truncate it abruptly to a finite length. Note that the sinc function is symmetric about the origin. Since we are looking for a linear-phase FIR filter, we can abruptly truncate the impulse response sequence in (7.10) to $\pm M$ samples about the origin. This will result in an impulse response sequence of length $N = 2M + 1$ samples. It is important to remember that even though we have obtained the impulse response of a linear-phase FIR filter corresponding to an ideal lowpass filter, the resulting frequency response will not be identical to that of the ideal lowpass filter. Why? Because we have limited the duration of the FIR filter to a finite length. There are a couple of observations. One, the abrupt truncation is equivalent to multiplying the ideal impulse response sequence by a rectangular window of length $N = 2M + 1$ samples centered at $n = 0$. Thus, we can write the impulse response of the desired FIR filter as

$$h[n] = h_I[n]w_R[n], \ -M \leq n \leq M, \tag{7.12}$$

where the rectangular window of length $2M + 1$ is described by

$$w_R[n] = \begin{cases} 1, \ -M \leq n \leq M \\ 0, otherwise \end{cases} \tag{7.13}$$

Because we multiply the ideal impulse response by the rectangular window, the design technique is called *windowing technique*. The second observation is that the truncated impulse response approaches the ideal impulse response as the window length increases. Is rectangular window the only window available for the design of an FIR filter, or are there other windows with desirable properties? We will deal with the types of windows, their properties, and their effects on the frequency response of an FIR filter in the sections to follow. One other thing in the abovementioned design is that the resulting FIR filter is non-causal because its impulse response is not zero for n < 0. However, since the impulse response sequence is of finite duration, we can shift it to the right by M samples so that the impulse response of the causal FIR filter can be described by

$$h[n] = h_I[n - M]w_R[n - M], 0 \leq n \leq 2M \tag{7.14}$$

Example 7.1 Design a lowpass, length-9 FIR filter that approximates an ideal lowpass filter whose normalized cutoff frequency is $\frac{\pi}{2}$ rad.

Solution The impulse response of the ideal lowpass filter from Eq. (7.10) is

$$h_I[n] = \frac{\Omega_c}{\pi} sinc\left(\frac{n\Omega_c}{\pi}\right) = \frac{1}{2} sinc\left(\frac{n}{2}\right) = \frac{1}{2}\frac{\sin\left(\frac{n\pi}{2}\right)}{\frac{n\pi}{2}}, \ -\infty \leq n \leq \infty \tag{7.15}$$

The impulse response of the length-9, causal FIR filter is then obtained by multiplying the ideal impulse response sequence by the rectangular window of length 9 and shifting the resulting sequence to the right by four samples. Thus, the desired impulse response is

$$h[n] = \frac{1}{2}\frac{\sin\left(\frac{(n-4)\pi}{2}\right)}{\frac{(n-4)\pi}{2}}, 0 \leq n \leq 8 \tag{7.16}$$

The length-9 impulse response of the lowpass filter is shown in Fig. 7.1. Its magnitude of frequency response in dB and phase response are shown in Fig. 7.2a, b, respectively. As expected, the phase response is linear due to the symmetry of the impulse response. However, we see ripples in both the passband and stopband of the frequency response. What if we increase the filter length to, say, 25? The impulse response of length-25 lowpass filter is shown in Fig. 7.3. Even if we increase the

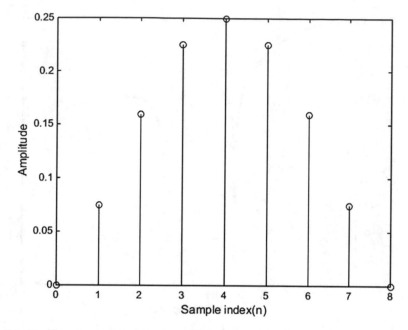

Fig. 7.1 Ideal impulse response of the lowpass FIR filter simply truncated to length 9

filter length to 25, we still see the ripples in the frequency response, as shown in Fig. 7.4a. The phase response is still linear, as can be seen from Fig. 7.4b. However, the ripple width in the magnitude response near zero frequency gets smaller but increases toward the passband edge. This is depicted in Fig. 7.5, which shows the magnitude of the frequency response for length 9 and length 25, respectively. The reason for this is explained in the next section. The M-file for this problem is named *Example7_1.m*.

7.2.2 Gibbs Phenomenon

In the last section, we introduced the design of a linear-phase lowpass FIR filter by simply truncating the impulse response of an ideal lowpass filter to a finite length. This simple or abrupt truncation is equivalent to multiplying the ideal, infinite-length impulse response by a rectangular window of finite-length sequence as in Eq. (7.12). From the multiplication property of the DTFT in the discrete-time domain of discrete-time signals, we know that the DTFT of the truncated impulse response is the convolution of the DTFTs of the ideal impulse response and that of the rectangular window. The DTFT of a rectangular window of finite length is a sinc function. As we saw in the previous example, the sinc function has a main lobe and an infinite number of side lobes as shown in Fig. 7.6. However, the amplitudes of the side lobes decrease as the frequency increases. Therefore, if we convolve the ideal frequency

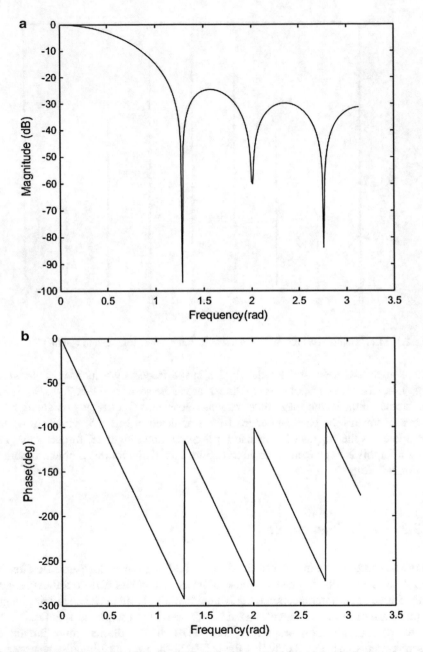

Fig. 7.2 Frequency response of length-9 FIR filter of Example 7.1. (**a**) Magnitude response in dB, (**b**) phase response

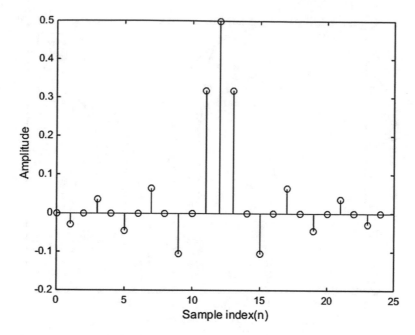

Fig. 7.3 Impulse response of the length-25 FIR filter of Example 7.1 with the same cutoff frequency

response of a lowpass filter with that of the rectangular window, which is a sinc function, we will get ripples in both the passband and stopband. This rippling effect in the frequency response is termed *Gibbs phenomenon*. No matter what the filter length is, there will always be ripples in both the passband and stopband. As the filter length increases, the ripple width gets smaller near the zero frequency but increases toward the passband edge, as depicted in Fig. 7.7. The question is how can we design a linear-phase FIR lowpass filter with a better approximation to the ideal brick wall response? The answer lies in the proper choice of a window. Remember that in using a rectangular window, the transition region narrows as the filter length is increased, but the minimum stopband attenuation does not change. Therefore, we must come up with a window whose frequency response has a main lobe only. This may not be a realizable window. However, if we can reduce the amplitudes of the side lobes, we can reduce the ripples in the resulting FIR filter. Fortunately, there exist several windows with each having a unique feature. Some windows have no independent control on the transition width and minimum stopband attenuation. These are called *fixed windows*. A variable window, on the other hand, has independent control on both the transition width and minimum stopband attenuation. We will look at both types of windows.

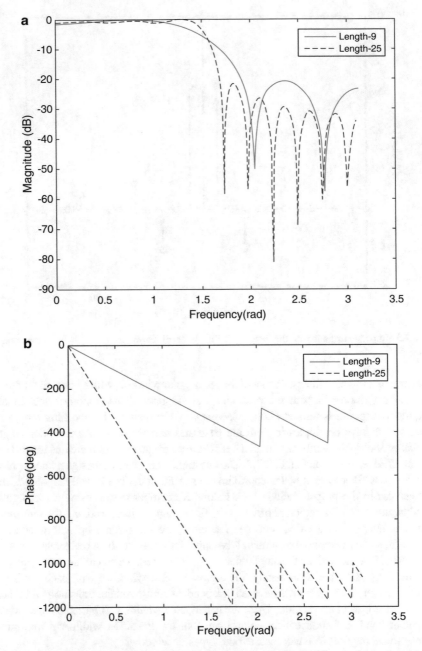

Fig. 7.4 Frequency response of length-25 FIR filter of Example 7.1. (**a**) Magnitude response in dB, (**b**) phase response

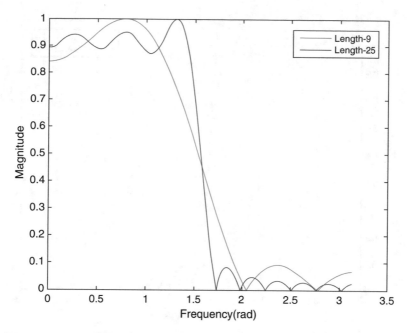

Fig. 7.5 Magnitude of the frequency responses of length-9 and length-25 FIR filters of Example 7.1

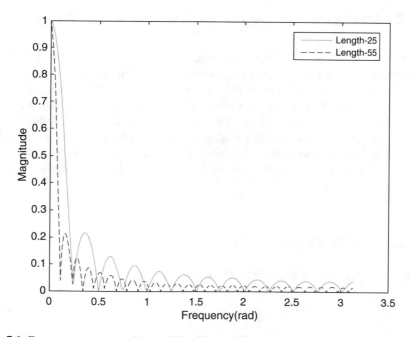

Fig. 7.6 Frequency responses of length-25 and length-55 rectangular windows

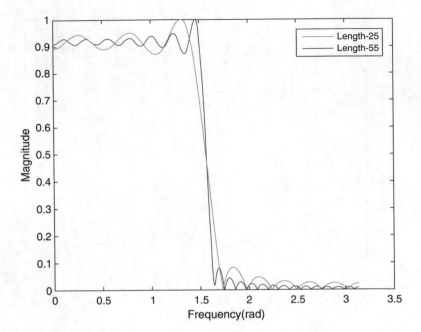

Fig. 7.7 Gibbs phenomenon

7.2.3 Windowed Lowpass Linear-Phase FIR Filter Design

With the length of the filter specified, a lowpass, linear-phase FIR filter corresponding to an ideal lowpass filter is obtained by multiplying the ideal impulse response by a suitable window of the same length as that of the desired filter and shifting the resulting impulse response by half of the filter order. Remember that we are dealing with type I linear-phase FIR filter. So, what is a suitable window? For a given filter length, we must get a minimum stopband attenuation larger than that obtainable from a rectangular window of the same length. This is possible by smoothly tapering the window. Let us look at some desirable windows in the following.

7.2.3.1 Bartlett Window

The Bartlett window of length $N = 2M + 1$ is described by

$$w[n] = \begin{cases} 1 - \dfrac{n}{M+1}, & -M \leq n \leq M \\ 0, otherwise \end{cases} \tag{7.17}$$

As can be seen from (7.17), the Bartlett window tapers off linearly.

7.2.3.2 Haan Window

A length $N = 2M + 1$ Haan window is described by

$$w[n] = \begin{cases} \frac{1}{2}\left(1 + \cos\left(\frac{n\pi}{M}\right)\right), & -M \leq n \leq M \\ 0, otherwise \end{cases} \tag{7.18}$$

Since the cosine function varies more smoothly than a linear function, the Haan window tapers off more smoothly than the Bartlett window.

7.2.3.3 Hamming Window

The Hamming window of length $N = 2M + 1$ is described by

$$w[n] = \begin{cases} 0.54 + 0.46\cos\left(\frac{n\pi}{M}\right), & -M \leq n \leq M \\ 0, otherwise \end{cases} \tag{7.19}$$

From Eq. (7.19), we notice that the Hamming window is not zero at $\pm M$; instead it is 0.08. This makes it even smoother than the Bartlett window.

7.2.3.4 Blackman Window

The Blackman window of length $2M + 1$ is defined as

$$w[n] = \begin{cases} 0.42 + 0.5\cos\left(\frac{n\pi}{M}\right) + 0.08\cos\left(\frac{2n\pi}{M}\right), & -M \leq n \leq M \\ 0, otherwise \end{cases} \tag{7.20}$$

Blackman window function is zero at $\pm M$ and is smoother than the Hamming window function. The four windows described above are shown in Figs. 7.8 and 7.9 for a length of 17. As can be seen from the figures, Blackman window is the smoothest. The corresponding DTFTs of the windows are shown in Figs. 7.10, 7.11, 7.12, 7.13, and 7.14. The Blackman window has the largest minimum stopband attenuation of 60 dB, while the rectangular window only offers 14 dB of minimum stopband attenuation. On the contrary, the rectangular window has the smallest transition width, while the Blackman window has the largest transition width. The windows described thus far are called *fixed windows*, because for a given filter length, the minimum obtainable stopband attenuation and the transition width are fixed. We will look at another type of window, which has the potential of trade-off between transition width and minimum stopband attenuation. Before we discuss further, let us revisit Example 7.1 and use fixed windows to see how they compare with the rectangular window.

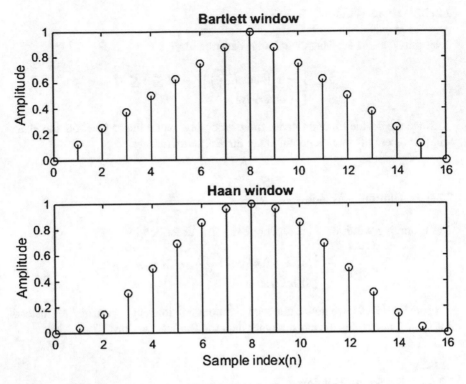

Fig. 7.8 Bartlett and Haan windows of length 17 in the discrete-time domain

The following table lists these window functions for readers' easy reference (Table 7.1).

Example 7.2 Repeat Example 7.1, but use the fixed windows and compare the resulting frequency responses.

Solution To recapitulate Example 7.1, we have to design a length-9 lowpass FIR digital filter by truncating the impulse response of an ideal lowpass filter with a cutoff frequency of $\frac{\pi}{2}$ rad. The length-9 impulse response sequence is then multiplied by the fixed windows mentioned above, which will result in windowed impulse response sequences. The DTFT of the windowed impulse response sequence will be the frequency response of the lowpass FIR filter. Depending on the chosen window, the minimum stopband attenuation and the transition width of the FIR filter will be different. As we saw earlier, the FIR filter using Blackman window will have the largest minimum stopband attenuation in dB and the widest transition width. The FIR filter using the rectangular window will have the narrowest transition width and smallest stopband attenuation in dB. The other fixed windows will have these values in between the rectangular and Blackman windows. These are clearly seen from Fig. 7.15. The M-file named *Example7_2.m* is used to solve this problem.

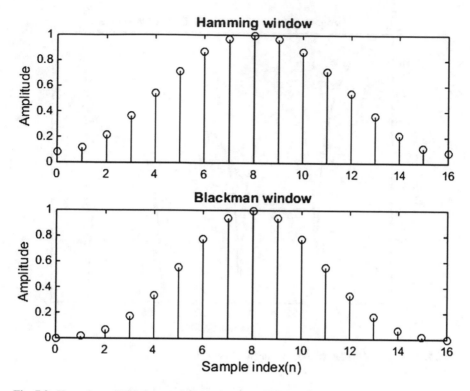

Fig. 7.9 Hamming and Blackman windows of length 17 in the discrete-time domain

7.2.3.5   Determination of FIR Filter Order

In the previous two examples, we assumed a value for the lowpass FIR filter order. In order to meet the given frequency specifications with as small a filter order as necessary, we will use a set of formulas. More specifically, we will use Kaiser's, Bellanger's, and Hermann's empirical formulas to determine the required FIR filter order. The frequency specifications of a lowpass filter used in these formulas are the (a) normalized passband edge frequency Ω_p, (b) normalized stopband edge frequency Ω_s, (c) peak passband ripple δ_p, and (d) peak stopband ripple δ_s. With these parameters being specified, we can use these formulas to calculate the filter order as follows.

Kaiser's Formula According to this formula, the required order N of a lowpass FIR filter with the above frequency specifications is calculated from

$$N \cong \frac{-20 log_{10}\left(\sqrt{\delta_p \delta_s}\right) - 13}{\frac{14.6\left(\Omega_s - \Omega_p\right)}{2\pi}}. \tag{7.21}$$

Since the filter order is an integer, one has to round the value in (7.21) up to the nearest integer.

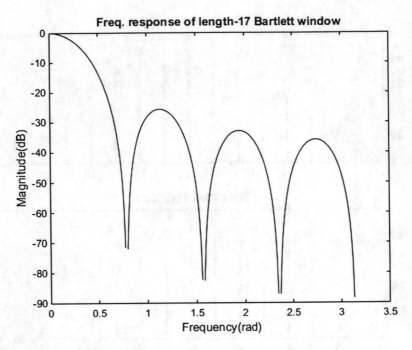

Fig. 7.10   DTFT of Bartlett window of length-17 samples

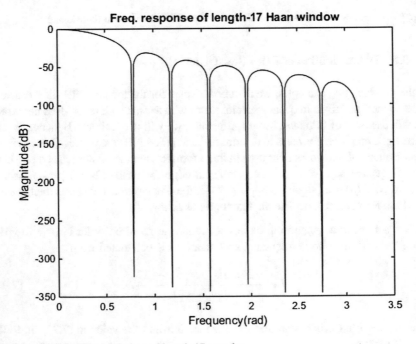

Fig. 7.11   DTFT of Haan window of length-17 samples

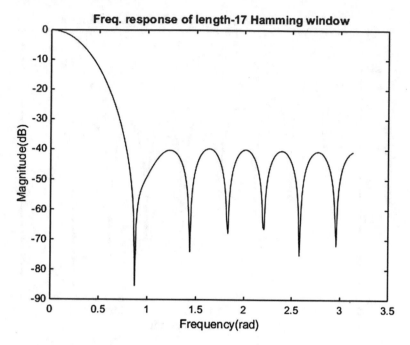

Fig. 7.12 DTFT of Hamming window of length-17 samples

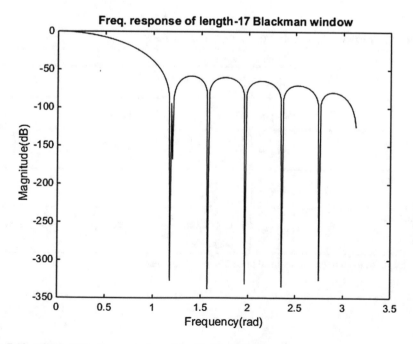

Fig. 7.13 DTFT of Blackman window of length-17 samples

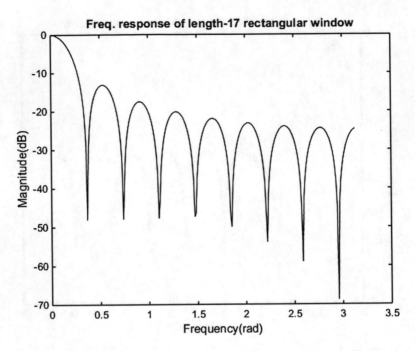

Fig. 7.14 DTFT of rectangular window of length-17 samples

Table 7.1 Window functions used for the design of FIR filters

Window name	Window function
Rectangular	$w[n] = \begin{cases} 1, & -M \le n \le M \\ 0, & otherwise \end{cases}$
Bartlett	$w[n] = \begin{cases} 1 - \dfrac{|n|}{M+1}, & -M \le n \le M \\ 0, & otherwise \end{cases}$
Haan	$w[n] = \begin{cases} \dfrac{1}{2}\left\{1 + \cos\left(\dfrac{n\pi}{M}\right)\right\}, & -M \le n \le M \\ 0, & otherwise \end{cases}$
Hamming	$w[n] = \begin{cases} 0.54 + 0.46\cos\left(\dfrac{n\pi}{M}\right), & -M \le n \le M \\ 0, & otherwise \end{cases}$
Blackman	$w[n] = \begin{cases} 0.42 + 0.5\cos\left(\dfrac{n\pi}{M}\right) + 0.08\cos\left(\dfrac{2n\pi}{M}\right), & -M \le n \le M \\ 0, & otherwise \end{cases}$

The length of the window is assumed to be 2M + 1, M an integer

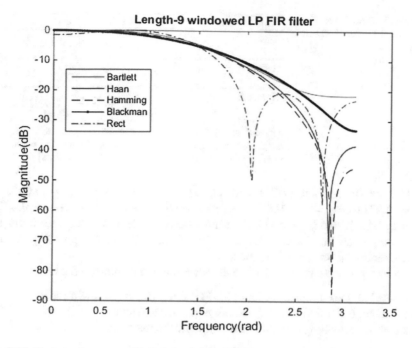

Fig. 7.15 Frequency response of length-9 windowed lowpass FIR filter of Example 7.2

Bellanger's Formula Bellanger's formula for the filter order takes the form

$$N \cong \frac{2log_{10}\left(10\delta_p\delta_s\right)}{\frac{3\left(\Omega_s-\Omega_p\right)}{2\pi}} - 1 \tag{7.22}$$

As we did in Kaiser's formula, the value in (7.22) must be rounded up to the nearest integer equal to or larger than the value in (7.22).

Hermann's Formula This formula for the filter order is given by

$$N \cong \frac{D_\infty\left(\delta_p,\delta_s\right) - F\left(\delta_p,\delta_s\right)\left[\frac{\left(\Omega_s-\Omega_p\right)}{2\pi}\right]^2}{\frac{\left(\Omega_s-\Omega_p\right)}{2\pi}}, \tag{7.23}$$

where the parameters are defined by

$$D_\infty\left(\delta_p,\delta_s\right) = \left\{a_1\left(log_{10}\delta_p\right)^2 + a_2\left(log_{10}\delta_p\right) + a_3\right\}log_{10}\delta_s$$
$$-\left\{a_4\left(log_{10}\delta_p\right)^2 + a_5\left(log_{10}\delta_p\right) + a_6\right\} \tag{7.24}$$
$$F\left(\delta_p,\delta_s\right) = b_1 + b_2\left(log_{10}\left(\delta_p\right) - log_{10}\left(\delta_s\right)\right) \tag{7.25}$$

Table 7.2 Formulas to calculate the order of an FIR filter

Window name	FIR filter order
Kaiser	$N \cong \dfrac{-20log_{10}\left(\sqrt{\delta_p\delta_s}\right)-13}{14.6\left(\Omega_s-\Omega_p\right)/2\pi}$
Bellanger	$N \cong \dfrac{2log_{10}\left(10\delta_p\delta_s\right)}{3\left(\Omega_s-\Omega_p\right)/2\pi} - 1$
Hermann	$N \cong \dfrac{D_\infty\left(\delta_p,\delta_s\right)-F\left(\delta_p,\delta_s\right)\left[\frac{(\Omega_s-\Omega_p)}{2\pi}\right]^2}{\left(\Omega_s-\Omega_p\right)/2\pi}$

δ_p and δ_s are, respectively, the passband and stopband ripples; Ω_p is the passband edge, and Ω_s is the stopband edge. The parameters in Hermann's formula are defined in (7.24) and (7.25)

The coefficients in (7.24) and (7.25) are as follows: $a_1 = 0.005309$, $a_2 = 0.07114$, $a_3 = -0.4761$, $a_4 = 0.00266$, $a_5 = 0.5941$, $a_6 = 0.4278$, $b_1 = 11.01217$, and $b_2 = 0.51244$. It must be mentioned that these formulas are only empirical. In Table 7.2, the FIR filter order corresponding to the abovementioned formulas is shown.

Let us try an example to calculate the filter order and compare the results.

Example 7.3 Determine the order of the FIR filter with the following specifications: passband edge at 1.5 kHz, stopband edge at 2 kHz, sampling frequency of 8 kHz, peak passband ripple $\alpha_p = 0.1\ dB$, and minimum stopband attenuation of $\alpha_s = 40\ dB$.

Solution We have to first convert the dBs to actual numbers to enter them in the filter order formulas. So,

$$\delta_p = 1 - 10^{-0.05\alpha_p} = 1 - 10^{-0.005} = 0.0114469, \qquad (7.26)$$

$$\delta_s = 10^{-0.05\alpha_s} = 10^{-2} = 0.01 \qquad (7.27)$$

From Kaiser's formula, we get

$$N \cong \left\lceil \frac{-20log_{10}\left(\sqrt{0.0114469*0.01}\right)-13}{\frac{14.6(2000-1500)}{8000}} \right\rceil = 28.9459 = 29 \qquad (7.28)$$

If we use the same values in Bellanger's formula, we get $N = 31$. Hermann's formula results in $N = 30$. So, Kaiser's formula gives the least value for the filter order. Note that $\Omega = \frac{2\pi f}{F_s}$, where F_s is the sampling frequency.

7.2.3.6   Adjustable Window Functions

The fixed windows we described earlier don't have the freedom to control both the minimum stopband attenuation and transition width for a specified filter length. There are a couple of window functions that can control both the minimum stopband attenuation and the transition width for a given filter length. These windows are Dolph-Chebyshev window and Kaiser window. Let us take a look at these two adjustable window functions.

7.2.3.6.1 Dolph-Chebyshev (DC) Window

The DC window function of length 2M + 1 is defined as

$$w[n] = \frac{1}{2M+1} \left\{ \frac{1}{\gamma} + 2 \sum_{k=1}^{M} T_{2M}\left(\beta \cos\left(\frac{k\pi}{2M+1} \right) \right) \cos\left(\frac{2nk\pi}{2M+1} \right) \right\}, \ -M \le n \le M,$$

(7.29)

where γ is the relative amplitude of the side lobe, that is,

$$\gamma = \frac{amplitude \ of \ side \ lobe}{amplitude \ of \ main \ lobe},$$

(7.30)

$$\beta = \cosh\left(\frac{1}{2M} \cosh^{-1}\left(\frac{1}{\gamma} \right) \right),$$

(7.31)

and the kth-order Chebyshev polynomial in the variable x is given by

$$T_k(x) = \begin{cases} \cos\left(k\cos^{-1}(x)\right), |x| \le 1 \\ \cosh\left(k\cosh^{-1}(x)\right), |x| > 1 \end{cases}$$

(7.32)

Unlike the fixed window functions, the order of the DC window is determined from

$$N = \frac{2.05\alpha_s - 16.4}{2.285(\Delta\Omega)},$$

(7.33)

where α_s is the minimum stopband attenuation in dB, and $\Delta\Omega = \Omega_s - \Omega_p$ is the transition width with Ω_p being the passband edge and Ω_s the stopband edge.

Example 7.4 Determine the lowpass FIR filter order with a passband edge at 0.23π, stopband edge at 0.5π, and a minimum stopband attenuation of 40 dB. Use the Dolph-Chebyshev window.

Solution From Eq. (7.33), we find

$$N = \frac{2.05 * 40 - 16.4}{2.285(0.5\pi - 0.23\pi)} \cong 33.85 = 34$$

(7.34)

Therefore, the filter length is 35.

Example 7.5 Compare the frequency responses of length-21 Dolph-Chebyshev windows for $\gamma = 10\ dB$ *and* $30\ dB$. Design a length-21 LP FIR filter with a cutoff frequency of 0.3π using Dolph-Chebyshev window. Compare the frequency responses of the LPFs for the two values of γ listed.

Solution A relative sidelobe amplitude of 10 dB is equal to an actual value of 3.1623. Similarly, $\gamma = 30\ dB \Rightarrow 31.6228$. Using these values, we can calculate β, which, for M = 10, are 0.9922 and 0.9882, respectively. We then calculate the length-21 Dolph-Chebyshev windows for the two cases. The frequency responses of the Dolph-Chebyshev windows for $\gamma = 10$ *and* $30\ dB$ are plotted as a function of the normalized frequency in rad and are shown in Fig. 7.16. From the figure we clearly notice the minimum sidelobe values to be 10 and 30 dB, respectively. The impulse response of the ideal LPF with a cutoff frequency of $\Omega_c = 0.23\pi$ is given in Eq. (7.10). After multiplying the impulse response by the respective windows, we obtain the windowed impulse response of the LPF. The Dolph-Chebyshev window and the corresponding windowed impulse response of the LP FIR filter for the two values of γ are shown in Figs. 7.17 and 7.18, respectively. In each figure, the top plot is the window function, and bottom plot is the windowed impulse response. Finally, the frequency response of the LP FIR filter for the two cases is shown in Fig. 7.19. The minimum stopband attenuation corresponding to $\gamma = 30\ dB$ is larger than that corresponding to the 10 dB case, while the transition width is larger for

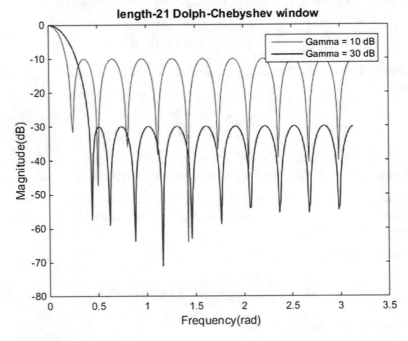

Fig. 7.16 Frequency response of length-21 Dolph-Chebyshev window for the two values of the relative amplitude of the sidelobe of 10 and 30 dB

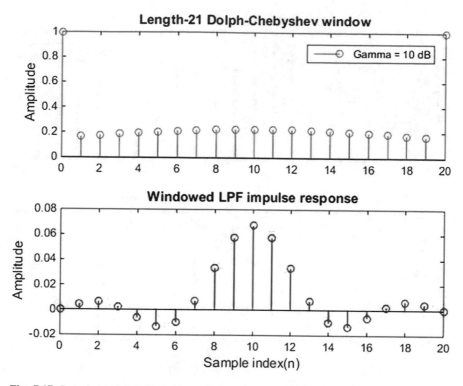

Fig. 7.17 Length-21 Dolph-Chebyshev window for $\gamma = 10$ dB. Top plot, Dolph-Chebyshev window; bottom plot, windowed impulse response of the LP FIR filter of Example 7.5

$\gamma = 30$ dB compared to the other case. The MATLAB M-file used to solve this problem is named *Example7_5.m*.

7.2.3.6.2 Kaiser Window

A length-$2M + 1$ Kaiser window function is defined as

$$w[n] = \frac{I_0\left(\beta\sqrt{1 - \left(\frac{n}{M}\right)^2}\right)}{I_0(\beta)}, \quad -M \le n \le M \tag{7.35}$$

In (7.35), β is an adjustable parameter and is given by

$$\beta = \begin{cases} 0.1102(\alpha_s - 8.7), \alpha_s > 50 \\ 0.5842(\alpha_s - 21)^{0.4} + 0.07886(\alpha_s - 21), 21 \le \alpha_s \le 50 \\ 0, \alpha_s < 21 \end{cases} \tag{7.36}$$

and $I_0(x)$ is the modified zeroth-order Bessel function, which is described by

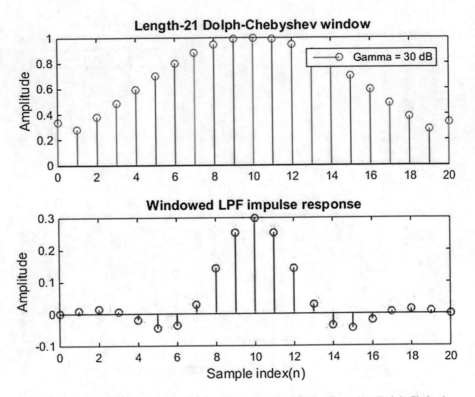

Fig. 7.18 Length-21 Dolph-Chebyshev window for $\gamma = 30$ *dB*. Top plot, Dolph-Chebyshev window; bottom plot, windowed impulse response of the LP FIR filter of Example 7.5

$$I_0(x) = 1 + \sum_{m=1}^{\infty} \left(\frac{\left(\frac{x}{2} \right)^m}{m!} \right)^2 \qquad (7.37)$$

For the Kaiser window, the FIR filter order is evaluated from

$$N = \frac{\alpha_s - 8}{2.285 \Delta \Omega} \qquad (7.38)$$

Example 7.6 For the same specifications as in Example 7.4, determine the order of the FIR filter if Kaiser window is used.

Solution Substituting the values for the relevant parameters in Eq. (7.38), we find

$$N = \frac{40 - 8}{2.285(0.5\pi - 0.23\pi)} = 16.5 = 17 \qquad (7.39)$$

Therefore, the filter length is 18. Note the difference in the filter order between Kaiser and DC window functions (Table 7.3)!

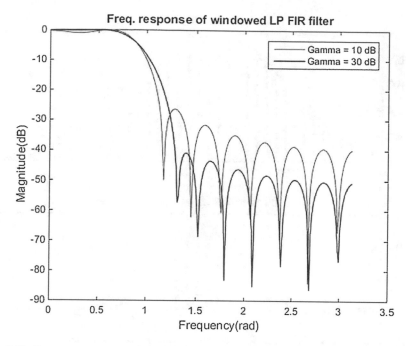

Fig. 7.19 Frequency response of the LP FIR filter of Example 7.5 using a length-21 Dolph-Chebyshev window

Table 7.3 Filter order for adjustable windows

Window type	Filter order
Kaiser	$N = \frac{\alpha_s - 8}{2.285\Delta\Omega}$
Dolph-Chebyshev	$N = \frac{2.056\alpha_s - 16.4}{2.285\Delta\Omega}$

α_s is the minimum stopband attenuation in dB and transition width is $\Delta\Omega = \Omega_s - \Omega_p$ and Ω_s is the stopband edge and Ω_p the passband edge, both in radians

Example 7.7 Design a LP FIR filter with a passband edge at $\Omega_p = 0.3\pi$, stopband edge at $\Omega_s = 0.5\pi$, and a minimum stopband attenuation of 60 dB using Kaiser window. Compare this with the LP FIR filter of the same length using Dolph-Chebyshev window.

Solution The transition width $\Delta\Omega = 0.5\pi - 0.3\pi = 0.2\pi$. Then from (7.38), we find the FIR filter order to be 37. Therefore, the FIR filter length is 38. From Eq. (7.36), we obtain the parameter $\beta = 5.6533$. Figure 7.20 shows the plot of the Kaiser window of length 38 along with that of the windowed impulse response of the LP FIR filter. Figure 7.21 plots the Dolph-Chebyshev window of length 38 as well as the corresponding windowed impulse response of the LP FIR filter. The relative sidelobe amplitude parameter for the Dolph-Chebyshev window is set at $\gamma = 30 \ dB$. The frequency response of the LP FIR filter using Kaiser window of length 38 is shown in Fig. 7.22 along with that using the same length Dolph-

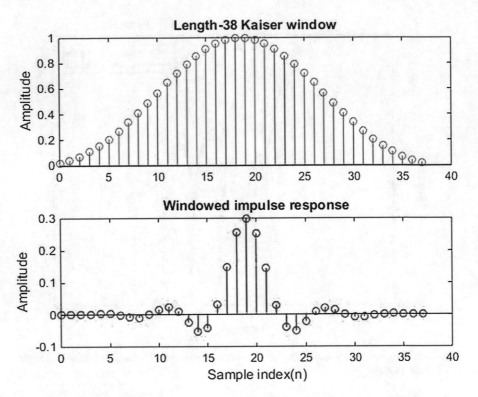

Fig. 7.20 A length-38 Kaiser window of Example 7.7. Top plot, Kaiser window; bottom plot, windowed impulse response

Chebyshev window with $\gamma = 30$ *dB*. As seen from the figure, the Kaiser window achieves a larger minimum stopband attenuation (60 dB, as specified) than that achieved by the Dolph-Chebyshev window of the same filter length at the expense of a larger transition width. We can increase the minimum stopband attenuation achievable by the Dolph-Chebyshev-windowed LP FIR filter by increasing the value of γ.

7.2.4   Design of a Highpass Linear-Phase FIR Filter

The frequency response of an ideal highpass filter with a cutoff frequency Ω_c is described by

$$H_{hp}\left(e^{j\Omega}\right) = \begin{cases} 0, & |\Omega| \leq \Omega_c \\ 1, \Omega_c \leq |\Omega| \leq \pi \end{cases} \tag{7.40}$$

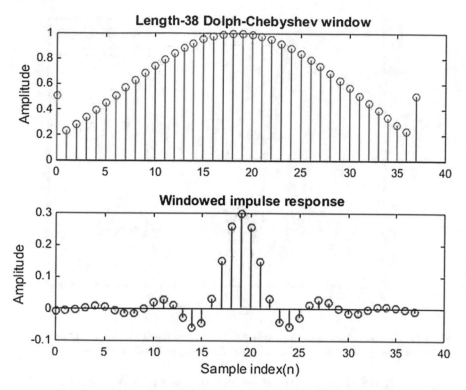

Fig. 7.21 A length-38 Dolph-Chebyshev window with $\gamma = 30\ dB$. Top plot, Dolph-Chebyshev window; bottom plot, windowed impulse response

Its impulse response is obtained from

$$
h_{hp}[n] = \frac{1}{2\pi}\left\{ \int\limits_{-\pi}^{-\Omega_c} e^{jn\Omega}d\Omega + \int\limits_{\Omega_c}^{\pi} e^{jn\Omega}d\Omega \right\} = \delta[n]
$$

$$
-\left(\frac{\Omega_c}{\pi}\right) sinc\left(\frac{n\Omega_c}{\pi}\right),\ -\infty \leq n \leq \infty
$$

(7.41)

As in the case of the lowpass FIR filter design, the impulse response of the desired highpass filter is obtained by multiplying the ideal impulse response by a suitable window of a given length and then shifting the resulting impulse response sequence by $M = \frac{N-1}{2}$ samples to the right. We have assumed the filter length to be odd. The following example illustrates the design of a windowed highpass FIR digital filter.

Example 7.8 Design a length-21 highpass FIR filter with a cutoff frequency of $\Omega_c = 0.5\pi$ using fixed windows, and plot the respective frequency responses.

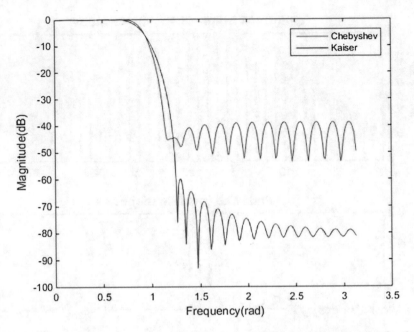

Fig. 7.22 Frequency response of length-38 LP FIR filter using Kaiser window. The plot also shows the frequency response of the LP filter using the same length Dolph-Chebyshev window

Solution The impulse response of the ideal highpass filter is given in Eq. (7.41) with the cutoff frequency specified in Example 7.8. We multiply the ideal impulse response by the four fixed windows of length-21 samples point by point to obtain the impulse response of the corresponding FIR filter. MATLAB has the function *window(@hamming, N)*, which will generate the Hamming window of length-N samples. Similarly, we can generate the other fixed windows by using the appropriate name for the windows. Figures 7.23 and 7.24 display the windowed impulse response of the highpass FIR filter. The magnitude in dB of the frequency response of the highpass filters is shown in Fig. 7.25. As a comparison, the frequency response of the highpass FIR filter using rectangular window is also plotted in Fig. 7.25. As expected, rectangular window has the smallest minimum stopband attenuation, while the Blackman window achieves the largest minimum stopband attenuation. The M-file named *Example7_8.m* is used to solve this problem.

7.2.5 Design of a Bandpass Linear-Phase FIR Filter

The frequency response of an ideal bandpass filter whose lower and upper cutoff frequencies are, respectively, Ω_1 and Ω_2 is defined as

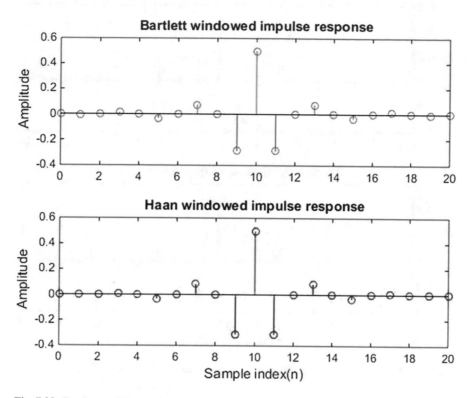

Fig. 7.23 Bartlett- and Haan-windowed impulse response of a length-21 highpass FIR filter

$$H_{bp}\left(e^{j\Omega}\right) = \begin{cases} 0, |\Omega| < \Omega_1 \\ 1, \Omega_1 \leq |\Omega| \leq \Omega_2 \\ 0, \Omega_2 < |\Omega| \leq \pi \end{cases} \tag{7.42}$$

The corresponding impulse response is found using the IDTFT and is given by

$$h_{bp}[n] = \frac{1}{2\pi} \int_{-\pi}^{\pi} H_{bp}\left(e^{j\Omega}\right) d\Omega = \left(\frac{\Omega_2}{\pi}\right) sinc\left(\frac{n\Omega_2}{\pi}\right)$$
$$- \left(\frac{\Omega_1}{\pi}\right) sinc\left(\frac{n\Omega_1}{\pi}\right), \ -\infty \leq n \leq \infty \tag{7.43}$$

Again, the impulse response of the desired bandpass FIR filter is derived by multiplying the ideal impulse response in (7.43) by a suitable window function of specified length (assumed odd) and then shifting the resulting sequence by $M = \frac{N-1}{2}$ samples to the right.

Example 7.9 Design a bandpass FIR filter with passband edges at $\Omega_1 = 0.4\pi$, $\Omega_2 = 0.6\pi$, transition width $\Delta\Omega = 0.1\pi$, and a minimum stopband attenuation of 30 dB. Use Kaiser window.

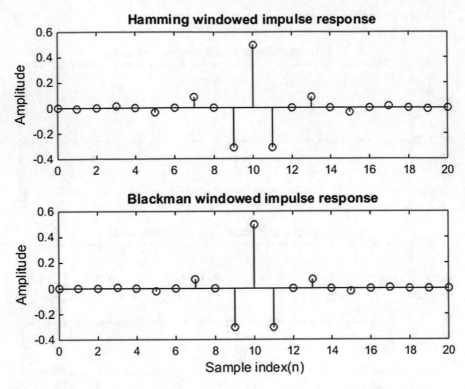

Fig. 7.24 Hamming- and Blackman-windowed impulse response of a length-21 highpass FIR filter

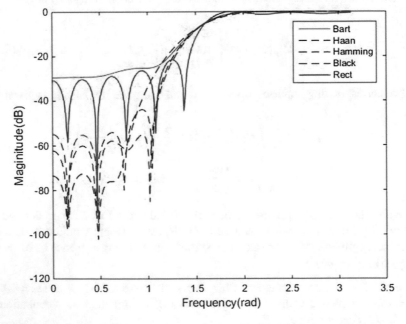

Fig. 7.25 Magnitude in dB of the frequency response of the FIR filter in Example 7.8 for the fixed window functions

Solution From Eq. (7.38) we find the FIR filter order to be 31 and the filter length to be 32. For the given specifications, we find $\beta = 2.1166$. The Kaiser window of length 32 is calculated using Eq. (7.35) and the impulse response of the ideal bandpass filter from Eq. (7.43). The impulse response of the desired bandpass FIR filter is then the point-by-point product of the window function and the ideal impulse response sequence. The length-32 ideal impulse response and the windowed impulse response are shown in the top and bottom plots in Fig. 7.26, respectively. Figure 7.27 shows the magnitude in dB of the frequency response of the bandpass FIR filter. It is seen that the minimum stopband attenuation is 30 dB, as specified. The MATLAB M-file to solve this problem is named *Example7_9.m*.

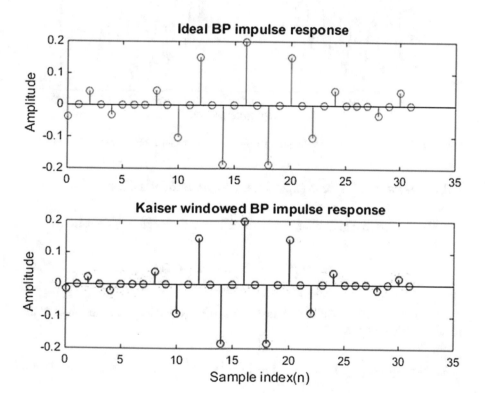

Fig. 7.26 Impulse response of length-32 FIR bandpass filter of Example 7.9. Top plot, ideal impulse response of the bandpass filter; bottom plot, windowed impulse response of the bandpass FIR filter using Kaiser window

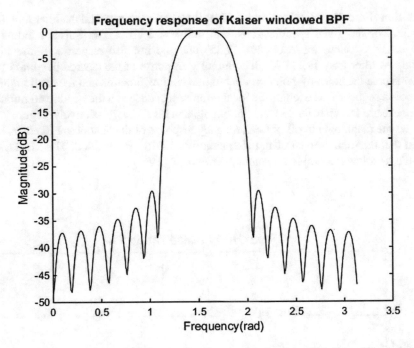

Fig. 7.27 Magnitude in dB of the frequency response of the bandpass FIR filter of Example 7.9

7.2.6   Design of a Bandstop Linear-Phase FIR Filter

The ideal bandstop digital filter can be described in the frequency domain by

$$H_{bs}\left(e^{j\Omega}\right) = \begin{cases} 1, |\Omega| \leq \Omega_1 \\ 0, \Omega_1 < |\Omega| < \Omega_2 \\ 1, \Omega_2 \leq |\Omega| \leq \pi \end{cases} \qquad (7.44)$$

Using (7.44) in the definition of the IDTFT, we obtain the impulse response of the ideal bandstop filter as

$$h_{bs}[n] = \delta[n] + \left(\frac{\Omega_1}{\pi}\right) sinc\left(\frac{n\Omega_1}{\pi}\right) - \left(\frac{\Omega_2}{\pi}\right) sinc\left(\frac{n\Omega_2}{\pi}\right), \; -\infty \leq n \leq \infty \quad (7.45)$$

The corresponding FIR filter's impulse response is obtained by multiplying the ideal impulse response in (7.45) by a suitable window of a specified length (odd) and shifting the resulting sequence by $M = \frac{N-1}{2}$ samples to the right. Let us consider the following example to design a bandstop linear-phase FIR filter using a window function.

Example 7.10 Design a bandstop FIR filter with passband edges at $\Omega_1 = 0.3\pi$ and $\Omega_2 = 0.7\pi$, transition width $\Delta\Omega = 0.1\pi$, and a minimum stopband attenuation of 30 dB. Use Kaiser window.

Solution The FIR filter order is found to be 31 and so the filter length is 32. The Kaiser window parameter is determined to be $\beta = 2.1166$. The ideal impulse response of the bandstop filter is given in Eq. (7.45). The length-32 Kaiser window with the above-listed parameter is obtained from (7.35), and the impulse response of the desired FIR bandstop filter is the point-by-point product of the ideal impulse response and the Kaiser window function. Figure 7.28 shows the ideal and windowed impulse responses in the top and bottom plots, respectively. The magnitude in dB of the frequency response of the bandstop FIR filter is shown in Fig. 7.29. The minimum stopband attenuation is seen to be 30 dB as specified in the problem. The M-file named *Example7_10.m* is used to solve this problem (Table 7.4).

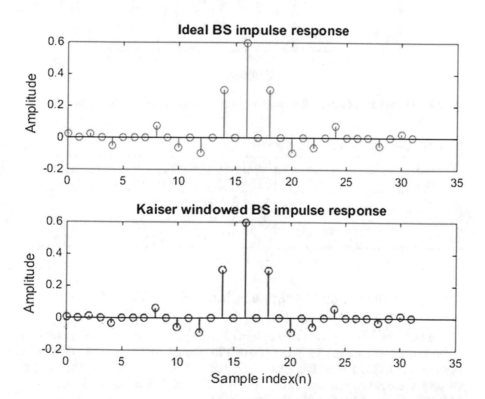

Fig. 7.28 Impulse response of length-32 FIR bandstop filter of Example 7.10. Top plot, ideal impulse response of the bandstop filter; bottom plot, windowed impulse response of the bandstop FIR filter using Kaiser window

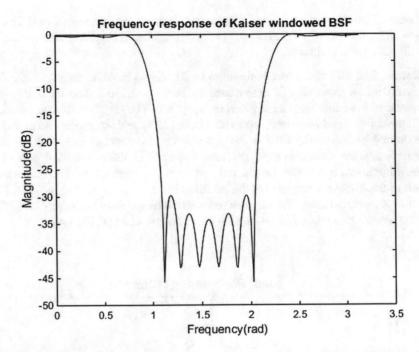

Fig. 7.29 Magnitude in dB of the frequency response of the bandstop FIR filter of Example 7.10

Table 7.4 Impulse responses of ideal filters

Filter type	Ideal impulse response
Lowpass	$\left(\frac{\Omega_c}{\pi}\right)sinc\left(\frac{n\Omega_c}{\pi}\right),\ -\infty < n < \infty$
Highpass	$\delta[n] - \left(\frac{\Omega_c}{\pi}\right)sinc\left(\frac{n\Omega_c}{\pi}\right),\ -\infty < n < \infty$
Bandpass	$\left(\frac{\Omega_2}{\pi}\right)sinc\left(\frac{n\Omega_2}{\pi}\right) - \left(\frac{\Omega_1}{\pi}\right)sinc\left(\frac{n\Omega_1}{\pi}\right),\ -\infty < n < \infty$
Bandstop	$\delta[n] + \left(\frac{\Omega_1}{\pi}\right)sinc\left(\frac{n\Omega_1}{\pi}\right) - \left(\frac{\Omega_2}{\pi}\right)sinc\left(\frac{n\Omega_2}{\pi}\right),\ -\infty < n < \infty$

7.3 Computer-Aided Design of Linear-Phase FIR Filters

As we learnt in the computer-aided design of IIR digital filters, there exists such an algorithm for the design of linear-phase FIR digital filters as well. One such algorithm is called Parks-McClellan algorithm and is an iterative procedure. Let the desired frequency response be denoted by $D(\Omega)$ and the frequency response of an Mth-order linear-phase FIR filter be described by

$$H\left(e^{j\Omega}\right) = e^{-j\frac{M}{2}\Omega}e^{j\varphi}\widehat{H}(\Omega), \tag{7.46}$$

where $\widehat{H}(\Omega)$ is a real function of Ω and $\varphi = \pm \pi$. Note that the frequency response in (7.46) corresponds to one of the four types of linear-phase FIR filters defined in Sect. 7.1. Define the weighted error function by

$$\varepsilon(\Omega) \equiv W(\Omega)\left(\widehat{H}(\Omega) - D(\Omega)\right) \qquad (7.47)$$

where $W(\Omega)$ is a positive weighting function, which is chosen to adjust the relative size of the peak errors in the specified bands. Parks-McClellan algorithm iteratively adjusts the impulse response of the FIR filter until the maximum of the absolute value of $\varepsilon(\Omega)$ is minimized. This iterative procedure known as *Remez exchange algorithm* is a highly efficient iterative procedure to arrive at the optimal result. MATLAB has the function *firpm*, which designs a linear-phase FIR filter of specified order, frequency bands, corresponding magnitudes, and a weighting function based on Parks-McClellan algorithm. Another computer-aided technique is called the *optimal least squares* technique, which approximates the desired frequency response by that of an Mth-order linear-phase FIR filter by minimizing the mean square error between the frequency responses of the FIR filter and the desired filter response. Again, MATLAB has a function called *firls*, which designs a linear-phase FIR filter of specified order, frequency bands of interest, and a weighting function based on the least mean square error procedure. We will use these two functions to design a linear-phase FIR filter as shown in the following example.

Example 7.11 Design a linear-phase lowpass FIR filter of order 20 with normalized frequency band edges at 0, 0.1, 0.3, and 1.0 with corresponding amplitudes of 1, 1, 0, and 0, respectively. Use a weighting factor of 0.25 in the passband and 1 in the stopband. Compare the frequency responses of the FIR filters designed with and without the weighting function.

Solution We will use MATLAB to solve this problem. As pointed out, we invoke the MATLAB function $h = firpm(M,F,A)$, where M is the filter order, which is 20, F is a vector of frequency band edges [0 0.1 0.3 1.0], A = [1 1 0 0], and h is a length-M + 1 impulse response of the linear-phase FIR filter. Note that 0.1 and 0.3 are the normalized passband and stopband frequency edges. These edges actually correspond to 0.1π and 0.3π, respectively. In order to use the specified weighting function, we have to call the function $h1 = firpm(M,F,A,W)$, where W = [0.25 1.0] is the weight vector and the rest of the arguments have the same meaning as in the previous invocation of the function. The weight vector W must be half the length of F. The output argument $h1$ is the impulse response of the linear-phase FIR filter designed using a weighting function in the Parks-McClellan iterative algorithm.

We next use the optimal least squares method of designing a length-21 linear-phase FIR filter to approximate the same specifications as used with Parks-McClellan method. Now we have to invoke the MATLAB function $hls = firls(M,F,A)$, where the output argument hls is the impulse response of the FIR filter corresponding to the least squares method. The other arguments are the same as used in the Parks-McClellan case. Similarly, the function call with a weighting vector W is $hls1 = firls(M,F,A,W)$.

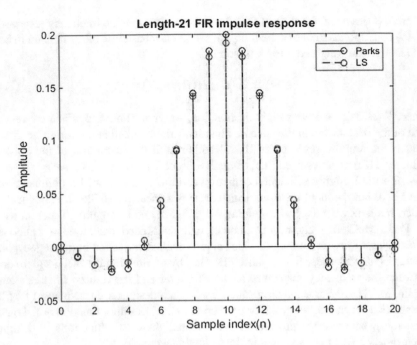

Fig. 7.30 Impulse response of length-21 linear-phase FIR filter of Example 7.11 designed by the two computer-aided design procedures without a weighting vector

Figure 7.30 shows the plots of the impulse response of the FIR filters for the two iterative design procedures without the weighting vectors. Though they appear identical, we can notice some differences. The corresponding frequency responses are shown in Fig. 7.31. We observe from Fig. 7.31 that the minimum stopband attenuation is about 40 dB for the Parks-McClellan case and about 50 dB for the least squares case. The impulse response and frequency response of the FIR filters for the two design procedures with the specified weighting vector are shown in Figs. 7.32 and 7.33, respectively. With the specified weighting vector, the minimum stopband attenuation is about the same for both design methods, which is seen to be about 50 dB.

Let us compare the results obtained from the computer-aided design of length-21 lowpass FIR filter with that obtained using Kaiser-windowed lowpass FIR filter. Since the FIR filter order is fixed at 20, we need to determine the parameter β in order to compute the Kaiser window. Since the minimum stopband attenuation for the Parks-McClellan case is 40 dB, let us specify the minimum stopband attenuation for the Kaiser window to be 40 dB also. Then we find $\beta = 3.3953$. The passband edge for the windowed filter is the same as that specified for the Parks-McClellan case, which is 0.1π. With the passband edge being known, we have the ideal impulse response of the lowpass filter, which, when multiplied by the Kaiser window, gives the windowed impulse response of length-21 lowpass FIR filter. The magnitude in dB of the frequency response of the Kaiser-windowed lowpass FIR filter is shown in Fig. 7.34 along with those of the FIR filter obtained using Parks-McClellan and least

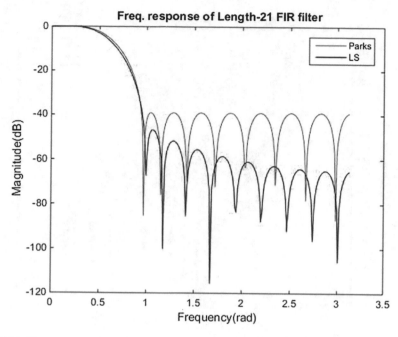

Fig. 7.31 Frequency response of length-21 linear-phase FIR filter of Example 7.11 corresponding to the impulse responses in Fig. 7.30

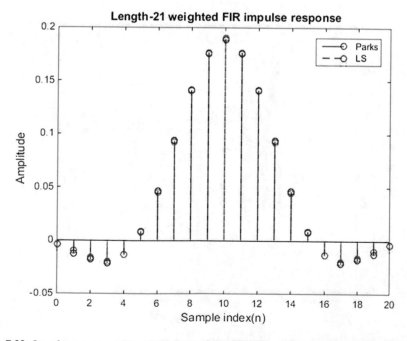

Fig. 7.32 Impulse response of length-21 linear-phase FIR filter of Example 7.11 designed by the two computer-aided design procedures with the weighting vector as specified in the problem

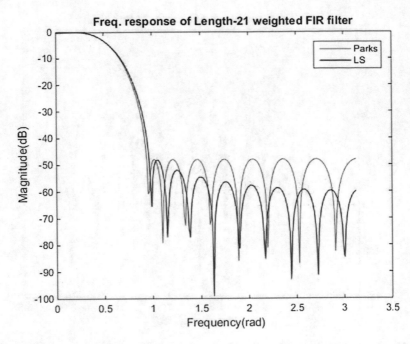

Fig. 7.33 Frequency response of length-21 linear-phase FIR filter of Example 7.11 corresponding to the impulse responses in Fig. 7.32

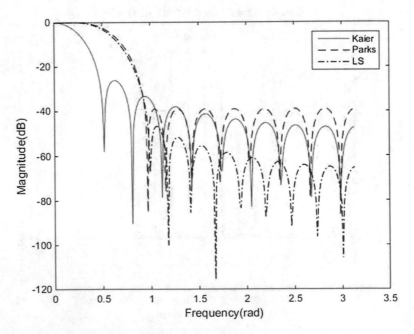

Fig. 7.34 Frequency response of length-21 FIR lowpass filter using Kaiser window. It is compared with those of the filters obtained using Parks-McClellan and least squares method

squares methods. The transition width is the smallest for the Kaiser-windowed filter, while the minimum stopband attenuation is the largest for the least squares case. Note that the filter length obtained using Eq. (7.38) is 24, which would have achieved minimum stopband attenuation greater than 40 dB. The MATLAB M-file for this problem is named *Example7_11.m*.

Example 7.12 Notch Filter A notch filter passes all frequencies except a single specified frequency. In other words, it creates a notch in its frequency response. Notch filters are useful in eliminating a specified frequency from the input. For instance, it can be used to filter out the 60 Hz AC power frequency from the input audio/speech signal. An ideal notch filter is specified in the digital filter frequency domain by

$$H\left(e^{j\Omega}\right) = \begin{cases} 0, |\Omega| = \Omega_0 \\ 1, otherwise \end{cases} \tag{7.48}$$

The corresponding impulse response can be obtained from the IDTFT of the frequency function in (7.48). To make the IDTFT process easier, we can rewrite (7.48) as

$$H\left(e^{j\Omega}\right) = 1 - \frac{1}{2}\delta(\Omega - \Omega_0) - \frac{1}{2}\delta(\Omega + \Omega_0), \ -\pi \le \Omega \le \pi \tag{7.49}$$

Then, the impulse response of the ideal digital notch filter with a notch frequency Ω_0 is found from

$$h[n] = \frac{1}{2\pi} \int\limits_{-\pi}^{\pi} H\left(e^{j\Omega}\right)e^{jn\Omega}d\Omega \tag{7.50}$$

From the DTFT pairs, we know that $\delta[n] \overset{DTFT}{\Leftrightarrow} 1$ and

$$IDTFT\{\delta(\Omega - \Omega_0)\} = \frac{1}{2\pi} \int\limits_{-\pi}^{\pi} \delta(\Omega - \Omega_0)e^{jn\Omega}d\Omega = \frac{e^{jn\Omega_0}}{2\pi} \tag{7.51a}$$

$$IDTFT\{\delta(\Omega + \Omega_0)\} = \frac{1}{2\pi} \int\limits_{-\pi}^{\pi} \delta(\Omega + \Omega_0)e^{jn\Omega}d\Omega = \frac{e^{-jn\Omega_0}}{2\pi} \tag{7.51b}$$

By using the above results, we can write the impulse response of the ideal notch filter as

$$h[n] = \begin{cases} 1 - \dfrac{1}{2\pi}, n = 0 \\ -\dfrac{\cos{(n\Omega_0)}}{2\pi}, |n| > 0 \end{cases} \tag{7.52}$$

Since the impulse response of the ideal digital notch filter is of infinite duration, it has to be truncated to ±2M samples by using an appropriate window function. For this example, let us choose the notch frequency to be 1500 Hz and the sampling frequency to be 10,000 Hz. Let us also fix the FIR filter length to 35. We can use either a fixed or an adjustable window to limit the impulse response to a length 35. It is found that a length-35 Blackman-Harris window results in a good attenuation at the chosen notch frequency. The Blackman-Harris window and the corresponding windowed impulse response of the notch filter are shown in the top and bottom plots in Fig. 7.35. The frequency response of the resulting notch filter is shown in Fig. 7.36, which indicates an attenuation of 30 dB at the notch frequency. To make it more practical, let us generate an input sequence consisting of two sinusoids at frequencies of 1500 and 2500 Hz and corresponding amplitudes of 2 and 1. We then filter this input sequence through the 35-point FIR notch filter that we just designed. We can use the MATLAB function *conv* to convolve the input and the impulse response sequences. As we know from the linear convolution that the output sequence will have a length equal to the sum of the lengths of the input and impulse response sequences minus one. To get the output length equal to the input length, we can state $y = conv(x,h,'same')$. The MATLAB code is in the M-file

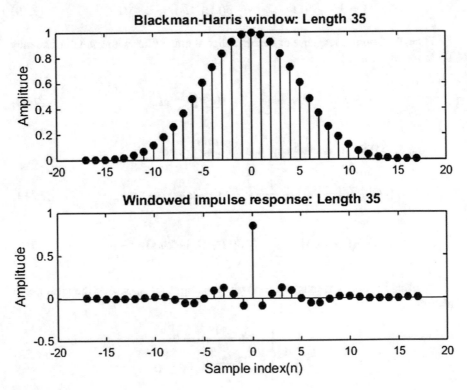

Fig. 7.35 Impulse response of the FIR notch filter of Example 7.12. Top plot, length-35 Blackman-Harris window function; bottom plot, windowed impulse response

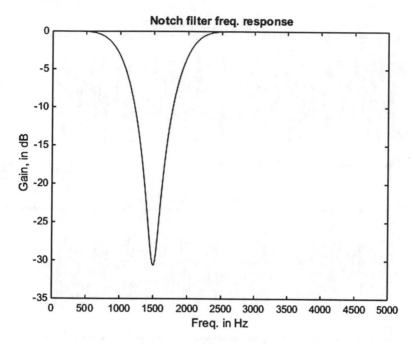

Fig. 7.36 Magnitude in dB of the frequency response of the notch filter of Example 7.12

named *Example7_12.m*. Figure 7.37 shows the input sequence and the filtered sequence in the top and bottom plots, respectively. The corresponding DTFTs are shown in Fig. 7.38. The top plot in Fig. 7.38 shows the two frequency components at 1500 and 2500 Hz, respectively. The bottom plot has only one component at 2500 Hz, implying that the notch filter has eliminated the 1500 Hz signal from the input. To exemplify further, we can play the sequences as sound using the MATLAB function *sound(x,Fs)*, where x is the audio sequence and Fs is the sampling frequency. It is found that the filtered sequence sounds as 2500 Hz signal.

Example 7.13 Audio Filtering Using Simulink In this example we will use MATLAB's Simulink to simulate filtering an audio signal with an FIR filter. We call it a simulation because we start with sampling an analog audio signal, then converting it to a digital signal, and then filtering it with an FIR filter, and finally dumping both the input and output signals in the workspace. The whole processing steps will be based on a block diagram. Each block can be configured to perform a specified task with the options for choosing the right parameters. Figure 7.39 shows the overall block diagram of the audio-filtering operation. It consists of two sine wave signal generators, whose frequencies are 4 kHz and 1 kHz, respectively. The parameters of the 4 kHz sine wave generator are shown in Fig. 7.40a, b. The amplitude of the 4 kHz sine wave is 0.5, the sampling frequency is chosen to be 10 kHz, and the output data type is real (floating point). Similar parameters are chosen for the second sine wave generator at a frequency of 1 kHz. Next, the two

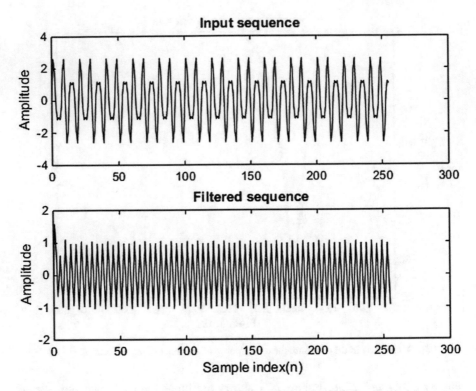

Fig. 7.37 Input and output sequences of the notch filter of Example 7.12. Top plot, input sequence consisting of 1500 and 2500 Hz sinusoids; bottom plot, filtered sequence

sine waves are added using an analog adder block. The output signal from the adder is then quantized by a uniform quantizer to 8-bit accuracy. Since we have not learnt quantization, we will not discuss the quantizer block further. Our objective in this example is to simulate an FIR filter. Therefore, we pass the signal from the quantizer through an FIR filter. Figure 7.41 lists the parameters of the FIR filter used in the example. The critical parameters are filter type is FIR, filter order is 30, passband edge is 1 kHz, stopband edge is 2 kHz, and the minimum stopband attenuation is 60 dB. The frequency response of the designed FIR filter is depicted in Fig. 7.42. Once the block parameters are specified and the blocks connected, we can run the simulation. The simulation time is chosen as 1 s. When the simulation is completed, the input and filtered output signals are stored in the workspace in the variables *input_from_AudioFiltering* and *output_from_AudioFiltering*, respectively.

In order to confirm that the filter has, indeed, filtered out the unwanted component at 4 kHz, we run another simulation, whose block diagram is shown in Fig. 7.43. As can be seen from the figure, the stored signals are read by the two blocks named *Signal From Workspace*. The DFT of the two signals are performed by the blocks named *Magnitude FFT of input* and *Magnitude FFT*

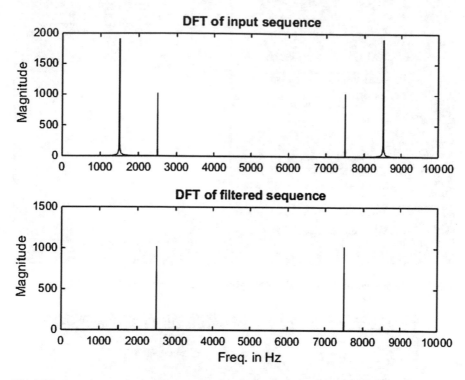

Fig. 7.38 Frequency response of the input and output sequences of the notch filter of Example 7.12. Top plot, DFT of the input sequence; bottom plot, DFT of the filtered sequence

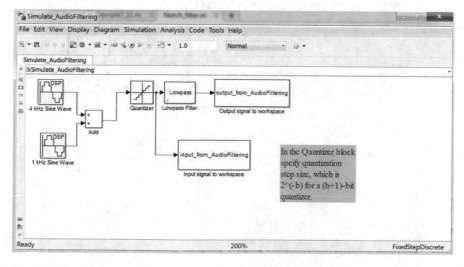

Fig. 7.39 Simulink block diagram of audio filtering in Example 7.13

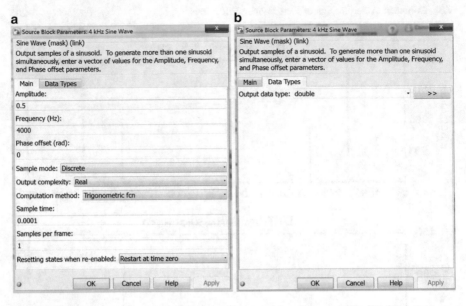

Fig. 7.40 Parameters of the 4 kHz sine wave generator: (**a**) main parameters, (**b**) data types

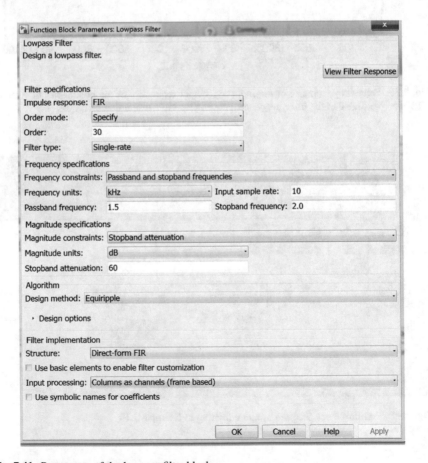

Fig. 7.41 Parameters of the lowpass filter block

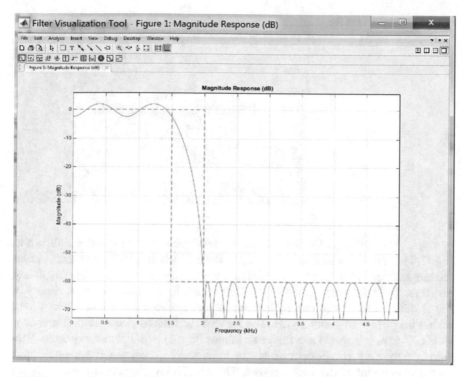

Fig. 7.42 Magnitude in dB of the frequency response of the lowpass FIR filter

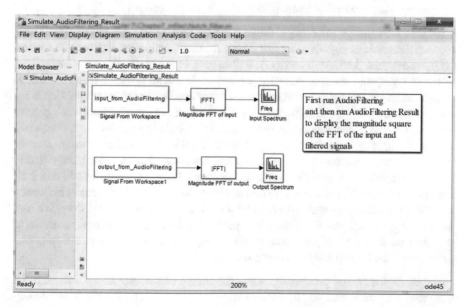

Fig. 7.43 Simulink block diagram to compute the DFT and to display the magnitude in dB on a scope

Fig. 7.44 Parameters of the FFT magnitude block

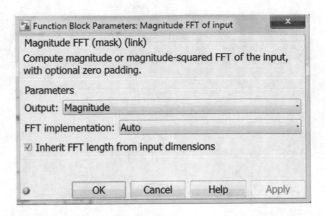

of output, respectively. The parameters of the Magnitude FFT block are shown in Fig. 7.44. These two blocks will perform the DFTs using FFTs and then output the magnitudes in dB. The two magnitudes are finally displayed on the frequency displays. The magnitude in dB of the DFT of the input audio signal is shown in Fig. 7.45a. It consists of two main components at frequencies 1 kHz and 4 kHz with a magnitude of about 35 dB. The DFT of the filtered output signal is shown in Fig. 7.45b, where we see the peak (about 35 dB) at 1 kHz as expected. The 4 kHz frequency component seems to be removed, because it appears about 34 dB below that of the 1 kHz signal. The MATLAB file that creates the input and filtered sequences is named *Example7_13a.slx*. The file to compute the DFTs of the sequences stored in the workspace is named *Example7_13b.slx*.

Example 7.14 Image Filtering Using a 2D FIR Filter So far we have played with filtering time-domain signals using FIR filters. A time-domain signal has one dimension (1D), namely, the time. An intensity (B&W or gray-scale) image is a two-dimensional signal in the spatial domain. Actually, we are dealing with a digital image, which is discrete in the two spatial axes and consists of discrete points in the 2D space. Therefore, we can consider a gray-scale image as a 2D array of picture elements, each picture element being an integer number. A picture element is called a *pixel* for short. Each pixel in a gray-scale image can range between 0 and 255 if it is an 8-bit digital picture. A color image has three-component images, all having the same size. A color picture that is viewable has red, green, and blue (RGB) components. In this example, we will deal with a gray-scale image. The problem is to design a lowpass FIR filter with a cutoff frequency of 0.1π in both the horizontal and vertical dimensions using a suitable window. We also want to design a highpass FIR filter with a cutoff frequency of 0.4π in both dimensions. We then want to filter a gray-scale image through these two filters and display the original and the filtered images to see the effect.

Since digital image processing is not the theme of this book, we will skip the description about it and use MATLAB to solve the problem. As we have already learnt the design of lowpass and highpass FIR filters with windows, we will assume that we have both lowpass and highpass 1D FIR filters of a specified order.

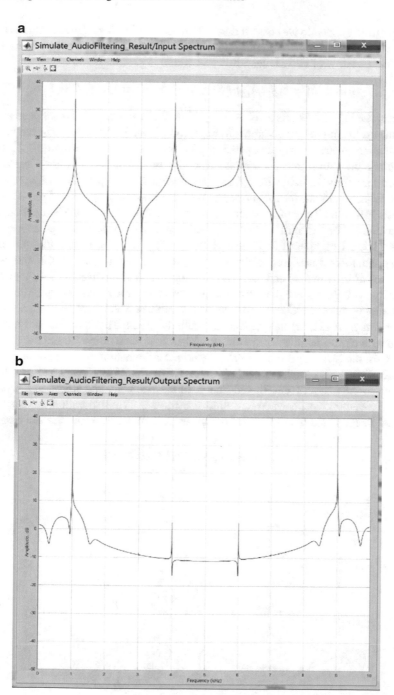

Fig. 7.45 Display on the scope: (**a**) magnitude in dB of the frequency response of the input signal consisting of 1 kHz and 4 kHz, (**b**) magnitude in dB of the frequency response of the filtered output, which only has the 1 kHz component

The question is how do we convert these 1D filters to 2D filters. One method is to do the outer product of the 1D impulse responses. Thus the impulse response of a 2D FIR filter in terms of two 1D length-N FIR filters is given by

$$h_{2D}[m,n] = h_{1D}[n]h_{1D}^{t}[n], 0 \leq m \leq N-1, 0 \leq n \leq N-1 \qquad (7.53)$$

Thus, the 2D FIR filter is an NxN array. Note that the number of rows need not be the same as the number of columns. In Eq. (7.53), $h_{1D}[n]$ is an N × 1 column vector, and t denotes matrix transpose operation. The other method of designing a 2D filter is to come up with a true 2D impulse response function. We will not discuss this method any further. Using Eq. (7.53), we design the lowpass and highpass FIR filters. Figure 7.46 shows the surface plot of the impulse response of the 21 × 21 lowpass FIR filter. To obtain a surface plot, we can call the MATLAB function *surf*. The details can be found in the M-file named *Example7_14.m*. The impulse response of the 21 × 21 highpass FIR filter is shown in Fig. 7.47. To ensure that the filters correspond to lowpass and highpass, we plot the magnitude of the frequency responses of the lowpass and highpass FIR filters and show in Figs. 7.48 and 7.49, respectively. Next we use these two filters to process a gray-scale image. The image we choose is called the *Cameraman* and is available in MATLAB. To read an image into an array, MATLAB has the function called *imread*. This function will read the image with a pre-specified image format and store it in an array. To filter the image, we invoke the MATLAB function $B = imfilter(A,h)$, where A is the input image and h is the 2D filter. So, we invoke this function twice to perform lowpass and highpass filtering of the image. Each filtered image, of course, has to be stored in a separate array. To display an image, we have to use the MATLAB function *imshow*. The original image is shown in Fig. 7.50. Figures 7.51 and 7.52 show the lowpass and

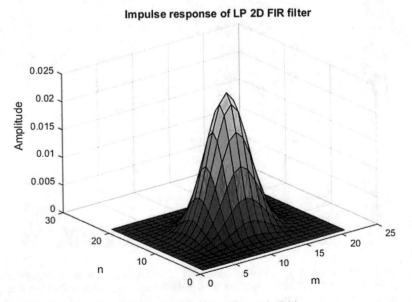

Impulse response of LP 2D FIR filter

Fig. 7.46 Surface plot of the 2D lowpass FIR filter of Example 7.14

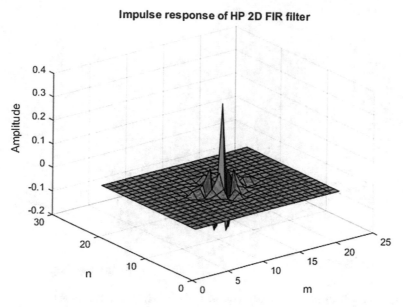

Fig. 7.47 Surface plot of the 2D highpass FIR filter of Example 7.14

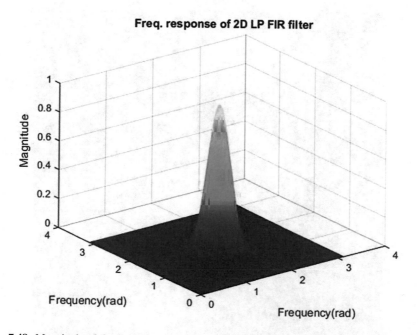

Fig. 7.48 Magnitude of the frequency response of the 2D lowpass FIR filter of Example 7.14

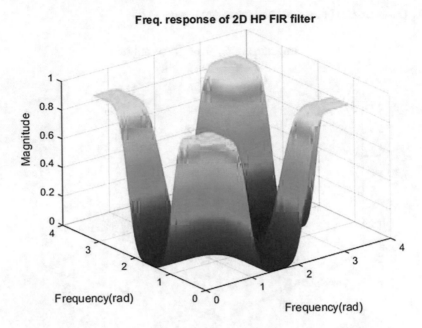

Fig. 7.49 Magnitude of the frequency response of the 2D highpass FIR filter of Example 7.14

Original image

Fig. 7.50 Original gray-scale Cameraman 8-bit image

Fig. 7.51 Lowpass filtered Cameraman image using the 2D lowpass FIR filter of Example 7.14

highpass filtered images, respectively. As can be seen from Fig. 7.51, the effect of lowpass filtering an image is to blur it – the smaller the cutoff frequency, the larger the blur. On the other hand, a highpass filter sharpens the image. It removes the DC values and retains only the edges.

Example 7.15 An FIR Differentiator A differentiator performs the operation of differentiation of an input signal. Its frequency response can be defined as

$$H\left(e^{j\Omega}\right) = j\Omega, \ -\pi \leq \Omega \leq \pi \tag{7.54}$$

Using the definition of the IDTFT, we can show the impulse response of an ideal differentiator to be

$$h[n] = \begin{cases} 0, n = 0 \\ \dfrac{(-1)^n}{n}, |n| > 0 \end{cases} \tag{7.55}$$

The impulse response of the corresponding FIR differentiator is the truncated version of the ideal impulse response in (7.55). In order to increase the minimum

highpass filtered image

Fig. 7.52 Highpass filtered Cameraman image using the 2D highpass FIR filter of Example 7.14

stopband attenuation, we will have to window the impulse response in (7.55) by a suitable window function. So, we will use a fixed window of length 21. Figure 7.53 plots the ideal and windowed impulse responses of the differentiator. The magnitude of the frequency response of the length-21 FIR differentiator is shown in Fig. 7.54. It corresponds to a linear function as specified. In order to verify that the filter is indeed a differentiator, let us filter a rectangular pulse of width-20 samples. When we differentiate a rectangular pulse of finite width, we will expect an impulse with positive amplitude at the rising edge and an impulse with negative amplitude at the falling edge. This is exactly what we see in Fig. 7.55, which shows the input rectangular pulse in the top plot and the output of the differentiator in the bottom plot. Consider another input, which is a triangular pulse. Since the two sides of the triangle have constant slopes – one positive and the other negative – the result of differentiating the triangular pulse is a positive pulse on the left side and a negative rectangular pulse on the right side of the triangle. This is what we see in Fig. 7.56. The MATLAB code for this example can be found in the M-file named *Example7_15.m*.

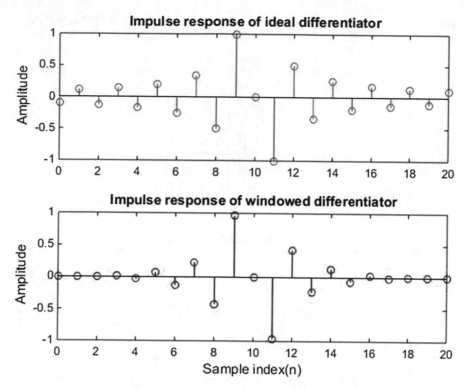

Fig. 7.53 Impulse response of a differentiator. Top plot, ideal impulse response; bottom plot, windowed impulse response

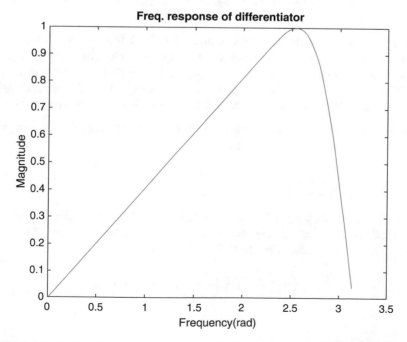

Fig. 7.54 Magnitude of the frequency response of a length-21 windowed FIR differentiator of Example 7.15

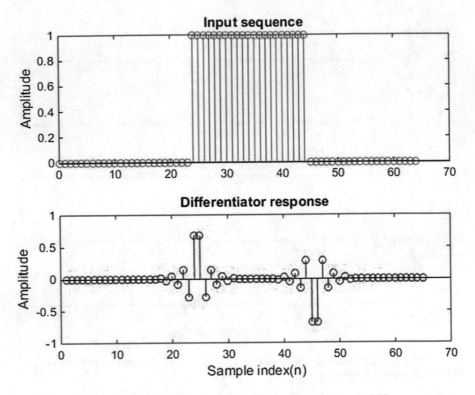

Fig. 7.55 Filtering a rectangular pulse through the differentiator of Example 7.15

Example 7.16 Comparison of SNR of an FIR Using Different Windows In this example we would like to filter a noisy input signal corrupted by a white Gaussian noise through a lowpass FIR filter and compute the resulting signal-to-noise ratio (SNR) at the filter output. We want to compare the SNR in dB of the FIR filter using Bartlett, Hann, Hamming, Blackman, and rectangular window functions.

Solution Let the filter order of the lowpass FIR filter be 16. Then the filter length is 17. Let the input signal be defined by

$$s[n] = 2\sin\left(\frac{2\pi f_1}{f_s}n\right) + \cos\left(\frac{2\pi f_2}{f_s}n\right) - 1.5\sin\left(\frac{2\pi f_3}{f_s}n\right), 0 \le n \le 1023$$

where $f_1 = 133\ Hz$, $f_2 = 205\ Hz$, $f_3 = 223\ Hz$, and the sampling frequency is $f_s = 1000\ Hz$. For a cutoff frequency of $\Omega_c\ rad$, the impulse response of the ideal lowpass filter is obtained using the IDTFT and is given by

$$h_I[n] = \left(\frac{\Omega_c}{\pi}\right) sinc\left(\frac{n\Omega_c}{\pi}\right), \quad -\infty < n < \infty$$

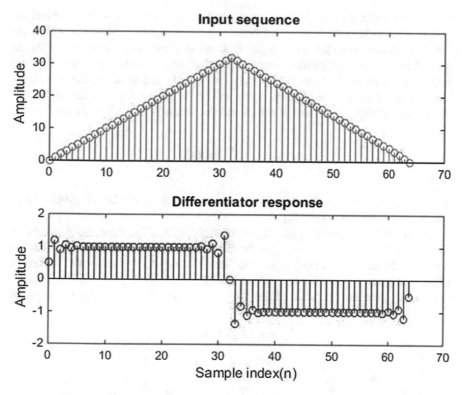

Fig. 7.56 Filtering a triangular pulse through the differentiator of Example 7.15

By multiplying the ideal impulse response sequence by a window function of length N = 17, we get the impulse response of the lowpass FIR filter of length 17 and is described by

$$h[n] = h_I[n]w[n], \ -8 \leq n \leq 8$$

To make the filter causal, we need to shift the impulse response of the FIR filter described above by eight samples to the right. Therefore, the desired impulse response of the lowpass FIR filter that approximates an ideal lowpass filter is found to be

$$h_d[n] = h_I[n-8]w[n-8], 0 \leq n \leq 16$$

As mentioned earlier, the window function is generated in MATLAB by the function *window(@Bartlett,17)*, which creates the Bartlett window of length 17. Similarly, other windows can be generated by using the appropriate window names in the function. The details of the program can be found in the M-file named *Example7_16.m*. In addition to the signal, we have to generate a white Gaussian noise sequence. This is achieved by the function call *sigma* × *randn(1,1024)*, which generates a Gaussian random vector of length 1024 with a standard deviation of sigma. The signal is added to the noise and the sum is input to the FIR filter. The filtering operation can be achieved by using the function *conv(x,h, 'same')*, where x is the input sequence and h is the impulse response of the FIR filter. By using the value 'same', the filtered output sequence will have the same length as that of the input sequence. One last thing is that by using the MATLAB function *rng('default')*, the noise sequence will be the same every time the program is run. Once the filtering operation is completed, the output SNR in dB is obtained by the equation

$$SNR = 10log_{10}\left(\frac{var(x)}{var(x-y)}\right), dB$$

where *var(x)* denotes the variance of the sequence x. The M-file for this problem also calculates the step responses for comparison.

Let us display the various plots and compare the results. The input signal plus noise sequence and the filtered sequence for the rectangular window is shown in Fig. 7.57 in the top and bottom plots, respectively. The filtered outputs corresponding to the Bartlett, Hann, Hamming, and Blackman windows are shown in Figs. 7.58 and 7.59. The names of the windows appear in the plot titles. Since it is difficult to compare the results in the

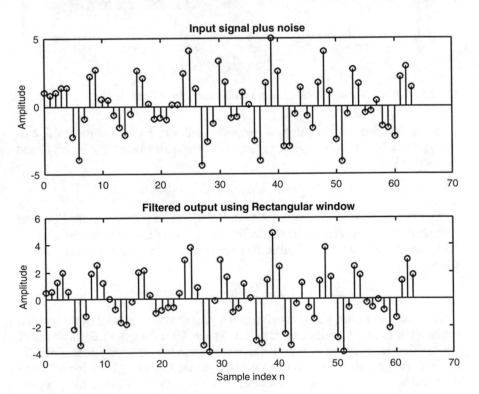

Fig. 7.57 Output of FIR filter using rectangular window: Top, input sequence; bottom, output sequence

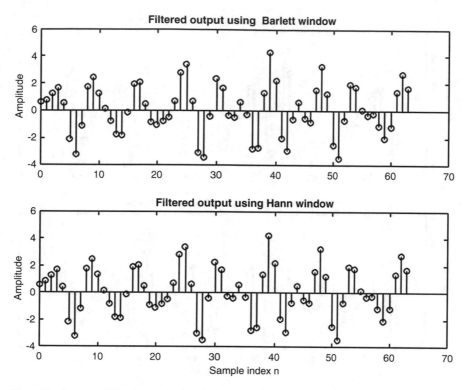

Fig. 7.58 Outputs of FIR filter using Bartlett and Hann windows: Top, Bartlett; bottom, Hann

discrete-time domain, we show the DTFTs of the filter outputs. Specifically, the DTFTs of the input signal plus noise and the filtered output corresponding to the rectangular window are shown in the top and bottom plots of Fig. 7.60, respectively. The DTFTs of the filter output for the other windows are shown in Figs. 7.61 and 7.62. As can be seen from Fig. 7.60, filtering removes the out-of-band noise. The SNRs are shown in the following table. Since the rectangular window has the smallest transition width, it rejects the maximum out-of-band noise at the filter output. The table also shows the rise time corresponding to the different windows. The step responses are displayed in Fig. 7.63 and the rise time plot in Fig. 7.64 and Table 7.5.

7.4 Discrete-Time Hilbert Transformer

Hilbert transformer is used in analog communications and speech processing. In amplitude modulation (AM), the message waveform modulates a carrier in its amplitude. The result is that the spectrum of the baseband message waveform is shifted to the carrier frequency. This produces what is called a double-sideband AM signal. The bandwidth of the double-sideband AM signal is twice the bandwidth of

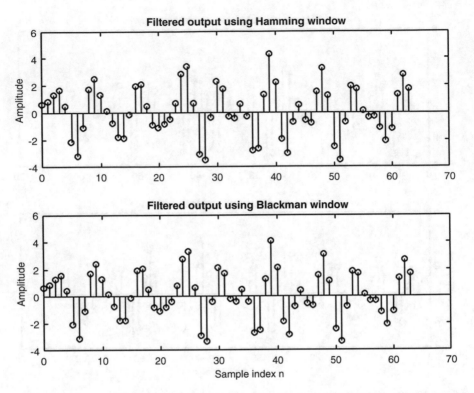

Fig. 7.59 Outputs of FIR filter using Hamming and Blackman windows: top, Hamming; bottom, Blackman

the message signal. If either the upper or lower sideband is filtered out, the bandwidth of the resulting AM signal is only half of the bandwidth of the double-sideband AM signal. This saves the transmitted spectrum of the AM signal. Incidentally, if only one of the sidebands is retained, the AM signal is called a single-sideband AM (SSB-AM) signal. An SSB-AM signal can be generated using the Hilbert transformer (HT). Let us consider the discrete-time version of the HT here.

The ideal HT in the frequency domain is characterized by the following DTFT function.

$$H\left(e^{j\Omega}\right) = \begin{cases} +j, \ -\pi < \Omega < 0 \\ -j, 0 < \Omega < \pi \end{cases} \tag{7.56}$$

From (7.56), it is observed that the magnitude of the DTFT of the HT is constant except at $\Omega = 0$ *and* π, where it is zero. The phase of the DTFT of the HT is $\pm\frac{\pi}{2}$. The corresponding impulse response or the discrete-time function of the HT is the IDTFT of the DTFT of the HT. Thus,

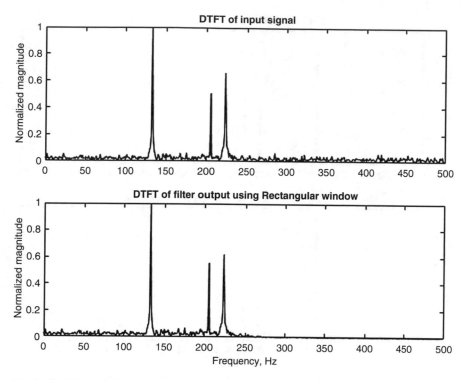

Fig. 7.60 DTFT of the output of FIR filter using rectangular window: top, DTFT of input sequence; bottom, DTFT of output sequence using rectangular window

$$h[n] = \frac{1}{2\pi} \int\limits_{-\pi}^{\pi} H\left(e^{j\Omega}\right) e^{jn\Omega} d\Omega = \frac{1}{2\pi} \int\limits_{-\pi}^{0} j e^{jn\Omega} d\Omega + \frac{1}{2\pi} \int\limits_{0}^{\pi} -j e^{jn\Omega} d\Omega \qquad (7.57)$$

After evaluating the two integrals and adding them, we obtain the discrete-time HT as

$$h[n] = \frac{1 - \cos(n\pi)}{n\pi} = \begin{cases} 0, n \ even, \\ \dfrac{2}{n\pi}, n \ odd \end{cases} \qquad (7.58)$$

The ideal HT in the discrete-time domain is of infinite duration and is, therefore, not physically realizable. However, by truncating it to $\pm M$ samples, the resulting HT is realizable by shifting it to the right by M samples. Instead of truncating it abruptly, we can use any of either fixed or adjustable windows. Once the HT of order $N = 2M + 1$, in the discrete-time domain, is determined, the HT of a given sequence can be found by convolving the given sequence with the HT. Let us get back to the SSB-AM for a second before we go further. The SSB-AM signal can be described by

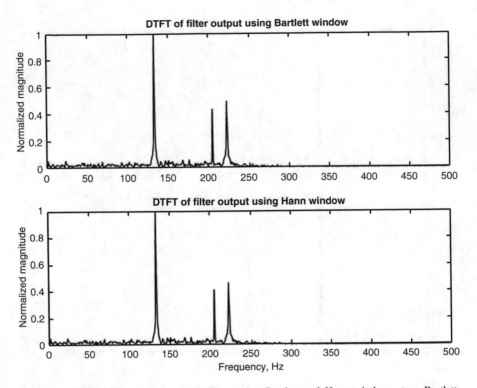

Fig. 7.61 DTFT of the outputs of FIR filter using Bartlett and Hann windows: top, Bartlett; bottom, Hann

$$x_{SSB}[n] = x[n] + j\widehat{x}[n] \tag{7.59}$$

The signal in (7.59) is also known as an *analytic signal*. In the above equation, both $x[n]$ and $\widehat{x}[n]$ are real signals. $\widehat{x}[n]$ is the Hilbert transform of the signal $x[n]$. The magnitude of the DTFT of the signal in (7.59) is odd, meaning, it is zero in the interval $[\pi, 2\pi]$. Remember that the magnitude of the DTFT of a discrete-time signal is periodic in period 2π and is symmetric about π. The reason for discussing HT here is to point out that it can be realized in the discrete-time domain using FIR filtering. We will illustrate the idea by an example.

Example of Hilbert Transformer in the Discrete-Time Domain Obtain a length-17 HT and compute its DTFT using MATLAB. Use a suitable window of length 17. Create a discrete-time sequence and find its HT. Plot the magnitude and phase of the input sequence, and compare it with that of the HT of the sequence.

Solution The length-17 HT is obtained from (7.58) by retaining 17 samples centered at zero. This amounts to multiplying the HT in (7.58) by a rectangular window. If we use another suitable window w[n], the HT will take the form

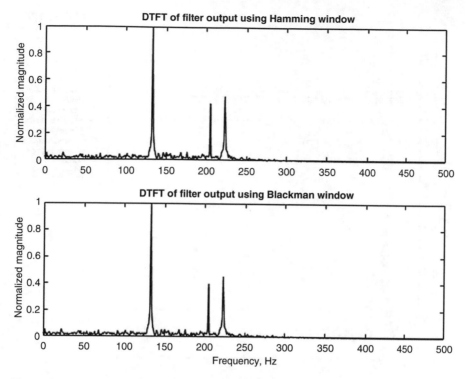

Fig. 7.62 DTFT of the outputs of FIR filter using Hamming and Blackman windows: top, Hamming; bottom, Blackman

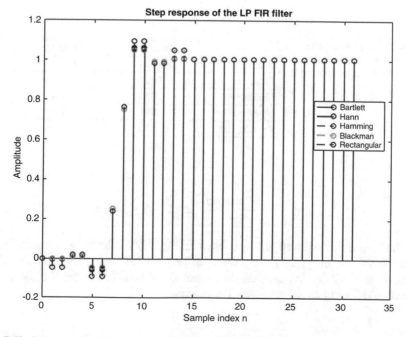

Fig. 7.63 Step response of the lowpass FIR filter of Example 7.16

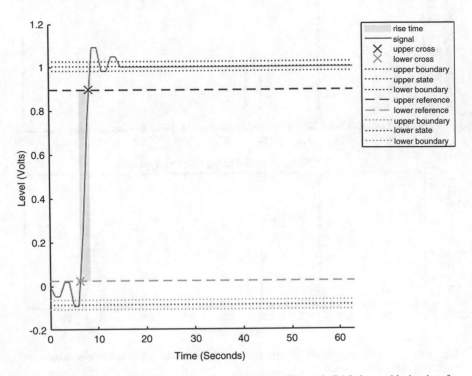

Fig. 7.64 Plot showing the step response of the FIR filter of Example 7.16 along with rise time for the rectangular window

Table 7.5 Signal-to-noise ratio and rise time of the lowpass FIR filter of Example 7.16

Window	SNR(dB)	Rise time (samples)
Bartlett	11.445	1.9677
Hann	10.891	2.0118
Hamming	10.058	2.1003
Blackman	10.611	2.0007
Rectangular	12.054	2.0573

$$h[n] = \left[\frac{1 - \cos{(n\pi)}}{n\pi} \right] w[n], \ -8 \le n \le 8$$

The ideal and windowed HTs are shown in Fig. 7.65 in the top and bottom plots, respectively. The magnitude of the DTFT of length-17 HT using Blackman window is shown in Fig. 7.66. As expected, the magnitude of the DTFT of the HT is wide but is zero at DC and half the sampling frequency.

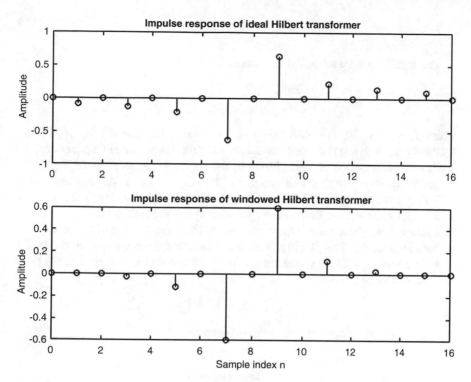

Fig. 7.65 Length-17 discrete-time HT: top, ideal HT; bottom, Blackman-windowed HT

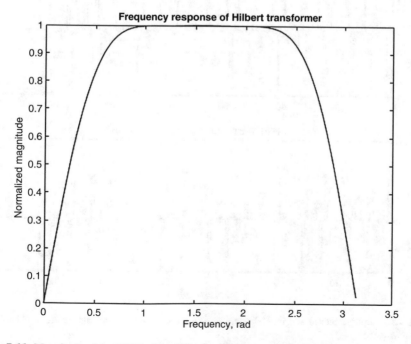

Fig. 7.66 Magnitude of the DTFT of the HT using Blackman window

The input sequence to the HT is described by

$$x[n] = \cos\left(\frac{2\pi f_m}{f_s}n\right)\cos\left(\frac{2\pi f_c}{f_s}n\right), 0 \le n \le 511$$

where $f_m = 123$ Hz, $f_c = 488$ Hz, and $f_s = 2000$ Hz. The HT of x[n] is then found by convolving the input sequence with the length-17 HT sequence. The input sequence and its HT are shown in the top and bottom plots of Fig. 7.67, respectively, over the first 128 samples. The magnitudes of the corresponding DTFTs are shown in Fig. 7.68. Both DTFTs show the double sidebands centered at the carrier frequency. Both spectra are also symmetric about half the sampling frequency. The phase functions of the two DTFTs are displayed in the top and bottom plots of Fig. 7.69. To verify the statement that the magnitude of the DTFT of an analytic signal is asymmetric about half the sampling frequency, we form the analytic signal

$$z[n] = x[n] + jy[n]$$

where x[n] is the input sequence as defined above and

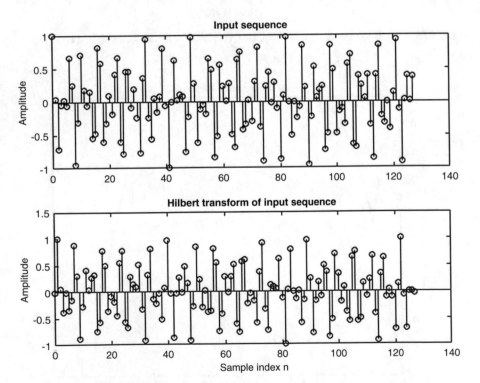

Fig. 7.67 Input sequence and its HT: top, input sequence x[n]; bottom, discrete-time HT of the input sequence

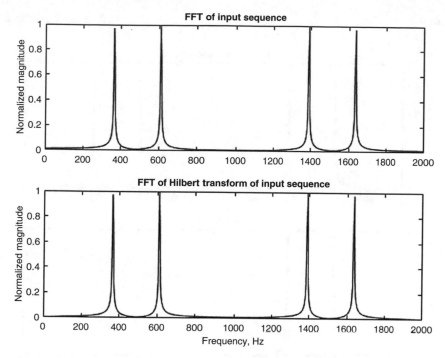

Fig. 7.68 Magnitude of the DTFT of the HT of the input sequence: top, magnitude of the DTFT of the input sequence; bottom, magnitude of the DTFT of its HT

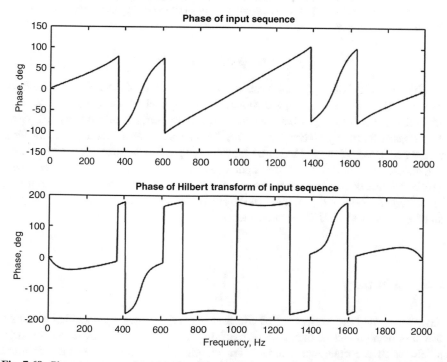

Fig. 7.69 Phase responses of the DTFTs of the input sequence and its HT: top, phase response of the DTFT of the input sequence; bottom, phase response of its HT

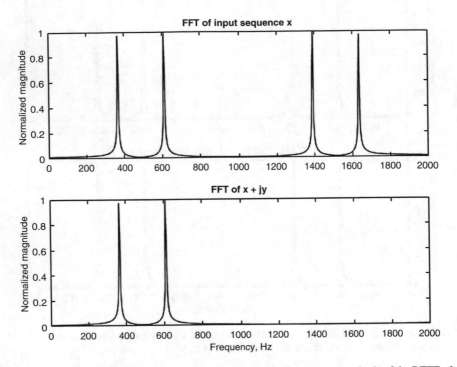

Fig. 7.70 Magnitudes of the input sequence and its analytic signal: top, magnitude of the DTFT of x[n]; bottom, magnitude of the DTFT of the analytic signal $z[n] = x[n] + jy[n]$, where y[n] is the HT of x[n]

$$y[n] = HT\{x[n]\} = x[n] \divideontimes h[n]$$

Then the DTFT of $\{z[n]\}$ is determined. The magnitudes of the DTFTs of the input sequence x[n] and its HT z[n] are shown in the top and bottom plots of Fig. 7.70, respectively. The bottom plot has only one side of the spectrum, and the other half above the sampling frequency is zero. This proves the statement. Finally, the Hilbert transform is also performed using the MATLAB function *hilbert*. For the sake of comparison, the magnitudes of the DTFTs of the input sequence and the analytic sequence using MATLAB function are shown in the top and bottom plots of Fig. 7.71, respectively. From Figs. 7.70 and 7.71, we find that the HT performed using the MATLAB function and the codes in M-file *Hil_tra.m* are identical.

7.5   Summary

Now we know how to design FIR filters to meet the frequency specifications of a lowpass or highpass or bandpass or bandstop filter. Due to the existence of large sidelobes in a rectangular window, the resulting FIR filter has small minimum

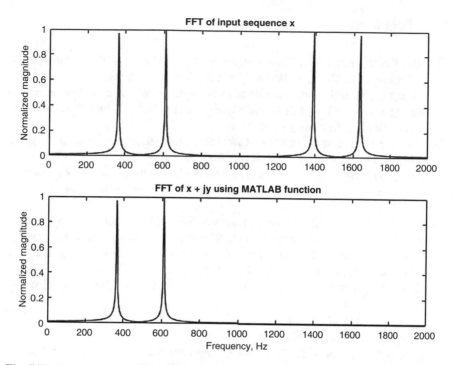

Fig. 7.71 Comparison of HT with MATLAB function: top, magnitude of the DTFT of x[n]; bottom, magnitude of the DTFT of the analytic signal using the MATLAB function

stopband attenuation. We circumvented this problem by choosing a proper window function. There are two types of windows – fixed and adjustable windows. Fixed windows, as the names imply, cannot control both the minimum stopband attenuation and transition width. For a given order, these two parameters are fixed. On the other hand, an adjustable window can trade off transition width for minimum stopband attenuation and vice versa. In addition to these two analytical window functions, we also described computer-aided windowed FIR filter design techniques. The computer-aided design arrives at the optimal impulse response of an FIR filter by iteratively minimizing an error function, which could be based on Parks-McClellan or least squares technique. Several examples, including MATLAB-based examples, were worked out to make the learning more efficient. The next thing for us to learn is how to realize the digital filters – IIR and FIR – in software or hardware. In the next chapter, we will describe these digital filters in terms of *signal flow graphs*, which are very useful in the realization process. Some of these structures may be more robust structures to coefficient quantization due to limited word length. We will also study the effect of limited coefficient word lengths on the resulting frequency response of the filters.

7.6 Problems

1. Find the smallest length of the lowpass FIR filter corresponding to the specifications $\omega_p = 0.42\pi$, $\omega_s = 0.58\pi$, $\delta_p = 0.002$, and $\delta_s = 0.008$.
2. Design a 12th-order lowpass FIR filter to approximate the ideal filter response having a passband edge at 1.5 kHz using rectangular and Hamming windows. Use a sampling frequency of 10 kHz.
3. Design a 15th-order highpass FIR filter to have a cutoff frequency of 0.4π rad using Blackman window, and plot its frequency response.
4. Design a bandpass FIR filter with the smallest length using the method based on fixed window to meet the following specifications: $\omega_1 = 0.4\pi$, $\omega_2 = 0.55\pi$, $\delta_p = 0.02$, and $\delta_s = 0.006$.
5. Design a lowpass FIR filter with a passband edge at 0.3π, stopband edge at 0.5π, and minimum stopband attenuation of 40 dB using Kaiser window. It is enough you show the ideal filter impulse response, the filter order, and the corresponding parameters of the Kaiser window.
6. Find the impulse response of the digital notch filter specified by
$$|H(e^{jw})| = \begin{cases} 0, w_1 \leq |w| \leq w_2 \\ 1, otherwise \end{cases}$$. Note that the frequency response of the digital filter is periodic with period 2π.
7. Design a bandstop FIR filter with the smallest length to meet the following specifications: lower passband edge $\Omega_{p1} = 0.25\pi$, $\Omega_{p2} = 0.6\pi$, $\Omega_{s1} = 0.45\pi$, $\Omega_{s2} = 0.8\pi$, $\delta_{p1} = 0.005$, $\delta_{p2} = 0.01$, and $\delta_s = 0.05$, where δ_{p1} and δ_{p2} are, respectively, the ripple in the lower and upper passbands.
8. Determine the filter order of the lowpass FIR filter having a passband edge of 0.3π, stopband edge of 0.6π, and a minimum stopband attenuation of 60 dB using Dolph-Chebyshev and Kaiser adjustable windows.
9. Show that an ideal bandstop digital filter can be realized as the sum of an ideal lowpass and highpass filters.
10. Show that an ideal bandpass digital filter can be realized as a cascade of lowpass and highpass digital filters.
11. Design a linear-phase lowpass FIR filter with the following specifications: passband edge at 500 Hz, stopband edge at 1200 Hz, maximum passband attenuation of 0.3 dB, minimum stopband attenuation of 45 dB, and a sampling frequency of 5000 Hz. Use Bartlett, Hamming, Haan, and Blackman window functions for the design. Plot the impulse responses and the magnitude responses in dB of the designed filters. Use MATLAB to solve the problem.
12. Repeat Problem 11 to design a linear-phase highpass FIR filter satisfying the following specifications: passband edge at 1200 Hz, stopband edge at 500 Hz, maximum passband attenuation of 0.3 dB, minimum stopband attenuation of 45 dB, and a sampling frequency of 5000 Hz.
13. Repeat Problem 11 using Kaiser window function.

14. Design a linear-phase FIR notch filter using a suitable window function to approximate an ideal notch filter of order 30 with a notch frequency of 60 Hz and a sampling frequency of 500 Hz. Plot the impulse response and magnitude of the frequency response in dB of the designed filter. Generate a signal consisting of 60 Hz and 130 Hz sinusoids, filter it through the notch filter, and plot the DTFT of the input and filtered signals. Verify that the notch filter rejects the 60 Hz component in the input signal. Use MATLAB to solve the problem.

15. Design a linear-phase lowpass FIR filter of lowest order based on windowing technique to meet the following specifications: passband edge at 0.25π rad, stopband edge at 0.4π rad, and a minimum stopband attenuation of 45 dB. Find out which window function will meet the given specifications.

References

1. Blackman RB (1965) Linear data smoothing and prediction in theory and practice. Addison-Wesley, Reading
2. Boite R, Leich H (1981) A new procedure for the design of high-order minimum-phase FIR digital or CCD filters. Sig Process 3:101–108
3. Crochiere RE, Rabiner LR (1976) On the properties of frequency transformations for variable cutoff linear phase digital filters. IEEE Trans Circ Syst CAS-23:684–686
4. Dutta Roy SC, Kumar B (1989) On digital differentiators, Hilbert transformers, and half-band lowpass filters. IEEE Trans Edu 32:314–318
5. Fatinopoulos I, Stathai T, Constantinides A(2001) A method for FIR filter design from joint amplitude and group delay characteristics. In: Proceedings of IEEE international conference on acoustics, speech, and signal processing, pp 621–625
6. Helms HD (1968) Nonrecursive digital filters: Design methods for achieving specifications on frequency response. IEEE Trans Audio Electroacoust AU-16:336–342
7. Herrmann O, Schussler HW (1970) Design of nonrecursive digital filters with minimum phase. Electron Lett 6:329–630
8. Herrmann O, Rabiner LR, Chan DSK (1973) Practical design rules for optimum finite impulse response lowpass digital filters. Bell Syst Tech J 52:769–799
9. Jarske P, Neuvo Y, Mitra SK (1988) A simple approach to the design of FIR digital filters with variable characteristics. Sig Process 14:313–326
10. Kaiser JF (1974) Nonrecursive digital filter design using the I_0-sinh window function. In: Proceedings of IEEE international symposium on circuits and systems, San Francisco, CA, pp 20–23
11. Kaiser JF, Hamming RW (1977) Sharpening the response of a symmetric nonrecursive filter by multiple use of the same filter. IEEE Trans Acoust Speech Sig Process ASSP-25:415–422
12. McClellan JH, Parks TW (1973) A unified approach to the design of optimum FIR linear phase digital filters. IEEE Trans Circ Theory CT-20:697–701
13. Mitra SK (2011) Digital signal processing: a computer-based approach, 4th edn. McGraw Hill, New York
14. Neuvo Y, Dong C-Y, Mitra SK (1984) Interpolated finite impulse response filters. IEEE Trans Acoust Speech Sig Process ASSP-32:563–570
15. Oetken G, Parks TW, Schussler HW (1975) New results in the design of digital interpolators. IEEE Trans Acoust Speech Sig Process ASSP-23:301–309
16. Oppenheim AV, Schafer RW (1975) Digital signal processing. Prentice Hall, Englewood Cliffs

17. Oppenheim AV, Mecklenbrauker WFG, Mersereau RM (1976) Variable cutoff linear phase digital filters. IEEE Trans Circ Syst CAS-23:199–203
18. Oppenheim AV, Schafer RW (1999) Discrete-time signal processing, 2nd edn. Prentice Hall, Englewood Cliffs
19. Parks TW, McClellan JH (1972) Chebyshev approximation for nonrecursive digital filters with linear phase. IEEE Trans Circ Theory CT-19:189–194
20. Rabiner LR (1973) Approximate design relationships for low-pass FIR digital filters. IEEE Trans Audio Electroacoust AU-21:456–460
21. Rabiner LR, Crochiere RE (1975) A novel implementation for narrow-band FIR digital filters. IEEE Trans Acoust Speech Sig Process 23(5):457–464
22. Rabiner LR, Gold B, McGonegal CA (1970) An approach to the approximation problem for nonrecursive digital filters. IEEE Trans Audio Electroacoust AU-18:83–106
23. Rabiner LR, McClellan JH, Parks TW (1975) FIR digital filter design techniques using weighted Chebyshev approximation. Proc IEEE 63(4):595–610
24. Saramaki T, Neuvo Y, Mitra SK (1988) Design of computationally efficient interpolated FIR filters. IEEE Trans Circ Syst CAS-35:70–88

Chapter 8
Digital Filter Structures

8.1 Signal Flow Graph

What we have learnt so far is how to design either an IIR or FIR digital filter to satisfy a given set of specifications in the frequency domain. We have also seen examples based on MATLAB wherein filtering operations are carried out by specific functions. We really don't know how these functions really work. If you are a S/W or H/W engineer and want to implement a digital filter in software or hardware, you should be able to describe the flow of signal from the input to the output. Thus, a digital filter structure describes the flow of signal as it propagates from the input to the output sample by sample. This filtering operation is described by a *signal flow graph*, which is a block diagram with blocks corresponding to the arithmetic operations of addition, multiplication, and unit delays. The blocks are connected by lines with arrows pointing in the direction of signal flow. In digital filter terminology, an adder has two inputs and one output, as shown in Fig. 8.1a. Similarly, a multiplier accepts an input signal and multiplies it by a coefficient a to produce an output, as shown in Fig. 8.1b. A unit delay block is a register, which can hold a sample from its input. The sample can be read from its output after one sample interval. Figure 8.1c illustrates a unit delay element. Note that the unit delay operation in the Z-domain is denoted by z^{-1}. Finally, Fig. 8.1d shows how a signal is tapped into. So, these are the basic building blocks of a digital filter structure. Let us look at a simple example.

Electronic supplementary material: The online version of this article (https://doi.org/10.1007/978-3-319-76029-2_8) contains supplementary material, which is available to authorized users.

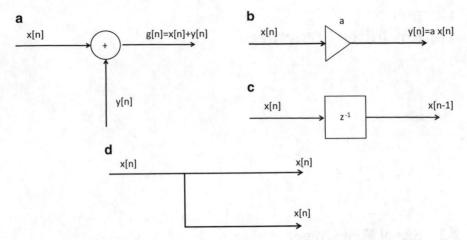

Fig. 8.1 Blocks in a signal flow graph: (**a**) adder, (**b**) multiplier, (**c**) unit delay, and (**d**) signal branch

Fig. 8.2 Signal flow graph of Example 8.1

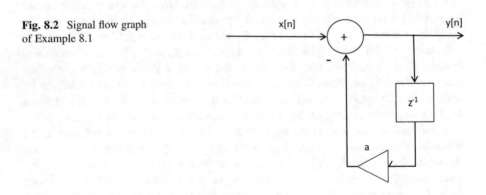

Example 8.1 Draw a signal flow graph to solve the LTI discrete-time system described by

$$y[n] = x[n] - ay[n-1], n \geq 0 \qquad (8.1)$$

Solution The above equation is familiar to us, which is a recursive linear difference equation. The signal flow graph consists of one adder, one unit delay, and a multiplier. The current output sample is obtained by multiplying the previous output sample by $-a$ and then adding it to the current input sample. Or the current output sample may be obtained by multiplying the previous output sample by the constant a and then subtracting it from the current input sample. Both are the same. Before the next sample arrives, the current output sample is fed to the unit delay. Figure 8.2 is the signal flow graph, which represents Eq. (8.1). At the time index $n = 0$, the sample value in the register corresponds to the initial condition. It must be pointed out that all arithmetic operations are performed with infinite accuracy. But in practice, the word lengths are finite, and so the arithmetic operations will not be exact. This inaccuracy in arithmetic operations will propagate from the input to the output and can be considered as noise. There are three sources of error in the filtering operation. The first is the error introduced in converting an analog signal into a digital signal. This error is actually outside of the signal flow graph. We have already established the error performance of an analog-to-digital converter in Chap. 2. The second source of error is in the word length used to represent the coefficients of the multipliers in the signal flow graph. The larger the word lengths of the multiplier coefficients, the smaller the inaccuracies in representing them. We will see the effect of finite word lengths of the coefficients on the filter later in this chapter. The third source of error is due to the inaccuracies of arithmetic operations because of finite word length.

8.2   IIR Digital Filter Structures

8.2.1   Direct Form I and II Structures

A given linear difference equation can be manipulated to form different signal flow graphs, each having specific features relating to filter coefficient sensitivity due to limited word length, efficiency of computations, etc. It is therefore important to study digital filter structures. For instance, consider the second-order transfer function described by

$$H(z) = \frac{Y(z)}{X(z)} = \frac{b_0 + b_1 z^{-1} + b_2 z^{-2}}{1 + a_1 z^{-1} + a_2 z^{-2}} = \frac{N(z)}{D(z)} \tag{8.2}$$

Define W(z) as

$$W(z) = X(z)N(z) = X(z)\left(b_0 + b_1 z^{-1} + b_2 z^{-2}\right) \tag{8.3}$$

In the discrete-time domain, Eq. (8.3) corresponds to

$$w[n] = b_0 x[n] + b_1 x[n-1] + b_2 x[n-2] \tag{8.4}$$

Figure 8.3a is the signal flow graph to compute w[n] from x[n] sample by sample. Note that it uses two unit delays, two adders, and three multipliers. Next the Z-transform of the output in terms of W(z) is

$$Y(z) = \frac{X(z)N(z)}{D(z)} = \frac{W(z)}{D(z)} \tag{8.5}$$

Therefore, the output in the discrete-time domain is expressed by

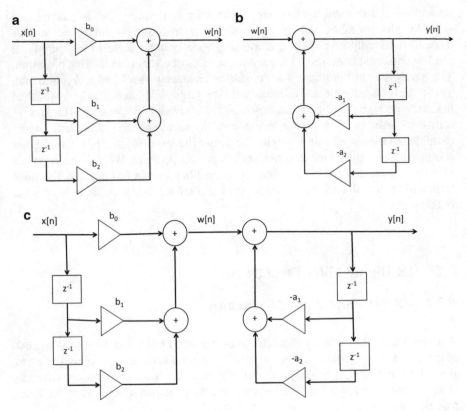

Fig. 8.3 A second-order Direct Form I IIR digital filter structure: (**a**) calculation of w[n] from the input x[n], (**b**) calculation of the output y[n] from w[n], and (**c**) calculation of y[n] from x[n]

$$y[n] = w[n] - a_1 y[n-1] - a_2 y[n-2] \tag{8.6}$$

In Fig. 8.3b the signal flow diagram for computing y[n] from w[n] is shown. As can be seen from the figure, it uses two adders, two multipliers, and two unit delays. Since our objective is to compute y[n] from x[n], we combine the two signal flow graphs in Fig. 8.3a, b, which is shown in Fig. 8.3c. This filter structure is known as *Direct Form I* structure and is *noncanonical* in delays meaning that it uses more number of delay elements than the order of the digital filter.

Though delay elements are not expensive from the point of view of hardware implementation, it is still a quest to find a structure that uses as many delays as the filter order is. In fact, it is feasible to realize such a *canonical* structure and is called *Direct Form II* digital filter structure. Let us see how we can achieve this. Using Eq. (8.2), define

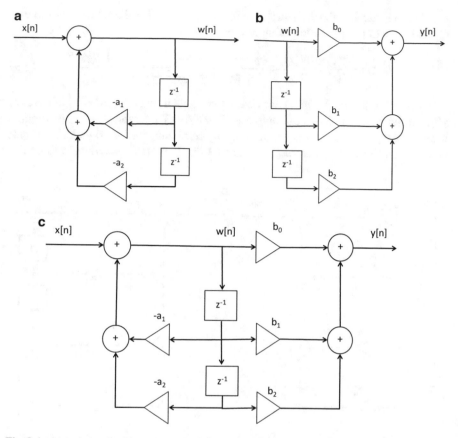

Fig. 8.4 A second-order Direct Form II IIR digital filter structure: (**a**) calculation of w[n] from the input x[n], (**b**) calculation of the output y[n] from w[n], and (**c**) calculation of y[n] from x[n]

$$W(z) = \frac{X(z)}{D(z)} \Rightarrow w[n] = x[n] - a_1 w[n-1] - a_2 w[n-2] \tag{8.7}$$

Figure 8.4a describes Eq. (8.7). Then, the Z-transform Y(z) of the output in terms of W(z) is described by

$$Y(z) = W(z)N(z) \Rightarrow y[n] = b_0 w[n] + b_1 w[n-1] + b_2 w[n-2] \tag{8.8}$$

The signal flow graph corresponding to Eq. (8.8) is depicted in Fig. 8.4b. By combining Fig. 8.4a, b, we get the overall Direct Form II filter structure of the second-order transfer function of (8.2) as shown in Fig. 8.4c. The only difference between Direct Form I and II structures is in the number of delay elements. Direct Form II structure uses two delays corresponding to the filter order, which is 2, whereas Direct Form I uses four delays.

We can generalize the filter structure for an Nth-order IIR digital filter as follows. Consider the transfer function of an Nth-order IIR digital filter described by

$$H(z) = \frac{Y(z)}{X(z)} = \frac{N(z)}{D(z)} = \frac{b_0 + b_1 z^{-1} + b_2 z^{-2} + \cdots\cdots + b_N z^{-N}}{1 + a_1 z^{-1} + a_2 z^{-2} + \cdots\cdots + a_N z^{-N}} \qquad (8.9)$$

By defining $W(z) = X(z)N(z)$, we obtain the signal flow graph for w[n] in terms of x[n], as shown in Fig. 8.5a. The output is $Y(z)D(z) = W(z)$, whose signal flow graph in terms of w[n] is shown in Fig. 8.5b. Then by combining the two signal flow graphs of W(z) and Y(z), we obtain the Direct Form I filter structure, as shown in Fig. 8.5c.

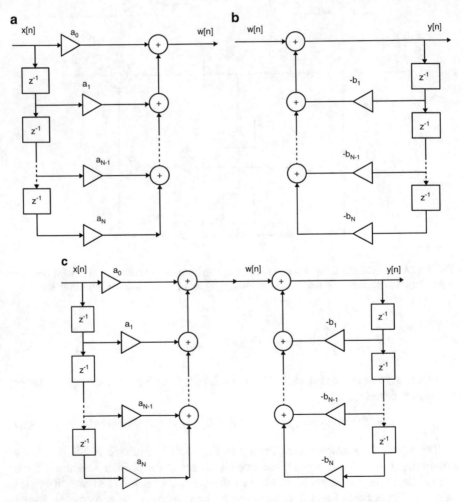

Fig. 8.5 An Nth-order Direct Form I IIR digital filter structure: (**a**) calculation of w[n] in terms of x [n], (**b**) calculation of y[n] in terms of w[n], and (**c**) overall input-output

To obtain the Direct Form II structure, we express

$$W(z)D(z) = X(z),$$ (8.10a)

and the output

$$Y(z) = W(z)N(z)$$ (8.10b)

Figures 8.6a, b, c show the signal flow graphs of the general Nth-order Direct Form II digital filter structure. As can be seen from the figure, the Direct Form II is canonic in the number of delays.

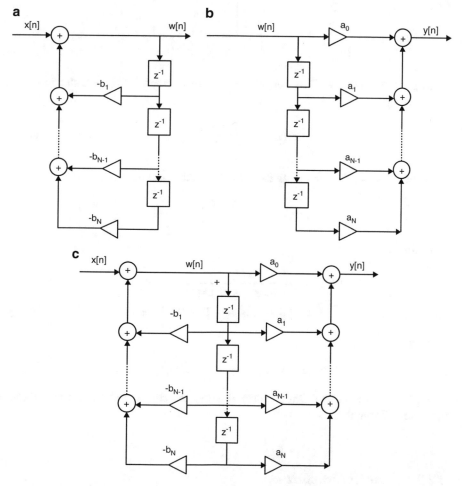

Fig. 8.6 An Nth-order Direct Form II IIR digital filter structure: (**a**) calculation of w[n] in terms of x[n], (**b**) calculation of y[n] in terms of w[n], and (**c**) overall input-output

Fig. 8.7 Direct Form II
structure of the transfer
function in Example 8.2

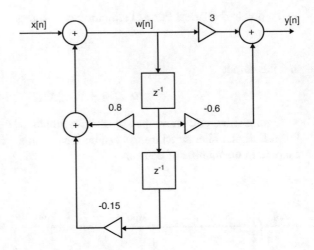

Example 8.2 Draw the Direct Form II signal flow graph for the transfer function described by

$$H(z) = \frac{3z^2 - 0.6z}{z^2 - 0.8z + 0.15} \qquad (8.11)$$

Solution First we rewrite the given transfer function in terms of z^{-1} as

$$H(z) = \frac{3 - 0.6z^{-1}}{1 - 0.8z^{-1} + 0.15z^{-2}} \qquad (8.12)$$

Define

$$W(z) = \frac{X(z)}{1 - 0.8z^{-1} + 0.15z^{-2}} \Rightarrow w[n]$$
$$= x[n] + 0.8w[n-1] - 0.15w[n-2] \qquad (8.13)$$

Then,

$$Y(z) = W(z)\left(3 - 0.6z^{-1}\right) \Rightarrow y[n] = 3w[n] - 0.6w[n-1] \qquad (8.14)$$

Figure 8.7 shows the Direct Form II structure corresponding to the transfer function in Example 8.2.

8.2.2   *Parallel Structure*

The transfer function of an Nth-order IIR digital filter can be expressed as a sum of the first- and/or second-order functions using partial fraction expansion. Thus the overall transfer function is written in the form

Fig. 8.8 Parallel form structure of an IIR digital filter

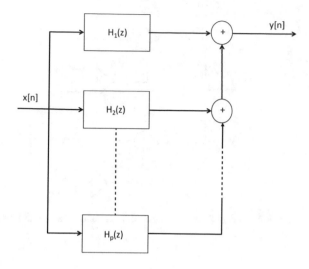

$$H(z) = \frac{N(z)}{D(z)} = H_1(z) + H_2(z) + \cdots + H_p(z) \qquad (8.15)$$

We then realize each transfer function in Direct Form I or II, where the input to each transfer function is the same and the outputs are added to obtain the overall response as shown in Fig. 8.8. One advantage in realizing a transfer function in parallel form is that the arithmetic error due to finite word length is restricted to each block. It must be pointed out that if a pole is complex, it must occur with its conjugate because the coefficients of the transfer function are real. So, after expressing the transfer function in partial fractions, the complex conjugate terms must be combined and expressed as a second-order function.

Example 8.3 Implement the following transfer function in parallel form:

$$H(z) = \frac{1 + z^{-1}}{(1 + 0.5z^{-1})(1 - 0.25z^{-1})} \qquad (8.16)$$

Solution We observe that the given transfer function is a proper function. Therefore, by expressing the above transfer function in partial fractions, we have

$$H(z) = \frac{A}{(1 + 0.5z^{-1})} + \frac{B}{(1 - 0.25z^{-1})}, \qquad (8.17)$$

where the residues are found from

$$A = H(z)(1 + 0.5z^{-1})\Big|_{z^{-1}=-2} = \frac{1 - 2}{1 + 0.5} = \frac{-2}{3}, \qquad (8.18a)$$

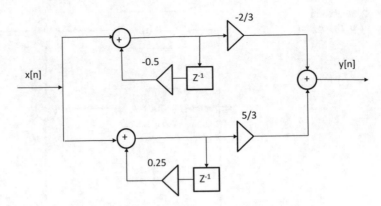

Fig. 8.9 Parallel form realization of the transfer function in Example 8.3

$$B = H(z)\left(1 - 0.25z^{-1}\right)\big|_{z^{-1}=4} = \frac{1+4}{1+2} = \frac{5}{3} \tag{8.18b}$$

Figure 8.9 shows the parallel form structure of the transfer function in Eq. (8.16), which consists of two first-order transfer functions.

Example 8.4 Realize the digital filter transfer function described by

$$H(z) = \frac{1 + 2z^{-1} + z^{-2}}{1 - z^{-1} + 0.5z^{-2} - 0.125z^{-3}} \tag{8.19}$$

in parallel form.

Solution The denominator is a third-order polynomial. Therefore, we have to factor it into first- and/or second-order polynomials. The poles are obtained by finding the roots of the denominator. To make the task easier, we can use the MATLAB function to find the residues, the poles, and the quotients of the transfer function in (8.19). Specifically, we call the function *[r,p,k] = residuez(B,A)*, where $B = [1\ 2\ 1\ 0]$ is the vector of coefficients of the numerator and $A = [1\quad -1\ 0.5\quad -0.125]$ is the vector of coefficients of the denominator in Eq. (8.19). With these input arguments, the MATLAB function returns the residues, the poles, and the quotients in the vectors r, p, and k, respectively. In this example, the poles are found to be 0.5, 0.25 + j0.433013, and 0.25–j0.433013. One of the poles is real and the other two are complex conjugate of each other. We now express the given transfer function in partial fraction as

$$H(z) = \frac{A}{1 - 0.5z^{-1}} + \frac{B}{1 - (0.25 + j0.433013)z^{-1}} \\ + \frac{B^*}{1 - (0.25 - j0.433013)z^{-1}} \tag{8.20}$$

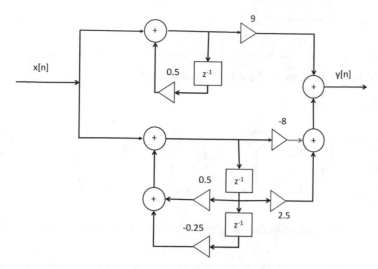

Fig. 8.10 Parallel form realization of the transfer function in Example 8.4 as first- and second-order sections

The residues are found to be $A = 9$, $B = -4 - j0.5774$, and $B^* = -4 + j0.5774$. The residues are returned in the vector r by the MATLAB function. Since two poles are complex conjugate of each other, we need to combine them into a single second-order function. Thus, the given transfer function in partial fractions is

$$H(z) = \frac{9}{1 - 0.5z^{-1}} + \frac{-8 + 2.5z^{-1}}{1 - 0.5z^{-1} + 0.25z^{-2}} \qquad (8.21)$$

We can now realize each term on the right-hand side of (8.21) in Direct Form II and then add the two outputs to obtain the overall response, as shown in Fig. 8.10

8.2.3 Cascade Structure

In parallel form realization, we expressed the given transfer function as a sum of first- and/or second-order functions, realized each function in Direct Form I or II and then added the individual outputs to obtain the overall response. The input to each section in the parallel structure is the same and is the input to the system. Another way of realizing the same given transfer function is to express it as a product of first-and/or second-order functions, realize each term in the product in Direct Form I or II, and then connect them in tandem. This implies that the input to the first block is the overall input, the input to the second block is the output of the first block, and so on. The output of the final block is the overall response of the system. More

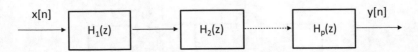

Fig. 8.11 Realization of an IIR digital filter in cascade form

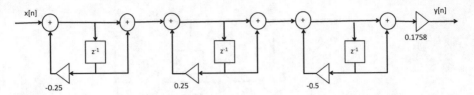

Fig. 8.12 Cascade realization of the transfer function of an IIR digital filter in Example 8.5

specifically, we express the given transfer function of an IIR digital filter as a product of p functions as given by

$$H(z) = H_1(z)H_2(z)\cdots\cdots H_p(z) \tag{8.22}$$

Each individual function on the right-hand side of (8.22) may be first- and/or second-order function. Figure 8.11 is a *cascade* realization of the transfer function in (8.22).

Example 8.5 Show a cascade structure to realize the transfer function as given below:

$$H(z) = \frac{K(1 + z^{-1})^3}{(1 + 0.25z^{-1})(1 - 0.25z^{-1})(1 + 0.5z^{-1})}, \tag{8.23}$$

where the constant K is chosen so that the transfer function has a value 1 at DC.

Solution We can express the given transfer function as a product of three first-order functions as

$$H(z) = K\left(\frac{1 + z^{-1}}{1 + 0.25z^{-1}}\right)\left(\frac{1 + z^{-1}}{1 - 0.25z^{-1}}\right)\left(\frac{1 + z^{-1}}{1 + 0.5z^{-1}}\right) \tag{8.24}$$

We can determine the value of the constant from

$$H(z)|_{z^{-1}=1} \Rightarrow K = 0.1758 \tag{8.25}$$

Figure 8.12 depicts a cascade realization of the transfer function in (8.23) as three first-order functions.

8.3   FIR Filter Structures

An FIR digital filter of order N has the transfer function described by

$$H(z) = \sum_{n=0}^{N} h[n]z^{-n} \tag{8.26}$$

There are N + 1 number of elements in its impulse response sequence. The FIR filter can be realized in two obvious forms, namely, direct form and cascade form. There is another possible structure known as polyphase structure for the realization of an FIR filter. We have seen before that an FIR filter, in general, has a linear-phase response. The impulse response of a linear-phase FIR filter has even or odd symmetry, which can be exploited to our advantage in reducing the number of multipliers. In this section, we will focus our effort in learning the possible structures of an FIR digital filter.

8.3.1   Direct Form Structure of an FIR Filter

The direct form structure of an FIR filter is a straightforward implementation of the transfer function in (8.26), where all the multipliers correspond to the elements of its impulse response sequence. To exemplify further, consider a fourth-order FIR filter, whose transfer function is given by

$$H(z) = \frac{Y(z)}{X(z)} = h[0] + h[1]z^{-1} + h[2]z^{-2} + h[3]z^{-3} + h[4]z^{-4} \tag{8.27}$$

The direct form structure of the FIR digital filter in Eq. (8.27) is shown in Fig. 8.13, where the multipliers are exactly the filter's coefficients or impulse response. The response to an input can be written from (8.27) as

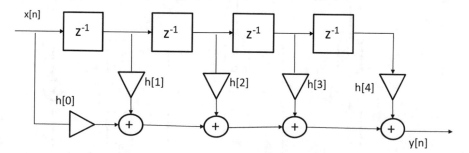

Fig. 8.13 Direct form structure of a fourth-order FIR digital filter

$$y[n] = h[0]x[n] + h[1]x[n-1] + h[2]x[n-2]$$
$$+ h[3]x[n-3] + h[4]x[n-4] \tag{8.28}$$

An FIR digital filter is non-recursive and so the output depends only on its current and past input samples, as seen from (8.28). This means that there is no feedback from its output, which is also seen from Fig. 8.13.

8.3.2   Cascade Structure of an FIR Digital Filter

A cascade structure corresponds to a transfer function expressed as a product of first- and/or second-order sections. The transfer function of an Nth-order FIR digital filter in Eq. (8.26) can also be written as

$$H(z) = h[0] \prod_{k=1}^{K} H_k(z), \tag{8.29}$$

where

$$H_k(z) = 1 + a_{1,k}z^{-1} \tag{8.30a}$$

if $H_k(z)$ is a first-order function or

$$H_k(z) = 1 + a_{1,k}z^{-1} + a_{2,k}z^{-2} \tag{8.30b}$$

if $H_k(z)$ is a second-order function. If the order N is even, then K = N/2 and all sections are of second-order. If N is odd, then K = (N + 1)/2 with one first-order function and the rest second-order functions. An example of a cascade structure of a sixth-order FIR digital filter is shown in Fig. 8.14. Since the filter order 6 is an even number, there are three second-order sections in the cascade structure.

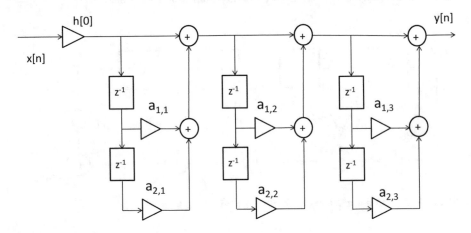

Fig. 8.14 Cascade structure of a sixth-order FIR digital filter

8.3.3   Linear-Phase FIR Filter Structure

Digital filters with linear-phase response are preferred in signal processing applications because they do not introduce phase distortions in the processed signal. Though human ears are tolerant to phase distortions in speech and music signals, digital images are susceptible to such distortions. A filter with nonlinear-phase response causes contour effect in image processing. Having said that, an FIR digital filter with a linear-phase response has either even or odd symmetry in its impulse response sequence. If the order is N, then due to symmetry, it only needs either $(N + 2)/2$ if N is even or $(N + 1)/2$ if N is odd, number of multipliers for the direct form structure. For example, consider a sixth-order linear-phase FIR digital filter. It has a length of 7. We can describe its transfer function as

$$
\begin{aligned}
H(z) = h[0] + h[1]z^{-1} + h[2]z^{-2} + h[3]z^{-3} + h[4]z^{-4} \\
+ h[5]z^{-5} + h[6]z^{-6}
\end{aligned}
\tag{8.31}
$$

Because of the even symmetry of the impulse response of the FIR digital filter in (8.31), we can rewrite H(z) as

$$
H(z) = h[0]\left(1 + z^{-6}\right) + h[1]\left(z^{-1} + z^{-5}\right) + h[2]\left(z^{-2} + z^{-4}\right) + h[3]z^{-3} \tag{8.32}
$$

From Eq. (8.32), we notice that there are only four multipliers instead of seven multipliers. What if the FIR filter's order was 7? We will still require four multipliers as seen from

$$
\begin{aligned}
H(z) = h[0](1 + z^{-7}) + h[1]\left(z^{-1} + z^{-6}\right) \\
+ h[2]\left(z^{-2} + z^{-5}\right) + h[3](z^{-3} + z^{-4})
\end{aligned}
\tag{8.33}
$$

The direct form structures corresponding to the filter orders 6 and 7 are shown in Fig. 8.15a, b, respectively.

8.3.4   Polyphase FIR Filter Structure

Before we describe a polyphase filter structure, it is important to mention the fact that polyphase filters are used in sample rate conversion. To increase the sample rate of an input sequence, we insert zero-valued samples between two consecutive input samples and then interpolate the in-between sample values. This process will upsample the input or increase the input sample rate. An advantage of a polyphase filter structure is that it avoids multiplying zero-valued samples in the interpolation process. To decrease the input sample rate, we have to decimate or throw away every few samples. This decimation process is called downsampling. Another advantage of a polyphase filter structure is to avoid multiplications of the samples that will

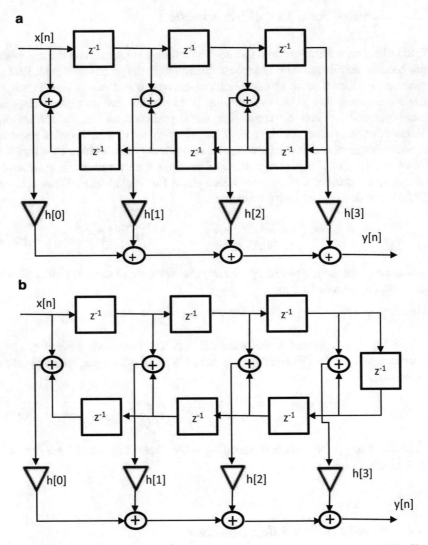

Fig. 8.15 Direct form structure of a linear-phase FIR filter with even symmetry: (**a**) FIR filter's order is 6; (**b**) FIR filter's order is 7

eventually be discarded at the output. Having said that, let us define polyphase decomposition of an FIR transfer function and then arrive at the structure. Consider a length-9 FIR digital filter function whose transfer function is described by

$$H(z) = \sum_{n=0}^{8} h[n]z^{-n} \tag{8.34}$$

We can rewrite Eq. (8.34) as

$$H(z) = \left(h[0] + h[3]z^{-3} + h[6]z^{-6}\right) + z^{-1}\left(h[1] + h[4]z^{-3} + h[7]z^{-6}\right) \\ + z^{-2}\left(h[2] + h[5]z^{-3} + h[8]z^{-6}\right) \tag{8.35}$$

Note that each bracketed factor is a sixth-order polynomial but with only three terms. Let us define the following polyphase components:

$$E_0(z) = h[0] + h[3]z^{-1} + h[6]z^{-2} \tag{8.36a}$$

$$E_1(z) = h[1] + h[4]z^{-1} + h[7]z^{-2} \tag{8.36b}$$

$$E_2(z) = h[2] + h[5]z^{-1} + h[8]z^{-2} \tag{8.36c}$$

We observe from the above three equations that

$$E_0\left(z^3\right) = h[0] + h[3]\left(z^{-1}\right)^3 + h[6]\left(z^{-2}\right)^3 = h[0] + h[3]z^{-3} + h[6]z^{-6} \tag{8.37a}$$

$$E_1\left(z^3\right) = h[1] + h[4]z^{-3} + h[7]z^{-6} \tag{8.37b}$$

$$E_2\left(z^3\right) = h[2] + h[5]z^{-3} + h[8]z^{-6} \tag{8.37c}$$

In terms of the three, polyphase components in Eqs. (8.36a), (8.36b), and (8.36c), we can express the eighth-order FIR filter transfer function in (8.34) as

$$H(z) = E_0\left(z^3\right) + z^{-1}E_1\left(z^3\right) + z^{-2}E_2\left(z^3\right) \tag{8.38}$$

Equation (8.38) is known as the *polyphase decomposition* of the FIR digital filter transfer function H(z) with three polyphase components. In general, the polyphase decomposition of an FIR digital filter of length N with L polyphase components is given by

$$H(z) = \sum_{m=0}^{L-1} z^{-m} E_m\left(z^L\right), \tag{8.39}$$

where

$$E_m(z) = \sum_{n=0}^{\left\lfloor \frac{N-1}{L} \right\rfloor} h[nL + m]z^{-n}, 0 \leq m \leq L - 1 \tag{8.40}$$

In Eq. (8.40), $\lfloor x \rfloor$ is the flooring operation. Figure 8.16 shows the polyphase structure of an eighth-order FIR filter with three parallel sections corresponding to three polyphase components.

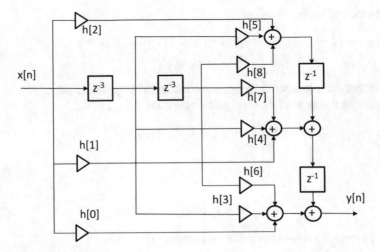

Fig. 8.16 Polyphase structure of an eighth-order FIR filter with three parallel sections corresponding to three polyphase components

Example 8.6 Design a lowpass FIR filter of order 8 with a cutoff frequency of 0.3π using Blackman window. Decompose the transfer function in a three-component polyphase sections. Upsample by a factor of 2 and downsample by a factor of 4 of a sinusoidal sequence of frequency 100 Hz. Use a sampling frequency of 4 kHz.

Solution We invoke the MATLAB function $w = blackman(9)$ to generate the Blackman window of length-9 samples. We then design the lowpass FIR filter by calling the MATLAB function $fir1(8,0.3,w)$, which returns the impulse response of the windowed lowpass FIR filter. Note that the *fir1* function accepts the cutoff frequency in the interval between 0 and 1 with 1 corresponding to half the sampling frequency. Since π corresponds to half the sampling frequency, we have to use 0.3 as the cutoff frequency in the MATLAB function. Once the impulse response of the lowpass FIR filter is determined, we can write the three polyphase components as given in Eq. (8.40) and the transfer function as in Eq. (8.39). We can evaluate the frequency response of the polyphase FIR filter as the sum of its polyphase components. In MATLAB, these are given, respectively, by

$$e_0 = [h[0]\ 0\ 0\ h[3]\ 0\ 0\ h[6]\ 0\ 0], \tag{8.41a}$$

$$e_1 = [0\ h[1]\ 0\ 0\ h[4]\ 0\ 0\ h[7]\ 0\ ], \tag{8.41b}$$

$$e_2 = [0\ 0\ h[2]\ 0\ 0\ h[5]\ 0\ 0\ h[8]] \tag{8.41c}$$

The frequency response of the lowpass FIR digital filter is the sum of the frequency responses of the polyphase components. Figure 8.17 shows the magnitude in dB of the frequency response of the FIR digital filter as a function of the frequency

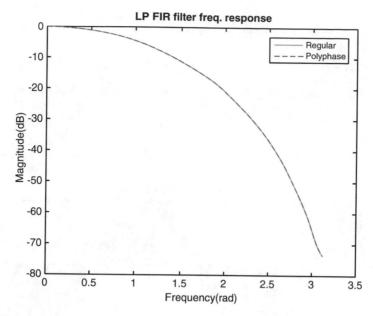

Fig. 8.17 Magnitude in dB of the frequency response of the polyphase FIR filter in Example 8.6

in rad. It also shows the magnitude in dB of the frequency response of the same FIR digital filter as a single unit. They are identical. Next we generate a length-64 sinusoidal signal of the specified frequency and sampling frequency. To increase its rate by 2, we first insert two zero-valued samples between consecutive samples of the input sequence and then filter it by the polyphase components. In MATLAB, we can use the function *conv(x,h, 'same')* to filter the sequence x by the filter h and the output sequence will have the same length as the input sequence. By filtering the zero-padded input sequence through the polyphase components and adding the outputs, we obtain the overall output with its rate doubled. The zero-padded input sequence and the filtered sequence are shown in Fig. 8.18 in the top and bottom plots, respectively. In order to down sample the upsampled sequence by a factor of 4, we discard every 4 samples between the filtered outputs. The input sequence and the upsampled-by-2/downsampled-by-4 are shown in the top and bottom plots of Fig. 8.19. MATLAB has the function *upfirdn(x,h,U,D)*, where x is the input sequence, h is the FIR filter impulse response, U is the upsampling factor, and D is the downsampling factor. In Fig. 8.20 the input sequence and upsampled-by-2/ downsampled-by-4 using *upfirdn* function in the top and bottom plots, respectively, are shown. They seem to match the results in Fig. 8.19.

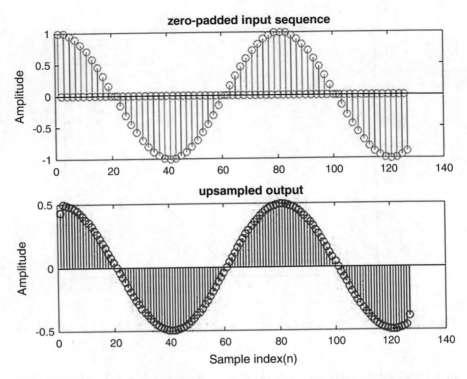

Fig. 8.18 Sampling rate increase by a factor of 2. Top plot, zero-padded input sequence; bottom plot, filtered signal with a sampling rate increase by a factor of 2

8.4   Finite Word Length Effect

In all our discussions so far, we have not paid any attention to the implementation of an IIR or FIR digital filter in hardware. We implicitly assumed that the coefficients of digital filters are represented exactly and that all arithmetic operations are performed with infinite accuracy. This is not a typical scenario in practice. Most real-time digital filters are implemented using either special-purpose hardware or DSP chips. These hardware systems have finite word lengths or bit widths for the representation of the filter coefficients as well as for carrying out arithmetic operations. That being the case, one must as a hardware engineer first ascertain that the designed digital filter will meet the given specifications with a chosen hardware platform. In other words, one must determine the minimum word length for the filter coefficients by simulation before implementing it in hardware. In a similar manner, one must determine the minimum word length used in the arithmetic operations so that the error or noise due to arithmetic inaccuracies in the output is acceptable. Before we delve into this task, let us familiarize ourselves with the binary representation of numbers since this is the number system used in computers.

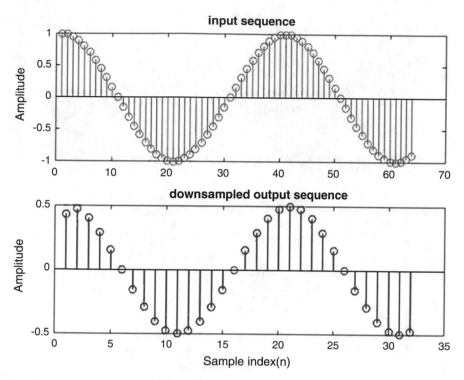

Fig. 8.19 Downsampling by a factor of 2: top plot, original input sequence; bottom plot, upsampled-by-2/downsampled-by-4 sequence

8.4.1   *Fixed-Point Binary Representation*

We are accustomed to decimal number system. It consists of ten symbols and an integer uses these symbols with *place value*. Thus, an integer decimal number 10875 is equivalent to

$$10875_{10} = 5*10^0 + 7*10^1 + 8*10^2 + 0*10^3 + 1*10^4 \qquad (8.42)$$

A decimal number such as 12.8432 is equivalent to

$$12.8432_{10} = 1*10^1 + 2*10^0 + 8*10^{-1} + 4*10^{-2} + 3*10^{-3} + 2*10^{-4} \quad (8.43)$$

That is, the digits to the right of the decimal point have the place values 10^{-d}, $d \in Z$. In a similar manner, the symbols used in the binary system are "0" and "1." An integer binary number 1101011 is equivalent to a decimal number given by

$$\begin{aligned}1101011_2 &= 1*2^0 + 1*2^1 + 0*2^2 + 1*2^3 + 0*2^4 + 1*2^5 + 1*2^6 \\ &= 107_{10}\end{aligned} \qquad (8.44)$$

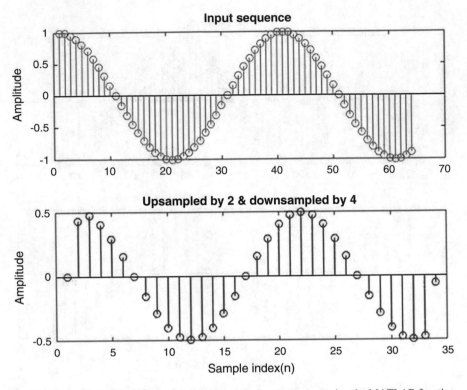

Fig. 8.20 Upsampled-by-2/downsampled-by-4 of the input sequence using the MATLAB function *upfirdn* with the filter in Example 8.6

Unlike the decimal system, a decimal number can be represented in the binary system in three different ways, namely, in one's complement or two's complement, or sign-magnitude form. A binary number can be in *fixed-point* or *floating-point* format. We will look into the fixed-point binary format in this section. As the name implies, a fixed-point binary number is a fraction with the binary point always fixed. The digit to the left of the binary point represents the sign of the fraction.

Sign-Magnitude Format   In the sign-magnitude format, a positive fraction has a "0" bit to the left of the binary point, while a negative fraction has a "1" bit to the left of the binary point. Thus,

$$0.11011_2 = 0.84375_{10} \tag{8.45a}$$

$$1.11011_2 = -0.84375_{10} \tag{8.45b}$$

One's Complement Format   In this fixed-point binary format, a positive fraction is represented in the same manner as in the sign-magnitude format. A negative fraction is represented as follows: First its magnitude is represented as a positive fraction. Then all the bits are complemented (1's to 0's and 0's to 1's) bit by bit to obtain the

1's complement. For example, we want to represent the decimal fraction -0.875_{10} in binary 1's complement format. First, the magnitude is $0.875_{10} = 0.111_2$. Then by complementing bit by bit the binary fraction of 0.875, we get $-0.875_{10} = 1.000_2$.

Two's Complement Format This format is similar to the one's complement format except that a "1" bit is added to the least significant bit (LSB) after performing one's complement. So, the decimal fraction -0.875_{10} in two's complement format will be

$$-0.875_{10} = 1.000_2 + 0.001_2 = 1.001_2 \tag{8.46}$$

8.4.2   Floating-Point Binary Representation

The range of values that can be represented using fixed-point binary format is limited as compared to that using the floating-point binary representation. In the normalized floating-point format, a positive number is represented by

$$x_{10} = M2^{\pm E}, \tag{8.47}$$

where M is called the mantissa and E the exponent. The mantissa is a binary fraction between $\frac{-1}{2} \leq M < 1$, and E is an integer. For instance, we want to represent the decimal number 67 in floating-point binary format. The exponent is determined by

$$E = \left\lceil \frac{log_{10}67}{log_2 2} \right\rceil = 7 = 111_2 \tag{8.48}$$

The mantissa is obtained by

$$M = \frac{67}{2^7} = 0.5234375 = 0.1000011_2 \tag{8.49}$$

Therefore, we have a 3-bit exponent of 111_2 and an 8-bit mantissa of 0.1000011_2. According to the 32-bit IEEE format, the most significant bit (MSB) is the sign bit, the next 8 bits are the exponent, and the last 23 bits form the mantissa. The exponent, however, is offset by 127. In this IEEE format, a decimal number x in binary floating-point format takes the form

$$x_{10} = (-1)^s E (M), \tag{8.50}$$

where s is the sign, 0 for positive value and 1 for negative value, and M is a fraction.

8.4.3   Filter Coefficient Sensitivity

When the coefficients of an IIR or FIR digital filter are realized in finite word length fixed-point representation, the realized coefficients are not exactly the same as when realized with full precision. These coefficient inaccuracies will alter the realized frequency response as well as the poles and zeros. Therefore, before implementing the digital filter in hardware, one must evaluate the effect of the bit width assigned to the coefficients. That is, one must determine if the frequency specifications of the filter are met for a given bit width and ensure that the resulting IIR digital filter is stable. We can use MATLAB to do the simulation. Specifically, the MATLAB function fi is used to represent a decimal fraction with a given number of bits. To demonstrate the finite word length effect on the filter frequency response, let us design a fifth-order Butterworth IIR digital filter and assign a total of 5 bits with 4 bits for the fraction to represent each of its coefficients. We can realize the designed filter either as a single section or as cascaded sections. In Fig. 8.21a the pole-zero plot of the fifth-order IIR Butterworth digital filter realized as a single section with full precision coefficients is shown. The pole-zero plot with 5-bit coefficients is shown in Fig. 8.21b. As can be seen from the figures, the five zeros corresponding to the full precision are split up with a pair lying outside the unit circle. As far as the poles are concerned, they are unchanged. Fortunately though, none of the poles is on the unit circle and so the filter is stable. Let us look at the frequency response of the Butterworth filter with coefficients in 5-bit fixed-point representation. Figure 8.22a shows the frequency response of the filter as a single section with 5-bit coefficients as well as with full precision. The frequency response seems to match the desired frequency response fairly closely. As a comparison, Fig. 8.22b shows the magnitude of the frequency response of the fifth-order Butterworth IIR digital filter realized as a single section with coefficient word lengths of 8 bits. It appears that 8 bits for the coefficients make a perfect match to the full precision case. The following table lists the values of the numerator and denominator coefficients of the fifth-order Butterworth IIR digital filter with full precision and 5-bits coefficients (Table 8.1).

When the same fifth-order Butterworth filter is realized as a cascade of first- and second-order sections, we get the numerator and denominator coefficients as shown in Table 8.2. Since some of the coefficients have magnitudes greater than unity, we assign 2 bits to the integer part and 4 bits to the fractional part of the coefficients. The resulting frequency response of the filter is shown in Fig. 8.23a. The frequency response corresponding to 6-bit coefficients nearly matches that with full precision coefficients. In Fig. 8.23b the magnitude of the frequency response of the Butterworth filter realized as cascaded sections with 8-bit precision coefficients, which is a perfect match to the full precision case is shown.

Next, let us try an elliptic IIR filter of order 5 with a passband ripple of 0.1 dB and a minimum stopband attenuation of 40 dB. The coefficients of the filter with full precision and 6-bit precision are listed in Table 8.3. The corresponding pole-zero plots of the elliptic IIR filter realized as a single section are shown in Fig. 8.24a, b, respectively, for full precision and 6-bit precision of the coefficients. The magnitude

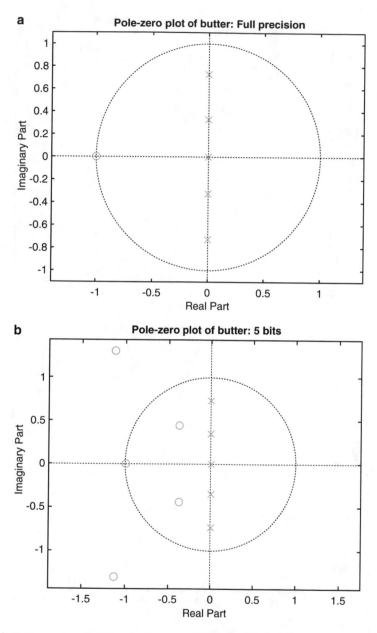

Fig. 8.21 Pole-zero plot of a fifth-order Butterworth IIR digital filter realized as a single section: (**a**) full precision for the coefficients; (**b**) 5-bit fixed-point representation of the coefficients

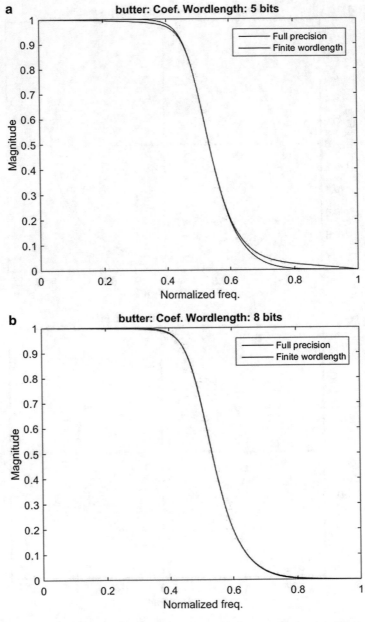

Fig. 8.22 Magnitude of the frequency response of the fifth-order Butterworth IIR digital filter realized as a single section: (**a**) 5-bit coefficients; (**b**) 8-bit coefficients

Table 8.1 Numerator and denominator coefficients of fifth-order Butterworth IIR digital filter as a single section

Numerator coefficients		Denominator coefficients	
Full precision	5-bit precision	Full precision	5-bit precision
0.0528	0.0625	1.0000	0.9375
0.2639	0.2500	0.0000	0.0000
0.5279	0.5000	0.6334	0.6250
0.5279	0.5000	0.0000	0.0000
0.2639	0.2500	0.0557	0.0625
0.0528	0.0625	0.0000	0.0000

Table 8.2 Numerator and denominator coefficients of fifth-order Butterworth IIR digital filter as cascaded sections

Section	Numerator coefficients		Denominator coefficients	
	Full	6-bit	Full	6-bit
1	1.0000	1.0000	1.0000	1.0000
	0.9989	1.0000	0.0000	0.0000
	0.0000	0.0000	0.0000	0.0000
2	1.0000	1.0000	1.0000	1.0000
	2.0018	1.9375	0.0000	0.0000
	1.0018	1.0000	0.1056	0.1250
3	1.0000	1.0000	1.0000	1.0000
	1.9993	1.9375	0.0000	0.0000
	0.9993	1.0000	0.5279	0.5000

of the frequency response of the IIR elliptic filter with 6-bit coefficients is shown in Fig. 8.25a. Figure 8.25b shows the magnitude response for coefficient word lengths of 8 bits. The same elliptic filter is also realized in cascaded form. The magnitude of the frequency response for 6-bit and 8-bit coefficient word lengths are shown in Fig. 8.26a, b, respectively. In this case, 6-bit coefficient word length meets the frequency specifications better than those of the single section realization.

Finally, let us see how the limited coefficient word length affects an FIR digital filter. To this end, let us design a 12th-order linear-phase FIR digital filter to approximate an ideal lowpass filter with a passband edge of $\frac{\pi}{2}$. We will use Blackman-Harris window. The MATLAB function $fir1(N,wn,wind)$ is invoked to design the FIR filter, where N is the filter order, wn is the normalized passband edge, and *wind* is the window function of length N + 1. The impulse response coefficients of the lowpass FIR filter are listed in Table 8.4 for 6-bit and 8-bit coefficient word lengths, respectively. The filter with 8-bit word length for the coefficients meets the desired specifications exactly, as can be seen from Fig. 8.27a, b. The MATLAB M-file for this example is named *Filter_coefft_sensitivity.m*.

Example 8.7 Simulation of Filtering Using Simulink   In this example we will do a simulation of digital filtering using fixed-point arithmetic with the aid of MATLAB's Simulink. The block diagram of the filtering operations is shown in Fig. 8.28.

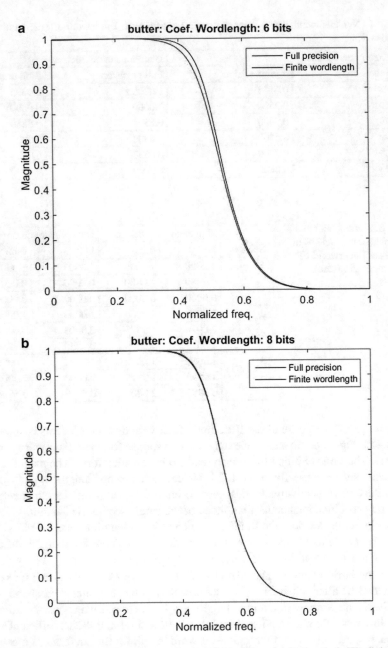

Fig. 8.23 Magnitude of the frequency response of the fifth-order Butterworth IIR digital filter realized as cascaded sections: (**a**) fixed-point 6-bit coefficients; (**b**) fixed-point 8-bit coefficients

Table 8.3 Numerator and denominator coefficients of fifth-order elliptic IIR digital filter as a single section

Numerator coefficients		Denominator coefficients	
Full precision	6-bit precision	Full precision	6-bit precision
0.1052	0.0938	1.0000	1.0000
0.3192	0.3125	−0.2291	−0.2500
0.5248	0.5313	1.1785	1.1875
0.5248	0.5313	−0.3114	−0.3125
0.3192	0.3125	0.3204	0.3125
0.1052	0.0938	−0.0600	−0.0625

It consists of a sinusoidal source of a specified frequency. The output of the sinusoidal source is in floating-point format. The block also allows us to specify the sampling rate. We can also choose a random signal source, which can output either a uniformly distributed random sequence or a Gaussian random sequence in floating-point format. In Fig. 8.28 we see a manual switch to switch between the input sources. To change the switch position, one has to double-click it and save the diagram before starting the simulation. The signal output is then converted to fixed-point format before it is filtered. This is achieved by the block named *Data Type Conversion*. Its parameters are shown in Fig. 8.29. In this example, we have chosen a word length of 10 bits with 7 bits for the fraction. This block allows other formats such as integer, double, etc. The converted data is then applied to the input of either an FIR or IIR digital filter via a switch, as shown in Fig. 8.28. The filter order, passband edge, etc. can be specified by double-clicking the blocks named *Lowpass FIR Filter* and *Lowpass IIR Filter* and then entering the values in the appropriate fields. In Fig. 8.30a, b the parameters chosen for the FIR and IIR filters, respectively, are shown. Finally, the time-domain input and output are displayed on the respective scopes. We have also included a power spectrum analyzer to display the power spectrum of the filtered signal. Now we are ready to start the simulation. Figure 8.31a shows the input sinusoidal signal with unit amplitude at a frequency of 1000 Hz and a sampling frequency of 8 kHz. The FIR filtered signal is shown in Fig. 8.31b, which is identical to the input except for a tiny difference in the amplitude. The power spectrum of the FIR filtered signal is depicted in Fig. 8.31c, which has a peak at 1 kHz corresponding to the input. Note that the spectrum is shown over the frequency range −4 kHz to +4 kHz. That is why we see two peaks at ±1 kHz. For the same sinusoidal input, the output of the IIR filter is shown in Fig. 8.31d. The corresponding power spectrum can be seen in Fig. 8.31e. Even though the IIR output power spectrum has a peak at 1 kHz, it has more peaks than those of the FIR output. The same filtering operations are repeated using the uniformly distributed random sequence as input. The input random sequence is shown in Fig. 8.32a. The FIR filtered output sequence and its power spectrum are shown in Fig. 8.32b, c, respectively. Since the FIR filter is a lowpass filter, we see the frequency components from DC to 1 kHz having much higher values than the rest of the frequency components. Similar observations are made from Fig. 8.32d, e, which correspond to the output from the IIR filter. The Simulink file to solve this example is named *Filter_fixed_point.slx*.

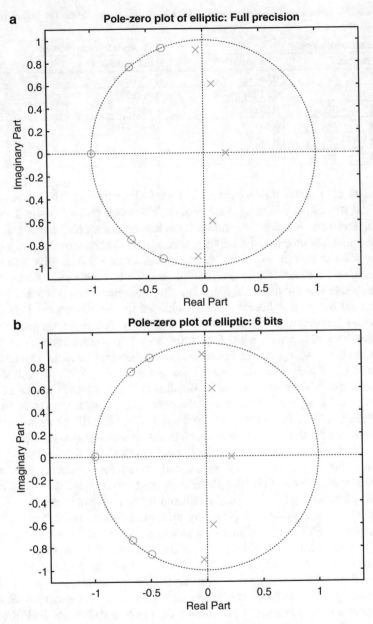

Fig. 8.24 Pole-zero plot of a fifth-order elliptic IIR digital filter realized as a single section: (**a**) full precision for the coefficients; (**b**) 6-bit fixed-point representation of the coefficients

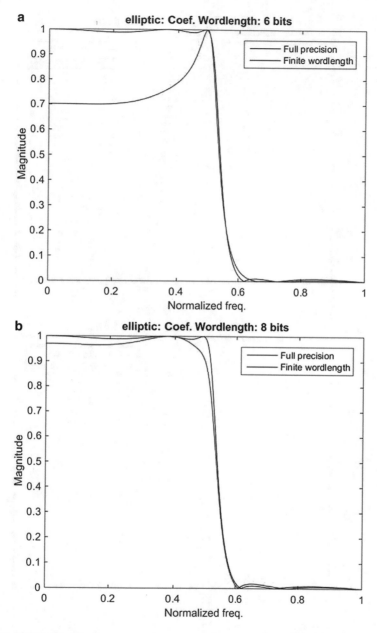

Fig. 8.25 Magnitude of the frequency response of the fifth-order elliptic IIR digital filter realized as a single section: (**a**) 6-bit coefficients; (**b**) 8-bit coefficients

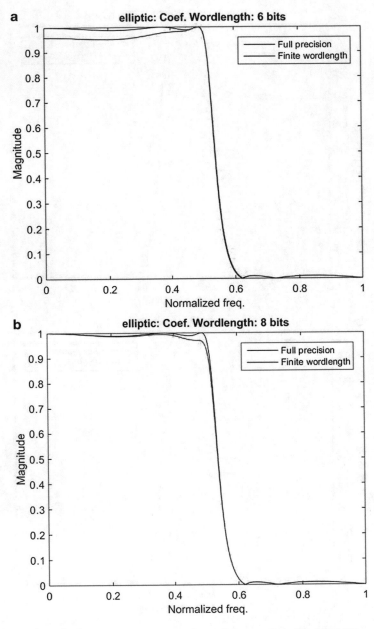

Fig. 8.26 Magnitude of the frequency response of the fifth-order elliptic IIR digital filter realized as cascaded sections: (**a**) fixed-point 6-bit coefficients; (**b**) fixed-point 8-bit coefficients

Table 8.4 FIR filter	Full precision	6-bit precision	8-bit precision
coefficients with 6-bit and 8-bit word lengths for the coefficients	0.0000	0.0000	0.0000
	0.0004	0.0000	0.0000
	0.0000	0.0000	0.0000
	−0.0231	−0.0313	−0.0234
	0.0000	0.0000	0.0000
	0.2720	0.2813	0.2734
	0.5014	0.5000	0.5000
	0.2720	0.2813	0.2734
	0.0000	0.0000	0.0000
	−0.0231	−0.0313	−0.0234
	0.0000	0.0000	0.0000
	0.0004	0.0000	0.0000
	0.0000	0.0000	0.0000

8.4.4   Error Due to Finite Word Length Arithmetic

In this section we will discuss the effect due to finite word length arithmetic on the output of both IIR and FIR digital filters. First let us analyze the first- and second-order sections of an IIR digital filter and then give an example based on MATLAB.

First-Order IIR Filter

Consider the first-order IIR digital filter shown in Fig. 8.33. Figure 8.33a shows the first-order filter with arithmetic performed with infinite precision, and Fig. 8.33b shows the same filter with finite precision arithmetic. The finite precision arithmetic results in an error at each sample and is modeled as a noise source denoted by e [n]. This error propagates through the filter and appears as additive noise at the output as indicated in the figure. This noise at the multiplier output is assumed to be uniformly distributed with a variance σ_e^2. It can be shown that the variance of the noise at the output with an impulse response sequence $h_e[n]$ is related to the noise due to the multiplier by

$$\sigma_{y_e}^2 = \sigma_e^2 \sum_{n=0}^{\infty} |h_e[n]|^2 \tag{8.51}$$

From Fig. 8.33b, we find that the transfer function of the noise due to the multiplier is the same as the transfer function of the first-order IIR digital filter. That is,

$$H_e(z) = H(z) = \frac{1}{1 - az^{-1}} \tag{8.52}$$

Therefore, the impulse response corresponding to the noise transfer function is given by

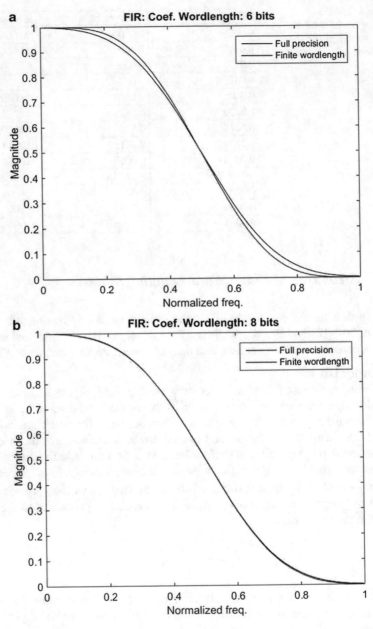

Fig. 8.27 Magnitude of the frequency response of the 12th-order linear-phase FIR digital filter realized as a single section: (**a**) fixed-point 6-bit coefficients; (**b**) fixed-point 8-bit coefficients

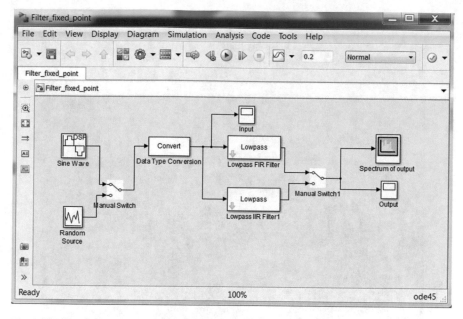

Fig. 8.28 Block diagram of digital filtering simulation

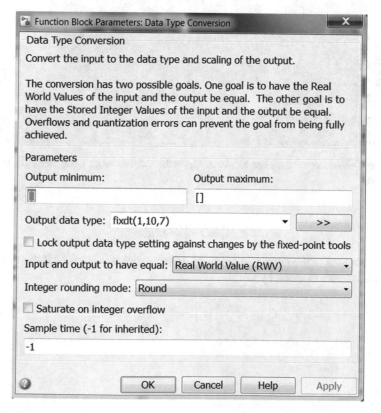

Fig. 8.29 Specification details of data type conversion block

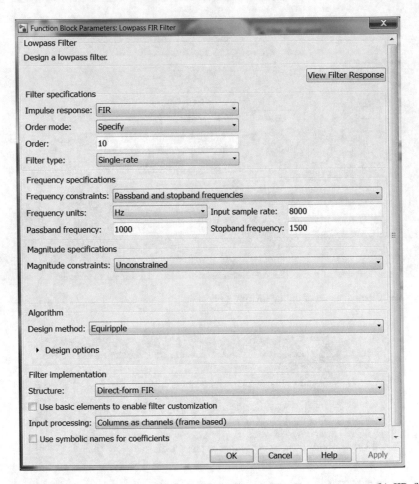

Fig. 8.30 Parameters of the filter blocks in Fig. 8.28: (**a**) FIR filter parameters; (**b**) IIR filter parameters

$$h_e[n] = h[n] = a^n u[n] \tag{8.53}$$

The variance of the output noise due to finite precision arithmetic is then obtained using (8.51) and is given by

$$\sigma_{y_e}^2 = \sigma_e^2 \sum_{n=0}^{\infty} |a^n|^2 = \frac{\sigma_e^2}{1 - a^2}, a < 1 \tag{8.54}$$

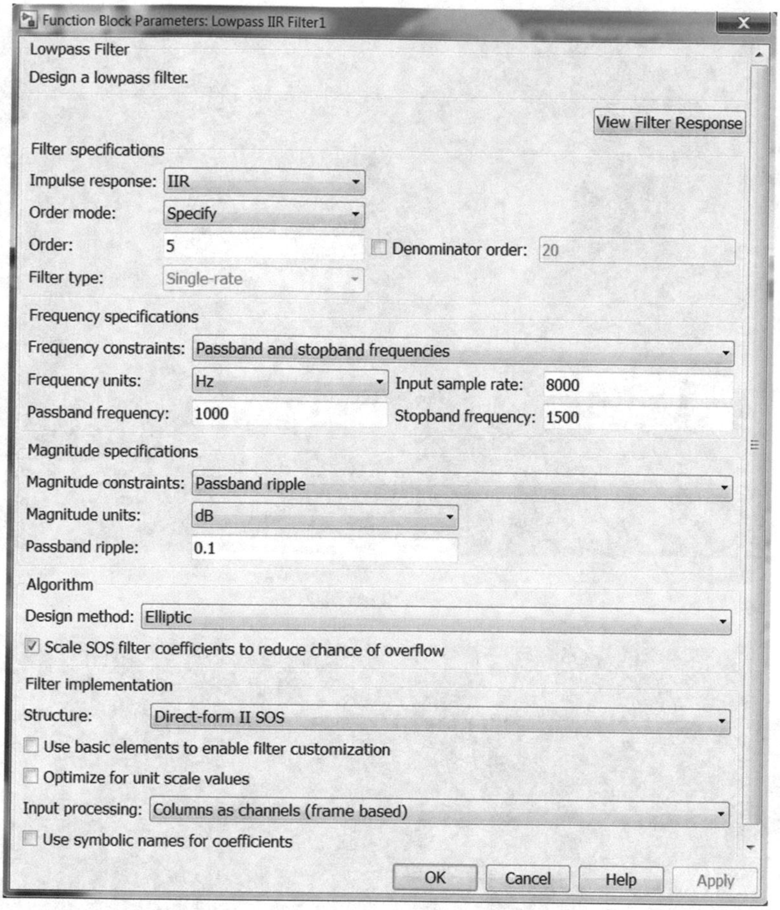

Fig. 8.30 (continued)

Second-Order IIR Filter

Next, let us consider the second-order IIR digital filter, which is shown in Fig. 8.34 along with the noise model. In this case also, the noise transfer functions for both multipliers are the same as the signal transfer function, which is

$$H_{e_a}(z) = H_{e_b}(z) = H(z) = \frac{1}{1 + az^{-1} + bz^{-2}} \tag{8.55}$$

Assuming both multiplier noise sources to have the same variance σ_e^2, the variance of the noise at the output due to both multipliers is found to be

$$\sigma_{y_e}^2 = 2\sigma_e^2 \sum_{n=0}^{\infty} |h[n]|^2 \tag{8.56}$$

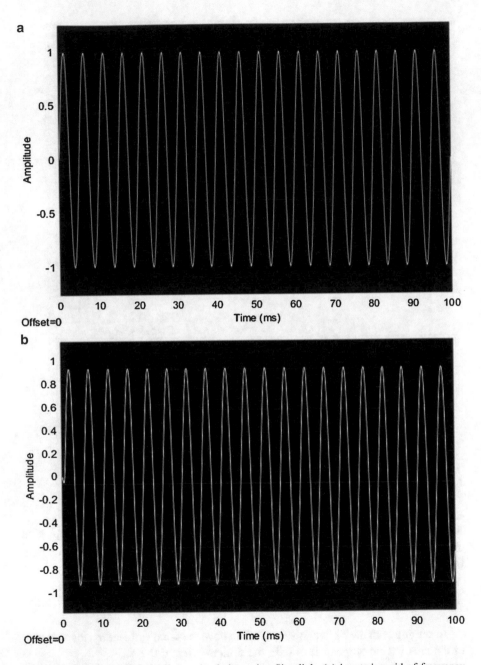

Fig. 8.31 Results of digital filtering simulation using Simulink: (**a**) input sinusoid of frequency 1 kHz; (**b**) FIR filtered output sequence; (**c**) power spectrum of the FIR filtered sequence; (**d**) IIR filtered output sequence; (**e**) power spectrum of the IIR filtered sequence

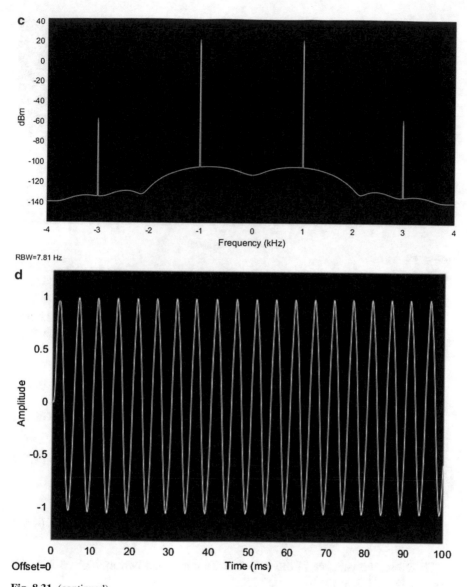

Fig. 8.31 (continued)

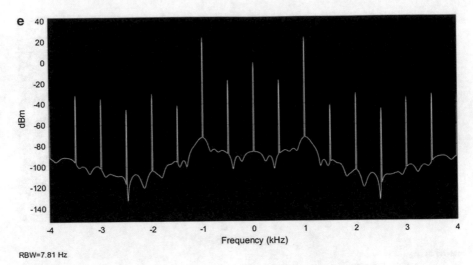

RBW=7.81 Hz

Fig. 8.31 (continued)

Note that the two noise sources are independent. As an example, if a = 0.6 and b = 0.0225, the transfer function in (8.55) can be expressed in partial fraction form as given by

$$H(z) = \frac{1.0774}{1 + 0.5598z^{-1}} - \frac{0.0774}{1 + 0.0402z^{-1}} \tag{8.57}$$

The corresponding impulse response is obtained by identifying each term on the right-hand side of (8.57) with a causal exponential sequence, which is

$$h[n] = 1.0774(-0.5598)^n u[n] - 0.0774(-0.0402)^n u[n] \tag{8.58}$$

Since the impulse response in (8.58) is convergent, we find that

$$\sum_{n=0}^{\infty} |h[n]|^2 \le 1.6966 \tag{8.59}$$

Therefore, the variance of the output noise due to the two multipliers is found from (8.56) and (8.59) to be

$$\sigma_{y_e}^2 \le 2\sigma_e^2 * 1.6966 \tag{8.60}$$

MATLAB Example: Fixed-Point Implementation of an IIR Digital Filter
Let us consider an example using MATLAB to compute the variance of the noise at the output of an Nth-order IIR digital filter using finite precision arithmetic operations. We will design Butterworth and elliptic IIR digital filters of order N and then compute the responses to a specified input. The filtering operation will be achieved

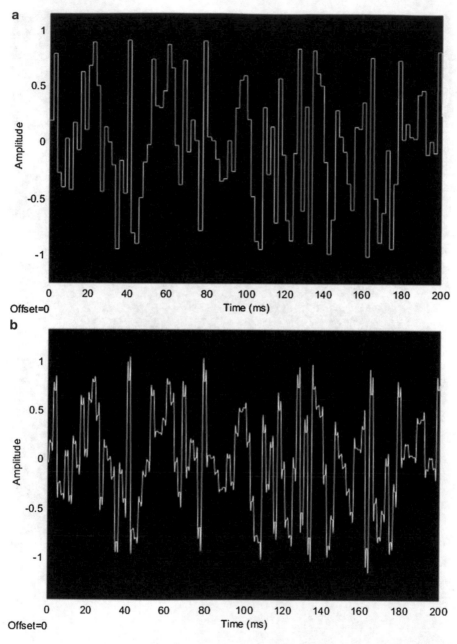

Fig. 8.32 Results of digital filtering simulation using Simulink: (**a**) uniformly distributed random sequence; (**b**) FIR filtered output sequence; (**c**) power spectrum of the FIR filtered sequence; (**d**) IIR filtered output sequence; (**e**) power spectrum of the IIR filtered sequence

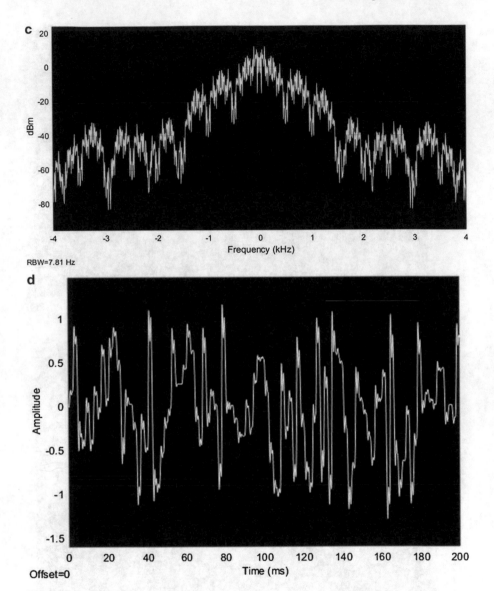

Fig. 8.32 (continued)

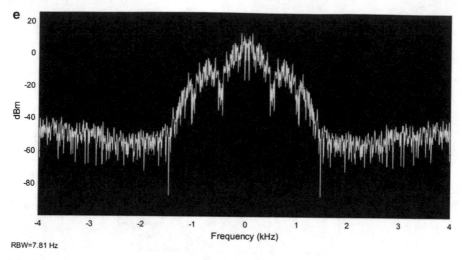

Fig. 8.32 (continued)

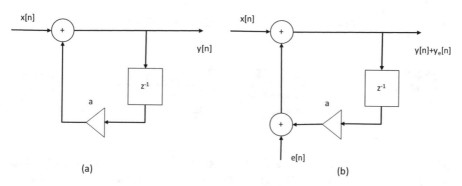

Fig. 8.33 Signal flow graph of a first-order IIR digital filter: (**a**) full precision arithmetic operation; (**b**) finite precision arithmetic operation

using fixed-point arithmetic with specified word length. Further, the designed filters will be implemented in either a single Direct Form II structure or a cascade form. The MATLAB tool to implement fixed-point arithmetic is *fimath*. The details for this example are given in the M-file "*Emulate_IIR_fixed_pt.m*". The signal-to-noise ratios (SNR) for the two types of filters are listed in the following Table. For all the cases, the word lengths for the filter coefficients are set to a total of signed 10 bits. The fractional bit lengths will depend on the maximum absolute value of the filter coefficients. The arithmetic operations are performed with a word length of 10 bits (Table 8.5).

Figure 8.35 shows the magnitude of the frequency response of the seventh-order Butterworth IIR digital filter. The filtered outputs with full precision and 10-bit

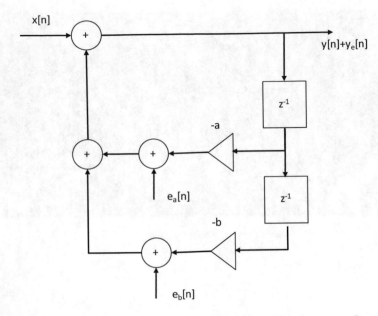

Fig. 8.34 Signal flow graph of a second-order IIR digital filter with noise sources due to the two multipliers

Table 8.5 SNR for IIR digital filters

Filter type	Single structure	Cascade structure
Butterworth	39.79 dB	24.56 dB
Elliptic	22.36 dB	40.46 dB

precision are shown in the top and bottom plots of Fig. 8.36. The corresponding spectra are shown in Fig. 8.37. As can be seen in the bottom plot, noise due to arithmetic errors is spread over the entire band of frequencies. The error between the full precision and finite precision outputs is shown in the top plot of Fig. 8.38. Its histogram and spectrum are shown in the middle and bottom plots, respectively. For the cascade realization of the seventh-order Butterworth filter, the arithmetic error, its histogram, and spectrum are shown in Fig. 8.39.

The same M-file is used to perform the filtering operation using a seventh-order elliptic IIR digital filter. The word lengths for the filter coefficients and arithmetic operations are the same as those used for the Butterworth filter. The results are shown in Figs. 8.40, 8.41, 8.42, 8.43, and 8.44. The same M-file also generates a zero-mean Gaussian noise with standard deviation of 0.5 and filters it using fixed-point arithmetic. For this case, all the word length specifications are the same as those used for the input signal. Figure 8.45 shows the spectrum of the filtered noise using a single structure, where the top plot corresponds to full precision arithmetic and the bottom plot corresponds to fixed-point implementation. The error sequence, its histogram, and the spectrum are shown in the top, middle, and bottom plots, respectively, in Fig. 8.46. The same quantities are shown in Figs. 8.47 and 8.48 for

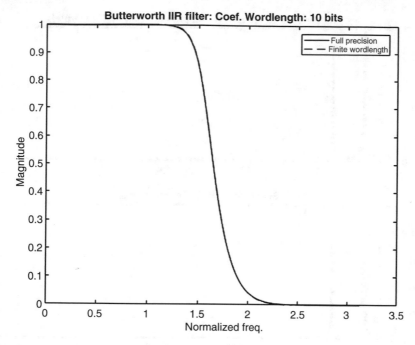

Fig. 8.35 Magnitude of frequency response of a seventh-order IIR Butterworth digital filter

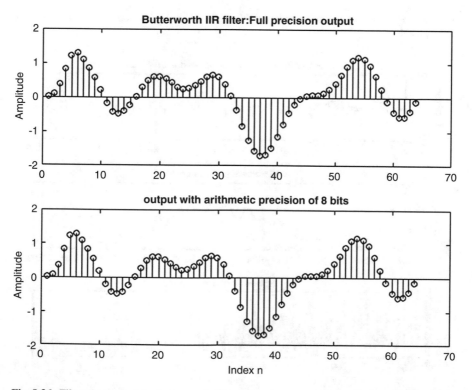

Fig. 8.36 Filtered output of the Butterworth filter whose frequency response is shown in Fig. 8.35. Top, full precision; bottom, finite precision

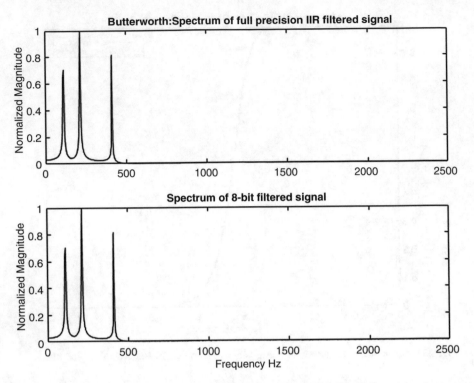

Fig. 8.37 Frequency spectrum of filtered output sequence in Fig. 8.36. Top, full precision; bottom, finite precision

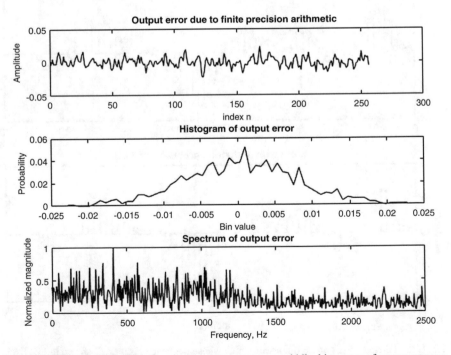

Fig. 8.38 Noise due to arithmetic error: top, error sequence; middle, histogram of error sequence; bottom, spectrum of error sequence

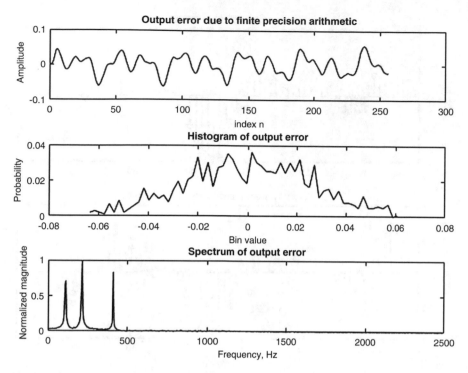

Fig. 8.39 Noise due to arithmetic error in the cascade realization of the seventh-order Butterworth IIR digital filter: top, error sequence; middle, histogram of error sequence; bottom, spectrum of error sequence

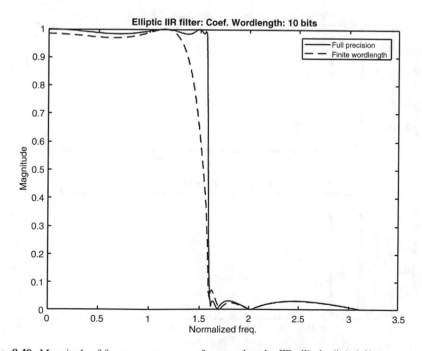

Fig. 8.40 Magnitude of frequency response of a seventh-order IIR elliptic digital filter

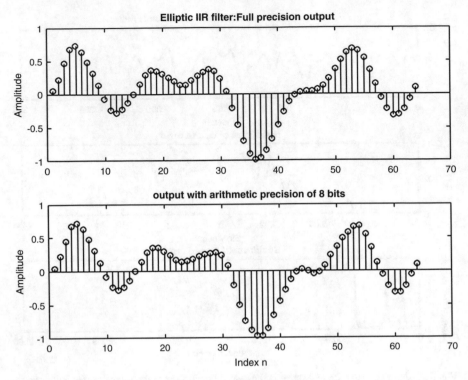

Fig. 8.41 Filtered output of the elliptic filter whose frequency response is shown in Fig. 8.40: top, full precision; bottom, finite precision

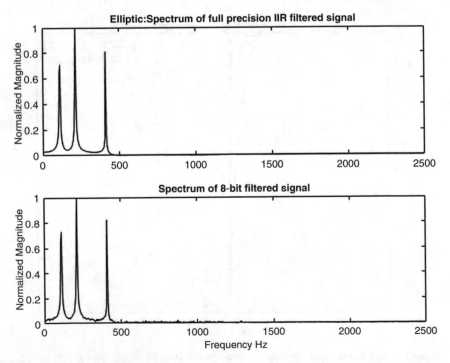

Fig. 8.42 Frequency spectrum of filtered sequence in Fig. 8.41. Top, full precision; bottom, finite precision

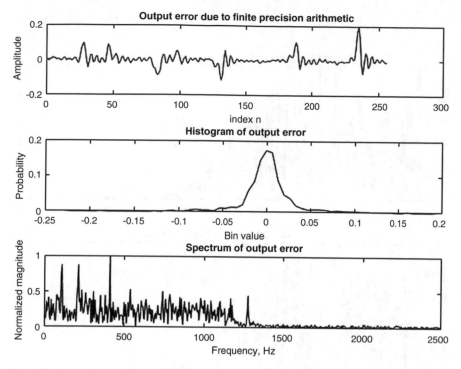

Fig. 8.43 Noise due to arithmetic error: top, error sequence; middle, histogram of error sequence; bottom, spectrum of error sequence

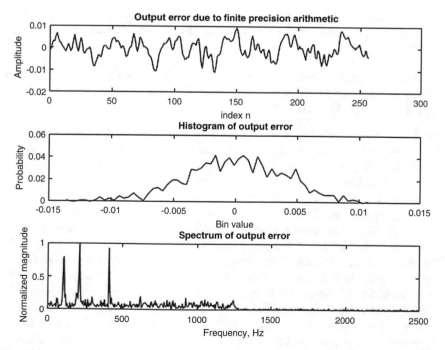

Fig. 8.44 Noise due to arithmetic error in the cascade realization of the seventh-order elliptic IIR digital filter: top, error sequence; middle, histogram of error sequence; bottom, spectrum of error sequence

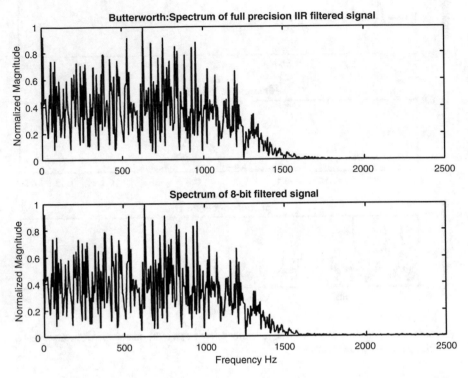

Fig. 8.45 Spectrum of filtered noise due to arithmetic error using a single structure Butterworth IIR digital filter: top, spectrum of full precision filtered noise; bottom, spectrum of finite precision filtered noise

the cascade structure. These results are obtained using a seventh-order Butterworth lowpass IIR digital filter. Similar results are shown for the elliptic IIR digital filter in Figs. 8.49, 8.50, 8.51, and 8.52.

MATLAB Example: Fixed-Point Implementation of an FIR Digital Filter

This example is similar to the previous example except that it deals with an FIR digital filter. In particular, we design a lowpass FIR digital filter of order 11 using Blackman window and then represent the filter coefficients in fixed-point representation with a word length of 8 bits. As before, the fractional word length will depend on the maximum magnitude of the filter coefficients. This filter is then used to filter either a signal or a zero-mean Gaussian random sequence. Both the signal and the noise are represented in fixed-point format with a word length of 8 bits. The magnitude of the frequency response of the lowpass FIR filter is shown in Fig. 8.53. The filtered signal is shown in Fig. 8.54, where the top plot represents the lowpass filtered signal with full precision and the bottom plot corresponds to the filtered signal using fixed-point arithmetic. The resulting signal-to-noise ratio due to finite precision arithmetic is found to be 26.94 dB. The spectra of the two filtered sequences are shown in Fig. 8.55. As can be seen from the figure, the arithmetic error

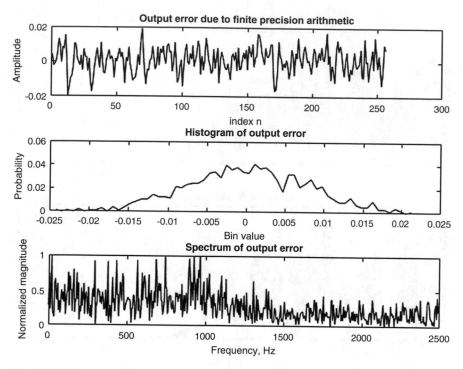

Fig. 8.46 Histogram of the error sequence due to arithmetic error using a single structure Butterworth IIR digital filter: top, error sequence; middle, histogram; bottom, spectrum of the error sequence

appears as noise over the whole frequency range. Figure 8.56 illustrates the error sequence, its histogram, and spectrum in the top, middle, and bottom plots, respectively. As a final step, a zero-mean Gaussian noise sequence in fixed-point format is input to the FIR filter, and the results are shown in Figs. 8.57, 8.58, and 8.59. These figures are self-explanatory. The details for this example are given in the M-file "*Emulate_FIR_fixed_pt.m*".

8.4.5 Limit Cycles in IIR Digital Filters

Finite precision arithmetic may cause unstable conditions in IIR digital filters. This will result in oscillation or periodic response for some specific inputs such as zero input with nonzero initial conditions or constant inputs. These oscillations are known as *limit cycles*. Limit cycles occur only in IIR digital filters and are due to the presence of feedbacks. There are two types of limit cycles possible in IIR digital filters, and they are called *granular* and *overflow* limit cycles. In granular limit cycle, the amplitude of oscillation is small, while the amplitude may be large in overflow limit cycle.

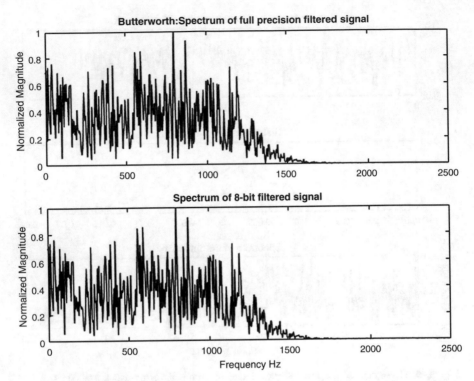

Fig. 8.47 Spectrum of filtered noise due to arithmetic error using a cascade structure Butterworth IIR digital filter: top, spectrum of full precision filtered noise; bottom, spectrum of finite precision filtered noise

We will illustrate the occurrence of limit cycles by an example using MATLAB. First we have to design an IIR digital filter. In this example, we will design either an elliptic or Butterworth lowpass IIR digital filter of order 4 with a cutoff frequency of 0.1π. For the elliptic filter, we will use a passband ripple of 0.1 dB and a minimum stopband attenuation of 50 dB. Once the specified filter is designed, we can convert it to second-order sections using the function *tf2sos*. Next, we convert the second-order sections to Direct Form II structures using the MATLAB function *dfilt.df2sos*. Now we have to represent the Direct Form II filter coefficients in fixed-point format. This is done by invoking the function *set(hd, 'arithmetic', 'fixed')* where, hd is the Direct Form II filter obtained in the previous step. Finally, we call the function *Res = limitcycle(hd)* to detect the presence of the limit cycle in the IIR digital filter. MATLAB detects the presence of limit cycles by performing Monte Carlo simulation using random initial states and zero input vectors of length twice that of the impulse response. One can set these numbers to any desired values. The M-file for this example is named *Test_limit_cycle.m*. The magnitude of the frequency response of the fourth-order elliptic IIR digital filter is shown in Fig. 8.60. When the M-file is run, it reports an overflow limit cycle, which is shown in Fig. 8.61. When the M-file is run again, we get a different overflow limit cycle shown in Fig. 8.62. When the

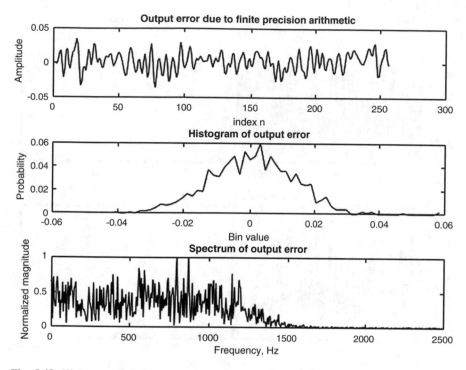

Fig. 8.48 Histogram of the error sequence due to arithmetic error using a cascade structure Butterworth IIR digital filter: top, error sequence; middle, histogram; bottom, spectrum of the error sequence

same M-file is run again, we get a granular limit cycle. Two different runs result in granular limit cycles and are shown in Figs. 8.63 and 8.64 for the elliptic IIR digital filter. When a fourth-order IIR Butterworth digital filter is used, the results are different from those of the elliptic filter. The magnitude of the frequency response of the lowpass IIR Butterworth digital filter is shown in Fig. 8.65, and the corresponding limit cycle obtained is shown in Fig. 8.66.

8.5 FIR Lattice Structure

Before we close this chapter, we will describe another structure to realize FIR digital filters. More specifically, we will learn how to realize an FIR filter in *lattice* form. To this end, consider the structure shown in Fig. 8.67, which is a second-order FIR filter in lattice form. The general form of the transfer function of a second-order FIR digital filter is given by

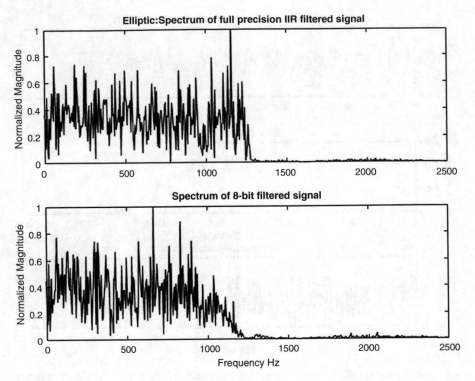

Fig. 8.49 Spectrum of filtered noise due to arithmetic error using a single structure elliptic IIR digital filter: top, spectrum of full precision filtered noise; bottom, spectrum of finite precision filtered noise

$$H(z) = b_0 + b_1 z^{-1} + b_2 z^{-2} \qquad (8.61)$$

The task is then to find the values of the multipliers g_1 and g_2 in terms of the coefficients of H(z) in (8.61). These multipliers are also called the *reflection coefficients*. From the lattice structure depicted in Fig. 8.67, we relate the variables in the Z-domain as follows.

$$X_0(z) = b_0 X(z) \qquad (8.62a)$$

$$Y_0(z) = X_0(z) \qquad (8.62b)$$

$$X_1(z) = X_0(z) + g_1 z^{-1} Y_0(z) \qquad (8.63a)$$

$$Y_1(z) = g_1 X_0(z) + z^{-1} Y_0(z) \qquad (8.63b)$$

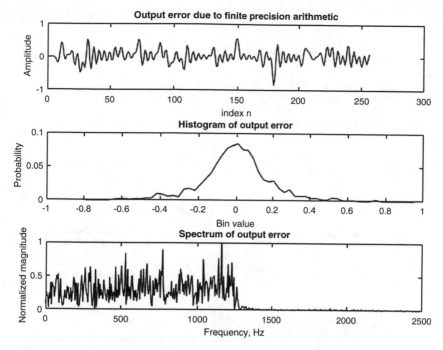

Fig. 8.50 Histogram of the error sequence due to arithmetic error using a single structure elliptic IIR digital filter: top, error sequence; middle, histogram; bottom, spectrum of the error sequence

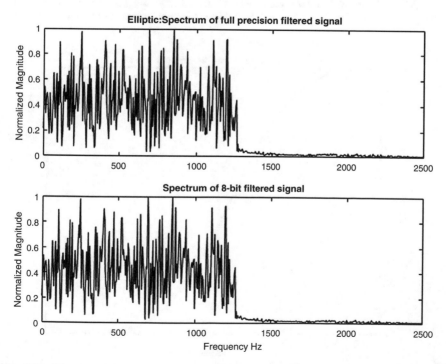

Fig. 8.51 Spectrum of filtered noise due to arithmetic error using a cascade structure elliptic IIR digital filter: top, spectrum of full precision filtered noise; bottom, spectrum of finite precision filtered noise

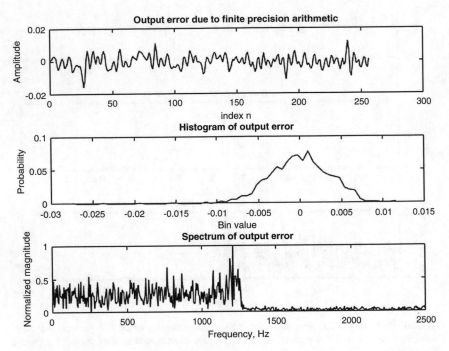

Fig. 8.52 Histogram of the error sequence due to arithmetic error using a cascade structure elliptic IIR digital filter: top, error sequence; middle, histogram; bottom, spectrum of the error sequence

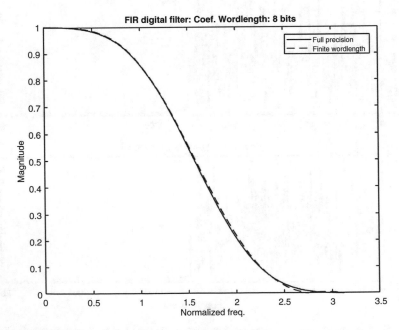

Fig. 8.53 Magnitude of frequency response of an 11th-order lowpass FIR digital filter

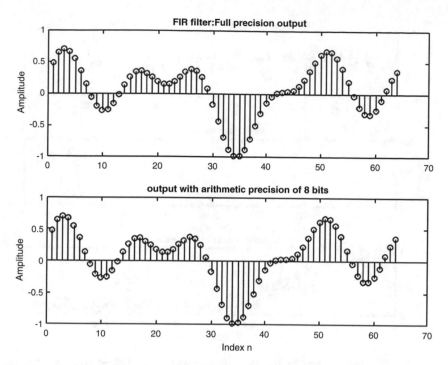

Fig. 8.54 Filtered output of the lowpass FIR filter whose frequency response is shown in Fig. 8.53: top, full precision; bottom, finite precision

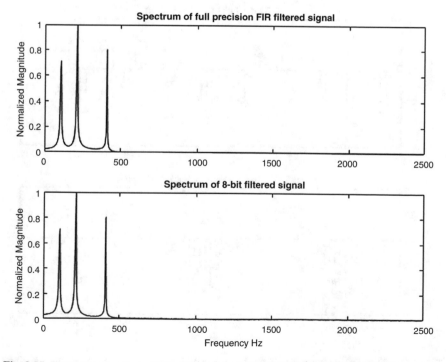

Fig. 8.55 Frequency spectrum of filtered sequence in Fig. 8.54. Top, full precision; bottom, finite precision

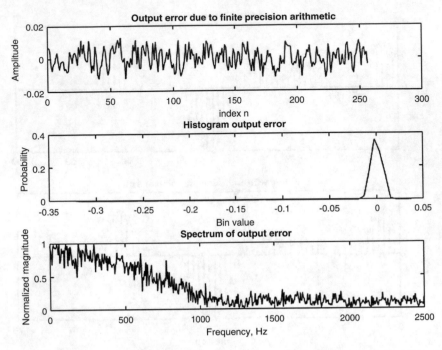

Fig. 8.56 Noise due to arithmetic error: top, error sequence; middle, histogram of error sequence; bottom, spectrum of error sequence

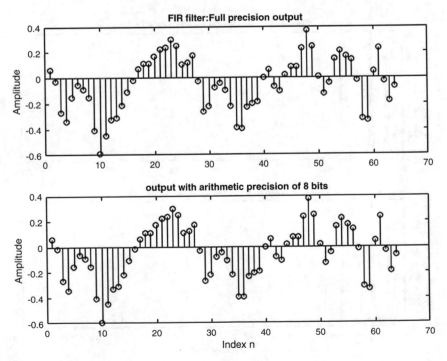

Fig. 8.57 Filtered Gaussian noise of the lowpass FIR filter: top, full precision; bottom, finite precision

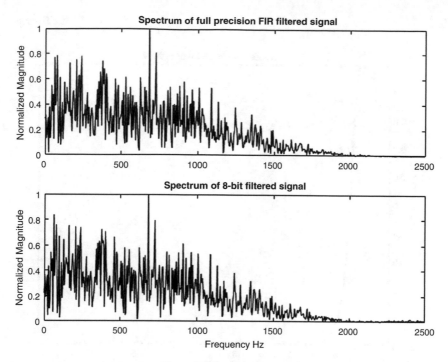

Fig. 8.58 Spectrum of filtered noise due to arithmetic error using the FIR digital filter: top: spectrum of full precision filtered noise; bottom, spectrum of finite precision filtered noise

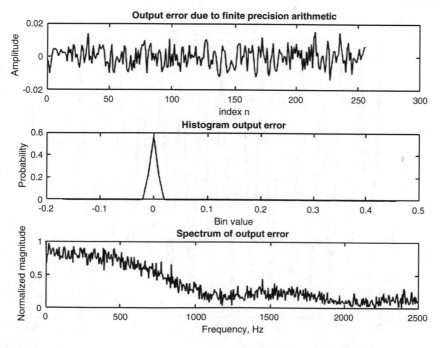

Fig. 8.59 Histogram of the error sequence due to arithmetic error using the FIR digital filter: top, error sequence; middle, histogram; bottom, spectrum of the error sequence

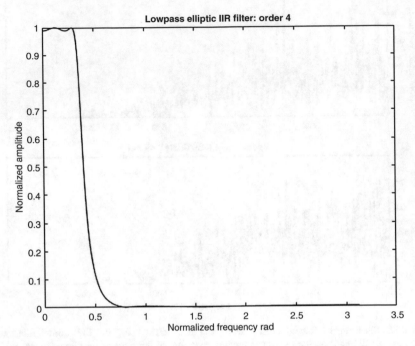

Fig. 8.60 Magnitude of the frequency response of the lowpass IIR elliptic digital filter used in the above example

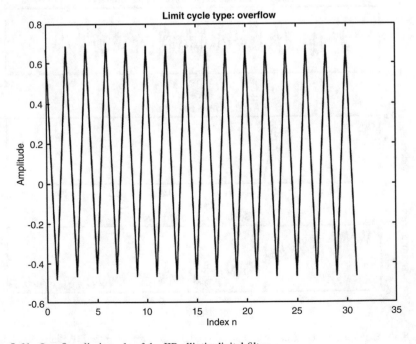

Fig. 8.61 Overflow limit cycle of the IIR elliptic digital filter

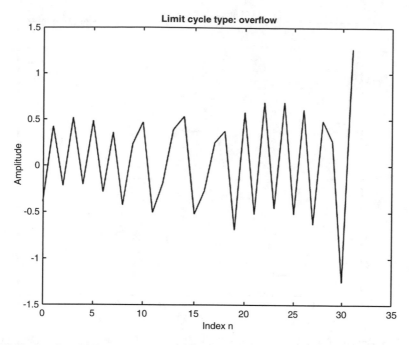

Fig. 8.62 Overflow limit cycle of the same elliptic filter when the M-file is run at a different time

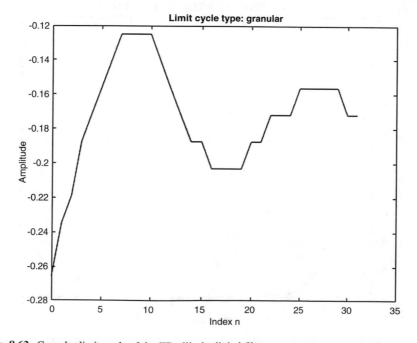

Fig. 8.63 Granular limit cycle of the IIR elliptic digital filter

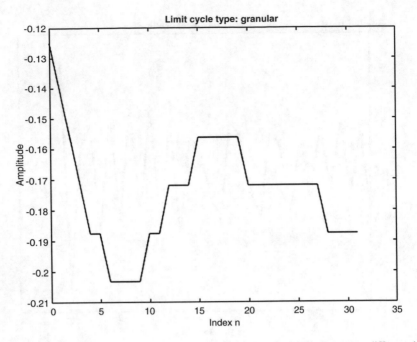

Fig. 8.64 Granular limit cycle of the same elliptic filter when the M-file is run at a different time

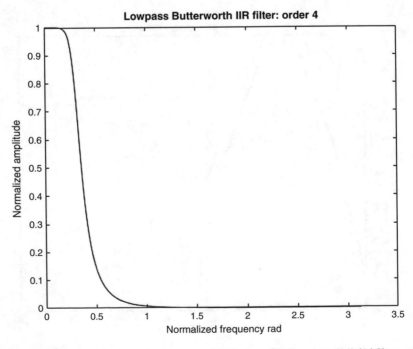

Fig. 8.65 Magnitude of the frequency response of the lowpass IIR Butterworth digital filter used in the above example

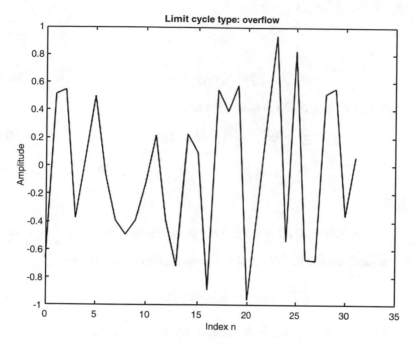

Fig. 8.66 Overflow limit cycle of the IIR Butterworth digital filter

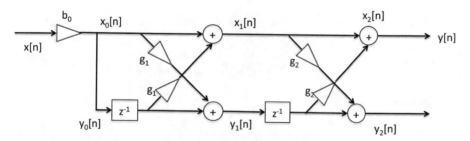

Fig. 8.67 Signal flow diagram of a second-order FIR lattice structure

$$X_2(z) = X_1(z) + g_2 z^{-1} Y_1(z) \tag{8.64a}$$

$$Y_2(z) = g_2 X_1(z) + z^{-1} Y_1(z) \tag{8.64b}$$

$$Y(z) = X_2(z) \tag{8.65}$$

Using (8.62a) and (8.62b) in (8.63), we have

$$X_1(z) = b_0 X(z)\left(1 + g_1 z^{-1}\right) \tag{8.66a}$$

$$Y_1(z) = b_0 X(z)\left(g_1 + z^{-1}\right) \tag{8.66b}$$

Similarly, by using (8.66) in (8.64), we get

$$X_2(z) = b_0 X(z)\left\{1 + g_1(1 + g_2)z^{-1} + g_2 z^{-2}\right\} \tag{8.67a}$$

$$Y_2(z) = b_0 X(z)\left\{g_2 + g_1 z^{-1}(1 + g_2) + z^{-2}\right\} \tag{8.67b}$$

and

$$Y(z) = X_2(z) = b_0 X(z)\left\{1 + g_1(1 + g_2)z^{-1} + g_2 z^{-2}\right\} \tag{8.68}$$

The overall transfer function is found from (8.68) and is given by

$$\frac{Y(z)}{X(z)} = b_0\left(1 + g_1[1 + g_2]z^{-1} + g_2 z^{-2}\right) \tag{8.69}$$

Since the transfer function in (8.69) must be the same as the transfer function in (8.61), we then find the reflection coefficients in the lattice structure in terms of the impulse response sequence of the FIR filter by solving the following equations.

$$b_0 g_1(1 + g_2) = b_1 \tag{8.70a}$$

$$b_0 g_2 = b_2 \Rightarrow g_2 = \frac{b_2}{b_0} \tag{8.70b}$$

Therefore,

$$g_1 = \frac{b_1}{b_0(1 + g_2)} = \frac{b_1}{b_0 + b_2} \tag{8.71}$$

By generalizing the lattice structure in Fig. 8.67 to a pth-order FIR filter, we can write the intermediate sequences in recursive equations as described by

$$X_0(z) = b_0 X(z) \tag{8.72a}$$

$$Y_0(z) = X_0(z) \tag{8.72b}$$

$$X_i(z) = X_{i-1}(z) + g_i z^{-1} Y_{i-1}(z) \tag{8.72c}$$

$$Y_i(z) = g_i X_{i-1}(z) + z^{-1} Y_{i-1}(z), 1 \le i \le p \tag{8.72d}$$

$$Y(z) = X_p(z) \tag{8.72e}$$

The general form of the transfer function of a pth-order FIR filter takes the form

$$H(z) = b_0 + b_1 z^{-1} + b_2 z^{-2} + \cdots + b_p z^{-p} \tag{8.73}$$

Then, there exists a procedure to determine the reflection coefficients $\{g_i, 1 \le i \le p\}$ in the lattice structure as described below.

1. By extracting the coefficient b_0, express $H(z)$ as $H(z) = b_0 A_p(z)$. Determine the pth multiplier by first determining $B_p(z) = z^{-p} A_p(z^{-1})$ and then evaluating

$$g_p = \lim_{z \to \infty} B_p(z)$$

2. For $i = p : -1 : 2$

$$\left\{ \begin{array}{l} \\ A_{i-1}(z) = \dfrac{A_i(z) - g_i B_i(z)}{1 - g_i^2} \\[2mm] B_{i-1}(z) = z^{-(i-1)} A_{i-1}(z^{-1}) \\[2mm] g_{i-1} = \lim_{z \to \infty} B_{i-1}(z) \\ \end{array} \right.$$

$$\}$$

Note that when the argument of A_i or B_i is z, then the corresponding polynomial is in descending power of z^{-1}. When the argument is z^{-1} in A_i or B_i, then the corresponding polynomial will be in ascending power of z. For this procedure to work properly, the reflection coefficients in the lattice structure should satisfy the condition $|g_k| \ne 1$, $1 \le k \le p$.

Example 8.8 Determine the reflection coefficients of the lattice structure realizing the FIR transfer function of (10.61) using the above procedure.

Solution From the given transfer function, we obtain

$$A_2(z) = 1 + \frac{b_1}{b_0} z^{-1} + \frac{b_2}{b_0} z^{-2}$$

Therefore,

$$B_2(z) = z^{-2} A_2(z^{-1}) = z^{-2} \left(1 + \frac{b_1}{b_0} z + \frac{b_2}{b_0} z^2 \right) = z^{-2} + \frac{b_1}{b_0} z^{-1} + \frac{b_2}{b_0}$$

And

$$g_2 = \lim_{z \to \infty} B_2(z) = \frac{b_2}{b_0}$$

Following the above procedure, we next find

$$A_1(z) = \frac{\left(1 - g_2^2\right) + \frac{b_1}{b_0}\left(1 - g_2\right)z^{-1}}{1 - g_2^2}$$

$$B_1(z) = z^{-1}A_1\left(z^{-1}\right) = \frac{z^{-1}\left(1 - g_2^2\right) + \frac{b_1}{b_0}\left(1 - g_2\right)}{1 - g_2^2}$$

Finally,

$$g_1 = \lim_{z \to \infty} B_1(z) = \frac{b_1}{b_0}\frac{1}{1 + g_2} = \frac{b_1}{b_0 + b_2}$$

Example 8.9 MATLAB Example

MATLAB has the function *tf2latc* to calculate the reflection coefficients of the lattice form of an FIR digital filter. It takes the coefficients of the FIR filter transfer function and returns the reflection coefficients of the corresponding lattice structure. The coefficients correspond to the descending powers of z^{-1}. The same MATLAB function can also be used to determine the reflection coefficients of either a maximum or minimum phase FIR filter. The transfer function of a fourth-order FIR filter is specified by

$$H(z) = 2 + 23z^{-1} + 73z^{-2} + 43z^{-3} - 15z^{-4}$$

We invoke the function $G = tf2latc(B)$, where $B = [2, \quad 23, \quad 73, \quad 43, \quad -15]$ is the vector of coefficients of H(z). The reflection coefficients returned by MATLAB are found to be 0.9698, 4.1782, -1.9502, and -7.5. The same function can also be used to calculate the reflection coefficients of an IIR digital lattice filter. The input arguments will be the coefficients of the numerator and denominator polynomials of the transfer function of the IIR filter. Again, the coefficients correspond to the respective polynomials in descending powers of z^{-1}. In this example, the IIR filter is a fourth-order Butterworth filter with a cutoff frequency of 0.3π. The resulting transfer function is found to be

$$H(z) = \frac{0.0186 + 0.0743z^{-1} + 0.1114z^{-2} + 0.0743z^{-3} + 0.0186z^{-4}}{1 - 1.5704z^{-1} + 1.2756z^{-2} - 0.4844z^{-3} + 0.0762z^{-4}}$$

The vectors of coefficients of the numerator and denominator polynomials are

$$B = [0.0186, 0.0743, 0.1114, 0.0743, 0.0186]$$

$$A = [1, -1.5704, 1.2756, -0.4844, 0.0762]$$

After the function call $K = tf2latc(B,A)$, the reflection coefficients are found to be

$$K = [-0.7459, 0.7158, -0.3669, 0.0762]$$

The M-file to solve this problem is named *FIR_lattice.m*.

8.6   Summary

We have described the signal flow graphs of both IIR and FIR digital filters as aids to their implementations. Signal flow graphs allow us to come up with different possible structures to realize the transfer function of a digital filter with each structure having its own advantages and disadvantages. Specifically, we defined three different structures for an IIR digital filter, namely, Direct Form I and II, cascade form, and parallel form. The Direct Form I structure uses more number of delay elements than the filter order and, so, is called a noncanonical structure. On the other hand, a direct form II structure is canonical in the delay elements. The cascade form involves the tandem connection of first- and/or second-order sections, which can be in either Direct Form I or II. In the cascade form, the output of the first section is the input to the second section and so on. The input to the first section is the overall input, and the output of the last section is the overall output. The parallel form consists of first- and/or second-order sections connected in parallel. The input to all the sections is the same, and the outputs are added to obtain the true overall output. An FIR filter can be realized in direct form or cascaded form. However, due to the linearity in phase of an FIR filter, one can take advantage of the symmetry/antisymmetry in its impulse response in reducing the number of multipliers. We showed several examples to illustrate the development of signal flow graphs for both FIR and IIR digital filters. We described fixed-point and floating-point representations of numbers used in implementing digital filters. Using fixed-point representation of numbers, we discussed the effect of finite word length coefficients on the frequency response of the digital filters. This was solidified by a couple of examples using MATLAB M-files and Simulink. Next we dealt with implementing digital filters using fixed-point arithmetic operations. Practical digital filters use finite word lengths to represent the filter coefficients and to perform arithmetic operations as well. It is, therefore, important to ascertain the performance of the designed digital filters before the actual implementation. To this end, we showed two examples based on MATLAB that involve IIR and FIR digital filters. Finally, we briefly described the limit cycle that may be present in IIR digital filters using finite word length arithmetic. In the next chapter, we will learn the efficient computation of the DFT through *fast Fourier transform* (FFT). Since DFT is used extensively in digital signal processing applications, it is a must for us to know how to compute it efficiently so that it can be used in real time.

8.7 Problems

1. Develop the canonic Direct Form II signal flow graph to realize the transfer function $H(z) = \frac{3+1.5z^{-1}}{1+1.2z^{-1}+0.35z^{-2}}$.

2. Realize the canonic Direct Form II structure of the transfer function $H(z) = \frac{2z^2+z}{z^2-1.0z+0.21}$.

3. Show the Direct Form II canonic realization of the transfer function $H(z) = \frac{3z^2-0.6z}{z^2-0.8z+0.15}$.

4. The three functions $H_1(z) = \frac{1-0.6z^{-1}}{1+0.25z^{-1}}$, $H_2(z) = \frac{0.2+z^{-1}}{1+0.3z^{-1}}$, and $H_3(z) = \frac{2}{1+0.25z^{-1}}$ are cascaded in that order. (a) Determine the transfer function of the overall system, (b) determine the difference equation characterizing the overall system, and (c) develop a parallel form realization of the overall system.

5. The transfer function of an IIR digital filter is expressed as the product of first- and/or second-order functions $H(z) = \prod_{m=1}^{M} \frac{B_m(z)}{A_m(z)}$, where the coefficients of the numerator and denominator polynomials are real. Determine the number of cascade realizations that are possible by pairing different poles and zeros as well as different orderings of the sections.

6. A fifth-order elliptic lowpass IIR digital filter has the transfer function described by $H(z) = \frac{0.0447+0.0547z^{-1}+0.0936z^{-2}+0.0936z^{-3}+0.0547z^{-4}+0.0447z^{-5}}{1-2.035z^{-1}+2.657z^{-2}-1.9297z^{-3}+0.8727z^{-4}-0.1791z^{-5}}$. (a) Express the transfer function in factored form. (b) Determine two different cascade forms of the transfer function. (c) Come up with two different parallel forms of the transfer function. (d) Use MATLAB to solve the problem. (e) Realize each second-order section in direct form II.

7. For the IIR filter in Problem 6, plot the magnitude of the frequency response in dB using 8 bits to represent the coefficients of the transfer function in fixed-point format. Compare the result with that using full precision for the coefficients. See what happens if the coefficient word length is reduced to 6 bits. Use MATLAB to solve the problem. Repeat the problem when the filter is realized in cascade and parallel forms.

8. Implement the IIR filter of Problem 6 using fixed-point arithmetic. Use an input signal $x[n] = \sin(0.15\pi n)u[n]$. Calculate the mean square error of the noise in dB at the output due to finite word length arithmetic. Determine the minimum bit width necessary for acceptable performance. Use full precision for the filter coefficients. You need to use MATLAB to solve the problem.

9. Compare the output noise due to arithmetic errors that you calculated in Problem 8 with that when the IIR filter in Problem 6 is implemented in (a) cascade form and (b) in parallel form.

10. Realize the FIR filter transfer function $H(z) = (1 + 0.4z^{-1})^4(1 - 0.2z^{-1})^2$ in (a) direct form and (b) cascade of three second-order sections.

11. In playing a DVD, it is determined that the time delay introduced by the FIR filter used in the reconstruction of the audio signal should be no greater than

6 ms for acceptable time synchronization with the video signal. If the sampling
frequency of the audio signal is 48 kHz, determine the maximum number of taps
in the FIR filter so as to satisfy the delay requirement.

12. The impulse response of a linear-phase lowpass FIR filter of order 12 has the
following sequence:

$$h[n] = \left\{ \begin{array}{l} 0.082, \, -0.2251, \, -0.3784, \, -0.083, 0.6438, 1.4181, 1.75, 1.4181, 0.6438, \\ \qquad\qquad -0.083, \, -0.3784, \, -0.2251, 0.082 \end{array} \right\}.$$

What is the value of the frequency response of the filter at DC? If the fourth and
the tenth coefficients are changed so that the new DC value is unity, what are the
values of those coefficients?

13. Design a 16th-order lowpass FIR filter with a passband edge at 0.3π, stopband
edge at 0.6π, and a minimum stopband attenuation of 40 dB based on Parks-
McClellan method using MATLAB.

14. Design a 32nd-order bandpass FIR digital filter using Parks-McClellan method
to satisfy the following specifications: passband edges at 0.3π and 0.5π and
stopband edges at 0.1π and 0.7π.

15. The following diagram depicts an FIR filter used in the video-processing chip
made by the Texas Instruments, Inc. The coefficient k controls the magnitude of
the frequency response of the filter and is user-defined. (a) Find the input-output
relation in the discrete-time domain corresponding to the filter structure shown.
(b) Is the phase response of the FIR filter linear? (c) What is the group delay of
the filter in samples?

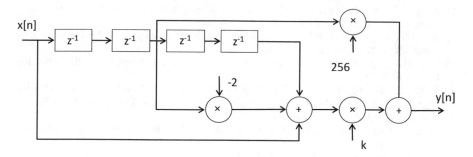

Signal flow graph of Problem 15

References

1. Agarwal RC, Burrus CS (1975) New recursive digital filter structures having very low sensitivity and roundoff noise. IEEE Trans Circ Syst, CAS-22(12): 921–927
2. Burrus CS (1972) Block realization of digital filters. IEEE Trans Audio Electroacoust
AU-20:230–235

3. Buttner M (1977) Elimination of limit cycles in digital filters with very low increase in the quantization noise. IEEE Trans Circ Syst CAS-24:300–304
4. Chan DSK, Rabiner LR (1973) Analysis of quantization errors in the direct form for finite impulse response digital filters. IEEE Trans Audio Electroacoust AU-21:354–366
5. Chang T-L, White SA (1981) An error cancellation digital-filter structure and its distributed-arithmetic implementation. IEEE Trans Circ Syst CAS-28:339–342
6. Classen TACM, Mecklenbrauker WFG, Peek JBH (1973) Some remarks on the classifications of limit cycles in digital filters. Philips Res Rep 28:297–305
7. Crochiere RE, Oppenheim AV (1975) Analysis of linear digital networks. Proc IEEE 62:581–595
8. Dutta Roy SC (2007) A new canonic lattice realization of arbitrary FIR transfer functions. IETE J Res 53:13–18
9. Dutta Roy SC (2008) A note on canonic lattice realization of arbitrary FIR transfer functions. IETE J Res 54:71–72
10. Ebert PM, Mazo JE, Taylor MG (1969) Overflow oscillations in digital filters. Bell Syst Tech J 48:2999–3020
11. Fettweis A (1971) Digital filter structures related to classical filter networks. Archiv fur Elektrotechnik und Ubertragungstechnik 25:79–81
12. Fettweis A (1975) On adapters for wave digital filters. IEEE Trans Acoust Speech Sig Process ASSP-23(6):516–525
13. Gray AH Jr, Markel JD (1973) Digital lattice and ladder filter synthesis. IEEE Trans Audio Electroacoust AU-21:491–500
14. Jackson LB (1969) An analysis of limit cycles due to multiplicative rounding in recursive digital filters. In: Proceedings, 7th Allerten conference on circuit and system theory, Monticello, IL, pp 69–78
15. Jackson LB (1970) On the interaction of roundoff noise and dynamic range in digital filters. Bell Syst Tech J 49:159–184
16. Jackson LB (1970) Roundoff-noise analysis for fixed-point digital filters realized in cascade or parallel form. IEEE Trans Audio Electroacoust AU-18:107–122
17. Jiang Z, Willson AN Jr (1997) Efficient digital filtering architectures using pipelining/interleaving. IEEE Trans Circuits Syst Part II 44:110–119
18. Kan EPF, Aggarwal JK (1971) Error analysis in digital filters employing floating-point arithmetic. IEEE Trans Circ Theory CT-18:678–686
19. Laroche L (1999) A modified lattice structure with pleasant scaling properties. IEEE Trans Sig Process 47:3423–3425
20. Lawrence VB, Mina KV (1978) Control of limit cycle oscillations in second-order recursive digital filters using constrained random quantization. IEEE Trans Acoust Speech Sig Process ASSP-26:127–134
21. Liu B, Kaneko T (1969) Error analysis of digital filters realized in floating-point arithmetic. Proc IEEE 57:1735–1747
22. Long JJ, Trick TN (1973) An absolute bound on limit cycles due to roundoff errors in digital filters. IEEE Trans Audio Electroacoust AU-21:27–30
23. Makhoul J (1978) A class of all-zero lattice digital filters: properties and applications. IEEE Trans Acoust Speech Sig Process 26:304–314
24. Mills WL, Mullis CT, Roberts RA (1978) Digital filter realizations without overflow oscillations. IEEE Trans Acoust Speech Sig Process ASSP-26:334–338
25. Mitra SK, Sherwood RJ (1973) Digital ladder networks. IEEE Trans Audio Electroacoust AU-21:30–36
26. Mitra SK, Hirano K, Sakaguchi H (1974) A simple method of computing the input quantization and the multiplication round off errors in digital filters. IEEE Trans Acoust Speech Sig Process ASSP-22:326–329
27. Mitra SK, Sherwood RJ (1974) Estimation of pole-zero displacements of a digital filter due to coefficient quantization. IEEE Trans Circ Syst CAS-21:116–124

28. Mitra SK, Mondal K, Szczupak J (1977) An alternate parallel realization of digital transfer functions. Proc IEEE (Lett) 65:577–578
29. Rabiner LR, Crochiere RE (1975) A novel implementation for narrow-band FIR digital filters. IEEE Trans Acoust Speech Sig Process 23(5):457–464
30. Renner K, Gupta SC (1973) On the design of wave digital filters with low sensitivity properties. IEEE Trans Circ Theory CT-20:555–567
31. Sandberg IW (1967) Floating-point-roundoff accumulation in digital filter realization. Bell Syst Tech J 46:1775–1791
32. Swamy MNS, Thyagarajan KS (1975) A new type of wave digital filter. J Frankl Inst 300 (1):41–58
33. Szczupak J, Mitra SK (1975) Digital filter realization using successive multiplier – extraction approach. IEEE Trans Acoust Speech Sig Process ASSP-23:235–239
34. Thyagarajan KS (1977) One and two-dimensional wave digital filters with low coefficient sensitivities. Ph.D. thesis, Concordia University, Montreal, Quebec, Canada

Chapter 9
Fast Fourier Transform

9.1 Brute-Force Computation of DFT

As was defined in a previous chapter, the discrete Fourier transform (DFT) is the sampled version of the discrete-time Fourier transform (DTFT), with a finite number of samples taken around the unit circle in the Z-domain. DFT is very useful in the analysis of discrete-time signals and linear time-invariant discrete-time systems. It is, therefore, necessary to determine the computational complexity in performing an N-point DFT of a sequence so that we may be able to come up with a more efficient computational algorithm. To this end, let us first evaluate the computational complexity of computing an N-point DFT using brute-force method. Consider an N-point discrete-time sequence $\{x[n]\}, 0 \le n \le N - 1, N \in Z$. Its DFT is given by

$$X[k] = \sum_{n=0}^{N-1} x[n] W_N^{nk}, 0 \le k \le N - 1, \tag{9.1}$$

where

$$W_N^{nk} \equiv e^{-j\frac{2\pi}{N}nk} \tag{9.2}$$

From Eq. (9.2), we notice that W_N^{nk} is complex, and so X[k] is also complex. In Eq. (9.1), there are N terms $x[n] W_N^{nk}$ for each value of k in the summation. Therefore, there are N number of complex multiplications for each k. Note that the sequence x[n] may be real or complex. The summation then has N-1 number of complex additions for each k. Therefore, to compute an N-point DFT, we need N^2 complex multiplications and $N(N - 1)$ complex additions. In practice, N is

Electronic supplementary material: The online version of this article (https://doi.org/10.1007/978-3-319-76029-2_9) contains supplementary material, which is available to authorized users.

very large. Therefore, the number of complex additions is very nearly equal to N^2
. It is customary to represent the computational complexity by $O(N^2)$. In other
words, the number of arithmetic operations increases as the square of the length
of the sequence if we use the brute-force technique. Each arithmetic operation is
equal to one complex multiplication and one complex addition. For instance, a
256-point DFT needs 2^{16} number of complex arithmetic operations. If the
sequence length is increased to 1024 points, then the computational load
becomes 2^{20}, which is more than a million! This becomes a problem for real-
time implementation of DFT. Fortunately there are algorithms, which require O
$\left(\frac{N}{2}\log_2 N\right)$ number of arithmetic operations to compute an N-point DFT. Such
algorithms are known as *fast algorithms*. One such fast algorithm is called the
fast Fourier transform (FFT) and is due to Cooley-Tukey fast algorithm.

9.2 Fast Fourier Transform

In this algorithm the sequence length N is assumed to be a positive integer power of
2. Hence the algorithm is known as radix-2 algorithm. There are two versions of the
fast algorithm, namely, *decimation-in-time* and *decimation-in-frequency* algorithms.
Both algorithms are fast algorithms meaning that they require $O\left(\frac{N}{2}\log_2 N\right)$ operations.
We will describe both algorithms and evaluate their computational complexities.

9.2.1 Decimation-in-Time FFT

In order to describe the algorithm with a signal flow graph, we will consider an eight-
point sequence. In general, N is a positive integer power of 2. If not, we will append
zeros to the sequence to make its length a power of 2. As the name implies, we first
divide the input discrete-time sequence into even-numbered and odd-numbered
sequences as defined by

$$x_1[n] = x[2n], 0 \le n \le \frac{N}{2} - 1 \tag{9.3a}$$

$$x_2[n] = x[2n+1], 0 \le n \le \frac{N}{2} - 1 \tag{9.3b}$$

The DFT of the input sequence $x[n]$ can be expressed as

$$X[k] = \sum_{n=0}^{N-1} x[n] W_N^{nk} = \sum_{n=0}^{\frac{N}{2}-1} x_1[n] W_N^{2nk} + \sum_{n=0}^{\frac{N}{2}-1} x_2[n] W_N^{(2n+1)k}, 0 \le k \le N-1 \tag{9.4}$$

However,

$$W_N^{2nk} = e^{-j\left(\frac{2\pi}{N}\right)2nk} = e^{-j\left(\frac{2\pi}{N/2}\right)nk} \equiv W_{\frac{N}{2}}^{nk}, \qquad (9.5a)$$

and

$$W_N^{(2n+1)k} = W_N^k W_N^{2nk} = W_N^k W_{\frac{N}{2}}^{nk} \qquad (9.5b)$$

Therefore, the N-point DFT of x[n] can be written as

$$X[k] = \sum_{n=0}^{\frac{N}{2}-1} x_1[n] W_{\frac{N}{2}}^{nk} + W_N^k \sum_{n=0}^{\frac{N}{2}-1} x_2[n] W_{\frac{N}{2}}^{nk} \qquad (9.6)$$

We see that each summation on the right-hand side of Eq. (9.6) corresponds to the DFT of an $\frac{N}{2}$-point sequence. By denoting

$$X_1[k] = \sum_{n=0}^{\frac{N}{2}-1} x_1[n] W_{\frac{N}{2}}^{nk}, \qquad (9.7a)$$

$$X_2[k] = \sum_{n=0}^{\frac{N}{2}-1} x_2[n] W_{\frac{N}{2}}^{nk} \qquad (9.7b)$$

we can write Eq. (9.4) as

$$X[k] = X_1[k] + W_N^k X_2[k], 0 \le k \le N - 1 \qquad (9.8)$$

In Eq. (9.8), there is something more than meets the eye. We notice that

$$W_N^{\left(k+\frac{N}{2}\right)} = W_N^k W_N^{\frac{N}{2}} = W_N^k e^{-j\left(\frac{2\pi}{N}\right)\frac{N}{2}} = W_N^k e^{-j\pi} = -W_N^k \qquad (9.9)$$

Therefore, the N-point DFT in terms of the two N-/2-point DFTs can be rewritten as

$$X[k] = X_1[k] + W_N^k X_2[k], 0 \le k \le \frac{N}{2} - 1 \qquad (9.10)$$

$$X\left[k + \frac{N}{2}\right] = X_1[k] - W_N^k X_2[k], 0 \le k \le \frac{N}{2} - 1$$

Assuming that both N-/2-point sequences $X_1[k]$ and $X_2[k]$ are available, we find from Eq. (9.10) that there are N/2 complex multiplications and N complex additions in computing the N-point DFT. The task is not over yet. We need to calculate the two N-/2-point DFTs. We can follow the same steps that we used in calculating X [k]. That is, we divide $x_1[n]$ and $x_2[n]$ into the respective even- and odd-numbered sequences of length N/4 and then compute the DFTs $X_1[k]$ and $X_2[k]$ in terms of the

N-/4-point DFTs as in (9.10). The number of complex multiplications and additions involved in computing the N-/2-point DFT $X_1[k]$ is N/4 and N/2, respectively. Similarly there are N/4 complex multiplications and N/2 complex additions in the computation of $X_2[k]$. We can continue this process of dividing the input sequence into two odd-numbered and even-numbered sequences and calculate the DFT in terms of the two half-length DFTs until we are left with two points. The DFT of a two-point sequence requires only additions and no multiplications. The conclusion is this: For $N = 2^m$, there are $m = \log_2 N$ stages; each stage has N/2 complex multiplications and N complex additions. Therefore, the total number of complex multiplications required in computing the N-point DFT using the decimation-in-time FFT is $\frac{N}{2}\log_2 N$ and the total number of complex additions is $N\log_2 N$. Note that a multiplication operation is costlier than an addition/subtraction operation. Thus the decimation-in-time FFT algorithm is a fast algorithm. We will illustrate the computational procedure by a signal flow graph using an eight-point sequence for simplicity.

Signal Flow Graph of an Eight-Point Decimation-in-Time FFT Algorithm

1. Divide the input sequence into even-numbered and odd-numbered sequences, each of length four points. The even-numbered sequence is $\{x[0], x[2], x[4], x[6]\}$, and the odd-numbered sequence is $\{x[1], x[3], x[5], x[7]\}$. These two sequences are the inputs to the two four-point DFT blocks, as shown in Fig. 9.1a.

2. Multiply the bottom half of the DFT $\{X[4], X[5], X[6], X[7]\}$ by the corresponding weights $\{W_8^0, W_8^1, W_8^2, W_8^3\}$. Add the top half of the DFT $\{X[0], X[1], X[2], X[3]\}$ to the weighted bottom half of the four-point DFT to obtain the first half of the eight-point DFT samples. Subtract the weighted bottom four DTF samples from the top four DFT samples to obtain the second half of the eight-point DFT samples as shown in Fig. 9.1a.

3. Next, we perform the four-point DFT. Divide the four-point even-numbered sequence into two two-point even- and odd-numbered sequences, which are, respectively, $\{x[0], x[4]\}$ and $\{x[2], x[6]\}$. Feed the two two-point sequences to the two-point DFT blocks, as shown in Fig. 9.1b. Similarly, divide the four-point odd-numbered sequence obtained in step 1 into two two-point even- and odd-numbered sequences, which are $\{x[1], x[5]\}$ and $\{x[3], x[7]\}$, respectively. Feed the two sequences to the bottom two two-point DFT blocks. Multiply the two DFT samples of the second DFT block by the weights $\{W_4^0, W_4^1\}$, and add them to the DFT samples of the first two-point DFT block to obtain the output points. The second two four-point DFT samples are obtained by subtracting the two weighted DFT samples from the DFT samples of the first two-point DFT block. Perform the same operations on the bottom two two-point DFT outputs. These operations are shown in Fig. 9.1b. Note that the outputs of the first two two-point DFT blocks are the inputs to the first four-point DFT block and the outputs of the bottom two two-point DFT blocks are the inputs to the second four-point DFT block.

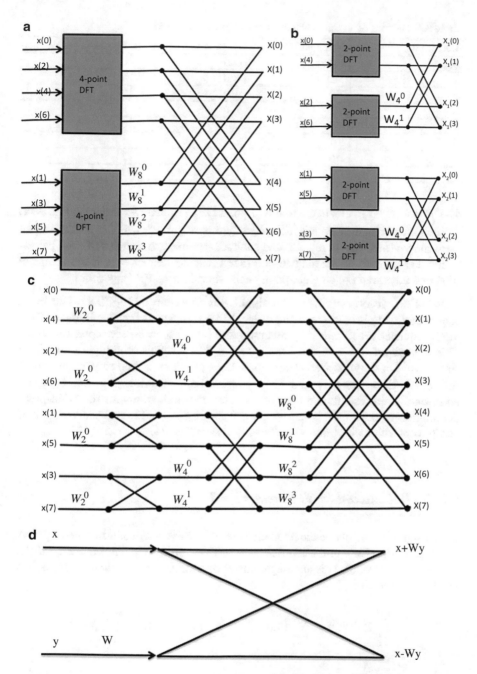

Fig. 9.1 Signal flow graph of an eight-point DFT using decimation-in-time FFT algorithm: (**a**) two four-point DFTs, (**b**) four two-point DFTs, (**c**) eight-point DFT, (**d**) butterfly

Table 9.1 Ordering of an eight-point input sequence using bit reversal

Decimal no.	Binary no.	Bit-reversed	Reordered decimal no.
0	000	000	0
1	001	100	4
2	010	010	2
3	011	110	6
4	100	001	1
5	101	101	5
6	110	011	3
7	111	111	7

4. The last step is to obtain the four two-point DFT samples. As shown in Fig. 9.1c, each second input sample is multiplied by the weight W_2^0, which is actually equal to unity. Thus, we simply add and subtract the first two samples of each pair of input samples to obtain the two-point DFT samples. Figure 9.1c is the complete signal flow graph of an eight-point decimation-in-time FFT algorithm.

What do we observe from the signal flow diagram of Fig. 9.1? The input sequence $\{x[n]\}$ is in bit-reversed order. What it means is that the indices of the input sequence are obtained by reversing the bits of their binary representations. This is depicted in Table 9.1. There are $log_2 8 = 3$ stages of computation. Each stage has $\frac{8}{2} = 4$ *butterflies*. The signal flow graph of a butterfly is shown in Fig. 9.1d. Each butterfly has one complex multiplication and two complex additions. Therefore, the number of complex multiplications for the eight-point decimation-in-time FFT is $\#stages * \# Butterflies = 12$, and the number of complex additions is $\#stages * \# Butterflies * 2 = 24$.

9.2.2 Decimation-in-Frequency FFT

In this algorithm we divide the frequency points into even- and odd-numbered points and perform the DFT of the input sequence. For a given N-point sequence $\{x[n]\}$, $0 \le n \le N - 1$, where N is an integer power of two, its even-numbered DFT points can be expressed as

$$X[2k] = \sum_{n=0}^{N-1} x[n] W_N^{n2k} = \sum_{n=0}^{\frac{N}{2}-1} x[n] W_N^{n2k} + \sum_{n=0}^{\frac{N}{2}-1} x\left[n + \frac{N}{2}\right] W_N^{(n+\frac{N}{2})2k}, 0 \le k$$

$$\le \frac{N}{2} - 1 \tag{9.11}$$

However, we observe that

$$W_N^{n2k} = e^{-j\left(\frac{2\pi}{N}\right)n2k} = e^{-j\left(\frac{2\pi}{\frac{N}{2}}\right)nk} \equiv W_{\frac{N}{2}}^{nk}, \qquad (9.12)$$

$$W_N^{\left(n+\frac{N}{2}\right)2k} = W_N^{n2k}W_N^{Nk} = W_{\frac{N}{2}}^{nk}e^{-j\left(\frac{2\pi}{N}\right)Nk} = W_{\frac{N}{2}}^{nk}e^{-j2\pi k} = W_{\frac{N}{2}}^{nk} \qquad (9.13)$$

We can, therefore, rewrite the even-numbered DFT points in Eq. (9.11) as

$$X[2k] = \sum_{n=0}^{\frac{N}{2}-1}\left(x[n] + x\left[n+\frac{N}{2}\right]\right)W_{\frac{N}{2}}^{nk}, 0 \le k \le \frac{N}{2} - 1 \qquad (9.14)$$

From Eq. (9.14), we deduce that the even-numbered DFT points are actually the N-/2-point DFT of the sequence $\left(x[n] + x\left[n+\frac{N}{2}\right]\right), 0 \le n \le \frac{N}{2} - 1$. We also find from (9.14) that the number of complex multiplications and additions required to calculate the even-numbered DFT points is N/2 and N, respectively. We have only found half the number of DFT points. The other half, namely, the odd-numbered points, can be found from

$$X[2k+1] = \sum_{n=0}^{\frac{N}{2}-1} x[n]W_N^{n(2k+1)} + \sum_{n=0}^{\frac{N}{2}-1} x\left[n+\frac{N}{2}\right]W_N^{\left(n+\frac{N}{2}\right)(2k+1)}, 0 \le k \le \frac{N}{2} - 1$$

$$(9.15)$$

However, since

$$W_N^{n(2k+1)} = W_N^n W_{\frac{N}{2}}^k, \qquad (9.16a)$$

$$W_N^{\left(n+\frac{N}{2}\right)(2k+1)} = W_N^{n(2k+1)}W_N^{Nk}W_N^{\frac{N}{2}} = -W_N^n W_{\frac{N}{2}}^{nk}, \qquad (9.16b)$$

we have

$$X[2k+1] = \sum_{n=0}^{\frac{N}{2}-1}\left\{\left(x[n] - x\left[n+\frac{N}{2}\right]\right)W_N^n\right\}W_{\frac{N}{2}}^{nk}, 0 \le k \le \frac{N}{2} - 1 \qquad (9.17)$$

From the above equation, it is clear that the odd-numbered DFT points correspond to the DFT of the N-/2-point sequence $\left\{x[n] - x\left[n+\frac{N}{2}\right]\right\}W_N^n, 0 \le n \le \frac{N}{2} - 1$. Similar to the even-numbered DFT points, the number of complex multiplications and additions required to compute the odd-numbered DFT points is N/2 and N, respectively. In the next step, we divide the N-/2-point DFTs into even- and odd-numbered points and calculate the respective DFTs. We can continue this process until we are left with two points. The total number of stages is log_2N with each stage having N/2 multiplications and N additions. Thus, the computational complexity of the decimation-in-frequency FFT algorithm is the same as that of the decimation-in-time FFT algorithm. The signal flow graph of an eight-point decimation-in-frequency FFT algorithm can be obtained directly from that of the decimation-in-time algorithm by simply transposing the signal flow graph of Fig. 9.1c. To transpose a signal flow graph, we need to change the directions of the arrows, adders will become branch-off points, and the branch-off points will become adders. The multiplier weights will remain the same.

Example 9.1 CPU Time for Calculating an N-point DFT   In this example, let us compare the CPU time required to calculate an N-point DFT using brute force against the FFT. In MATLAB, we can start and stop a stopwatch to determine how long the CPU takes to compute an N-point DFT. Specifically the MATLAB functions *tic* starts the stopwatch and *toc* stops the stopwatch. So, we start the stopwatch just before starting to calculate the DFT and stop the stopwatch just after finishing the calculation. The time taken by the CPU to calculate the DFT by FFT and brute-force methods is plotted with respect to the DFT length N and is shown in Fig. 9.2a. Here N is a positive integer power of 2. Note that the CPU time is in microseconds for FFT computation and is in milliseconds for the brute-force method. We also note that the data in Fig. 9.2a is for a sinusoidal input sequence. In Fig. 9.2b, the CPU time when the input is a Gaussian random sequence is shown. As expected, the CPU time varies as the square of the DFT length N for the brute-force case and is approximately linear for the FFT case. The MATLAB M-file *fft_proc_time.m* calculates the CPU time required to compute the DFT of an input sequence using the brute-force method and FFT technique.

9.2.3   Inverse FFT

We have seen a fast algorithm to compute an N-point DFT that takes only $O\left(\frac{N}{2}log_2N\right)$ number of multiplications instead of N^2 number of multiplications required by the brute-force method. Is there such a fast algorithm to compute the inverse DFT? After all, what is the use of DFT without its inverse? First, let us look at the inverse DFT of an N-point DFT, which is defined by

$$x[n] = \frac{1}{N} \sum_{k=0}^{N-1} X[k] W_N^{-nk}, 0 \leq n \leq N-1 \qquad (9.18)$$

What we observe from Eq. (9.18) is that the right-hand side of the equation without the scaling factor 1/N is identical to the DFT in Eq. (9.1) with the following differences: the sequence is X[k] instead of x[n] and the multiplying factor is the complex conjugate of that in (9.1). Therefore, we should be able to use the same flow graphs, namely, decimation-in-time and decimation-in-frequency, to calculate the inverse DFT. Since we are using N as a positive integer power of 2, the scaling factor 1/N can be incorporated as 1/2 at each stage of the signal flow graph. Further,

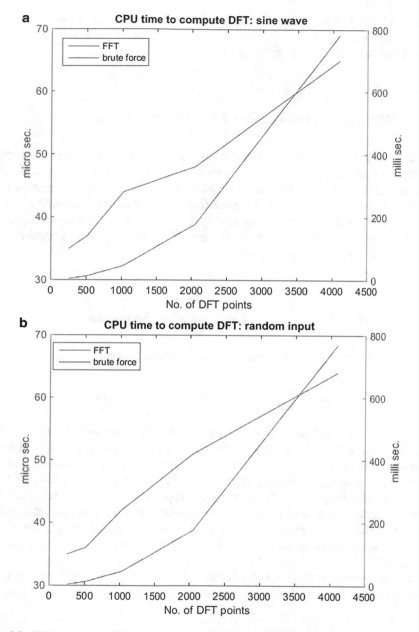

Fig. 9.2 CPU time versus DFT length N: (**a**) input sequence is deterministic; (**b**) input sequence is random

division by 2 is equivalent to 1-bit right shift; there is really no multiplication involved in incorporating the 1/N scale factor. Thus, the inverse DFT (IDFT) can also be accomplished by the FFT algorithm.

9.3 Spectral Analysis of Discrete-Time Sequences

In this section we will discuss the power spectral density of discrete-time sequences. Power spectral density is widely used in speech processing. It enables one to identify the characteristics of speech signals such as voiced and fricative sounds and silence. Speech compression is achieved by extracting these features and transmitting them instead of transmitting the entire speech signal. At the receiver, the speech signal is reconstructed using the transmitted features. Essentially, power spectrum of a discrete-time sequence is its signature.

9.3.1 Autocorrelation of a Discrete-Time Sequence

The autocorrelation of a discrete-time sequence $x[n]$ is defined as

$$c_{xx}[k] = \sum_{n=-\infty}^{\infty} x[n]x[n-k], k = 0, \pm 1, \pm 2, \ldots \qquad (9.19)$$

The interpretation from the above equation is that the autocorrelation at lag index k is the sum of the product of the sequence and the same sequence shifted by k samples to the right. We also notice that the autocorrelation is a maximum at zero lag and equals the energy in the sequence. If the autocorrelation function falls off very slowly with respect to the lag k, it means that adjacent samples of the sequence are highly correlated. On the other hand, if the autocorrelation function falls off rapidly, then the correlation between adjacent samples of the sequence is very low. That is to say that adjacent samples of the sequence are uncorrelated. This is typical in noise sequences.

It is interesting to see from (9.19) that the autocorrelation function looks very similar to the linear convolution of a sequence with itself. In fact, by rewriting (9.19) as

$$c_{xx}[k] = \sum_{n=-\infty}^{\infty} x[n]x[-(k-n)] = x[n]*x[-n], \qquad (9.20)$$

we find that the autocorrelation of a sequence is, in fact, the linear convolution of the sequence with its time-reversed version.

9.3.2   Relation Between Autocorrelation and DTFT of a Discrete-Time Sequence

The DTFT of the autocorrelation function in (9.20) denoted $C_{xx}(e^{j\Omega})$ is

$$DTFT\{c_{xx}[n]\} \equiv C_{xx}(e^{j\Omega}) = DTFT\{x[n]*x[-n]\} = X(e^{j\Omega})X(e^{-j\Omega}) \qquad (9.21)$$

In the above equation we have used the fact that the DTFT of a time-reversed discrete-time sequence is the complex conjugate of the DTFT of the original sequence. Therefore, the DTFT of the autocorrelation function of a discrete-time sequence is related to the DTFT of the sequence by

$$C_{xx}(e^{j\Omega}) = |X(e^{j\Omega})|^2 \qquad (9.22)$$

The power spectral density (PSD) of a discrete-time sequence is the square of the magnitude of its DTFT. Therefore, the PSD of a discrete-time sequence is the DTFT of the autocorrelation function of the sequence. Note that the autocorrelation is an even function as seen from (9.19). Therefore, its DTFT is a real function.

9.3.3   Autocorrelation of Periodic Sequences

In the previous section, we dealt with aperiodic sequences. If a discrete-time sequence is periodic, then its autocorrelation function is defined as follows. Let $\tilde{x}[n]$ be a periodic sequence with period N. Then,

$$c_{\tilde{x}\tilde{x}}[k] = \frac{1}{N} \sum_{n=0}^{N-1} \tilde{x}[n]\tilde{x}[n-k] \qquad (9.23)$$

In the above equation, the shift is circular as opposed to linear shift used for aperiodic sequences. It can be shown that the DFT of the autocorrelation function of a periodic sequence to be

$$C_{\tilde{x}\tilde{x}}[k] = DFT\{c_{\tilde{x}\tilde{x}}[n]\} = \frac{|X[k]|^2}{N} \qquad (9.24)$$

where $X[k]$ is the DFT of $\tilde{x}[n]$. If a sequence is of finite duration, then it can be considered to be periodic with period equal to the length of the sequence.

Example 9.2 Calculation of PSD Using MATLAB   Let us demonstrate the idea of the power spectral density using MATLAB. We can use the MATLAB function *cconv* to calculate the circular convolution of two periodic sequences. Further, as we found earlier, the autocorrelation of a periodic sequence can be computed as the circular convolution of the sequence with its time-reversed version. Thus, the autocorrelation using the MATLAB function is obtained by using the statement

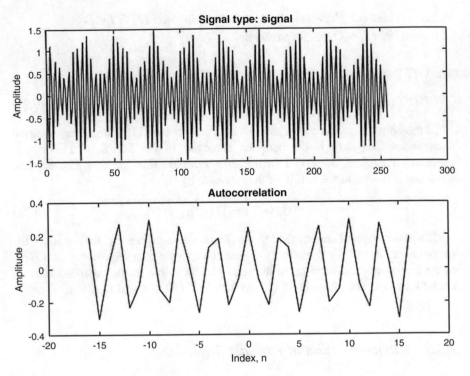

Fig. 9.3 Autocorrelation of an AM sequence: top, input sequence; bottom, its autocorrelation sequence

$c = cconv(x,fliplr(x),N)$. The rest of the statements are found in the MATLAB M-file named *Example 9_2.m*. In this example, we generate either an amplitude-modulated signal or random noise. We then calculate the corresponding autocorrelation function and its DFT. We also compute the DFT of the input sequence. The input signal and its autocorrelation function are plotted as shown in Fig. 9.3 in the top and bottom plots, respectively. In Fig. 9.4, the magnitude of the DFT of the autocorrelation in the top plot and the square of the magnitude of the DFT of the sequence in the bottom plot are shown. As seen in the figure, they are identical, proving the statement in (9.24). Similar quantities are shown in Figs. 9.5 and 9.6 when the input is a Gaussian random sequence.

9.3.4   Short-Time Fourier Transform

The Fourier transform of a signal gives the frequency contents of the signal. In order to achieve this goal, it has to examine the signal in the time domain over the entire time interval. A signal in real time evolves as time progresses. Consider an

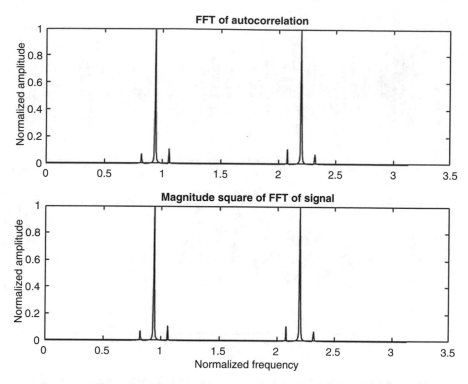

Fig. 9.4 Power spectrum of the input sequence in Fig. 9.3: top, normalized magnitude of the DFT of the autocorrelation sequence of the AM sequence; bottom, normalized square of the magnitude of the DFT of the AM sequence

orchestra performing in front of an audience. Different instruments are played at different time instants. These instruments produce different frequencies. If we apply the Fourier transform to the entire program, all we will know is what frequencies were present with what magnitudes. We will not know when the different frequencies appeared and disappeared. In order to be able to pinpoint the time instants at which different frequencies appear, we need to perform the Fourier transform over short intervals of time. This is accomplished by what is known as the *short-time Fourier transform* (STFT). The STFT is computed by first segmenting an input sequence into overlapping blocks and then performing the DFT on each block. The block lengths must be small so that the signal is stationary over the block time intervals. That is, the signal frequencies do not change over the block interval. The formal definition of STFT of a discrete-time sequence is as follows: Let $\{x[n]\}$ be the sequence whose STFT is to be computed. Divide the sequence into blocks of lengths N samples. Then, the STFT of the sequence is given by

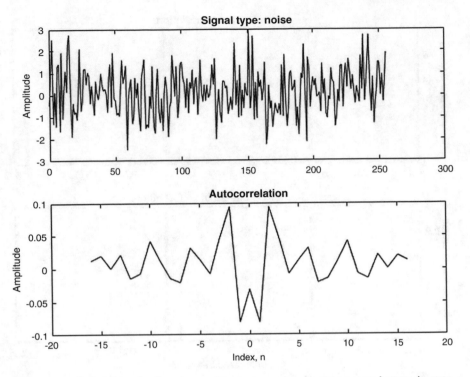

Fig. 9.5 Autocorrelation of a Gaussian random sequence: top, input sequence; bottom, its autocorrelation sequence

$$X\left(e^{j\Omega}, n\right) = \sum_{k=n}^{n+N-1} x[k]w[n-k]e^{-jk\Omega} \tag{9.25}$$

In (9.25), $w[n]$ is a suitable window of length N. As we see from (9.25), the STFT is a function of both the frequency and time. Since the DFT is computed over short intervals of time, the corresponding Fourier transform is called short-time Fourier transform!

STFT is useful in characterizing speech signals. It enables one to recognize the type of sound – voiced, fricative, etc. – using the frequency contents in each segment of the speech signal. Let us work out an example based on MATLAB.

Example 9.3 Calculation of STFT Using MATLAB The MATLAB function to compute the STFT of a discrete-time sequence is *spectrogram*. It accepts a sequence and returns its STFT. Invoking just the function *spectrogram* with proper input arguments will calculate the STFT of the input sequence and plot it with axes time and frequency. The magnitude of the DFT will be shown using color map. In this example, we will consider three types of signals: (1) an amplitude-modulated (AM) signal, (2) a frequency-modulated (FM) signal, and (3) a chirp signal.

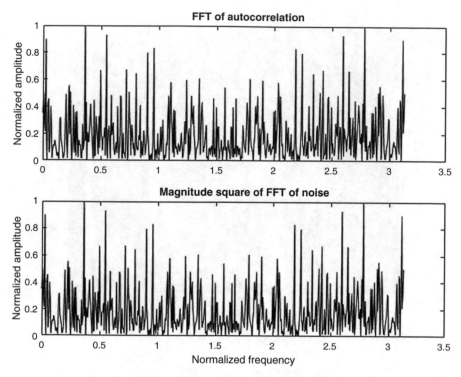

Fig. 9.6 Power spectrum of the noise sequence in Fig. 9.5: top, normalized magnitude of the DFT of the autocorrelation sequence of the noise; bottom, normalized square of the magnitude of the DFT of the noise sequence

A chirp signal is one in which the frequency varies with time. In this example, we consider the frequency of the chirp signal to vary linearly with time. In particular, the chirp signal in this example is described by

$$x[n] = \cos\left(\Omega_0 n^2\right), n \in Z \tag{9.26}$$

Since frequency is the time derivative of phase of a sinusoid, the instantaneous frequency of the chirp signal in (9.26) varies linearly with time. The STFT of the chirp signal obtained by running the M-file *Example 9_3.m* is shown in Fig. 9.7. The power level of the frequencies present is indicated by varying colors. The same M-file is used to compute the STFT of AM and FM signals as well. The STFTs of AM and FM signals are shown in Figs. 9.8 and 9.9, respectively. We see from Figs. 9.8 and 9.9 that the frequencies do not change with time, as we expected.

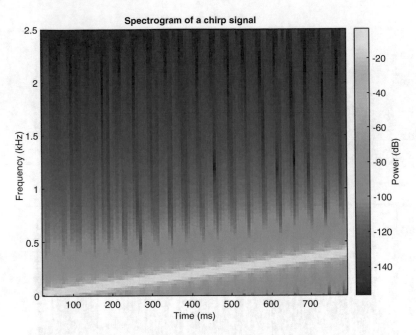

Fig. 9.7 STFT of a chirp signal with frequency varying linearly with time

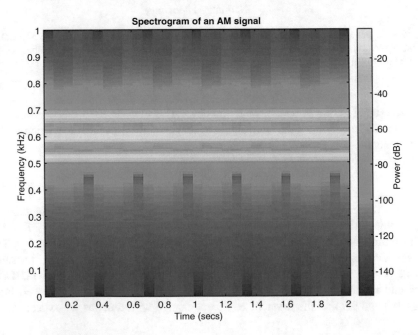

Fig. 9.8 STFT of an AM signal: carrier frequency is 600 Hz, modulating signal frequency is 75 Hz, and the sampling frequency is 2000 Hz

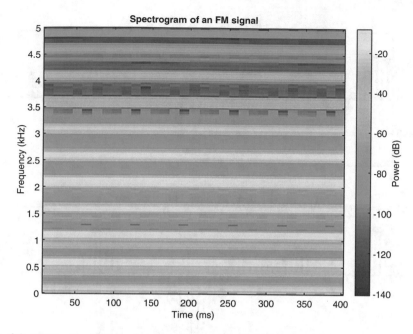

Fig. 9.9 STFT of an FM signal: carrier frequency is 3000 Hz, modulating signal frequency is 500 Hz, sampling frequency is 10,000 Hz, and modulation index is 2

9.4   Fixed-Point Implementation of FFT

So far we have used full precision arithmetic in calculating the DFT of discrete-time sequences. In practice, especially in real-time processing, the arithmetic operations are performed with finite precision. In addition, the DFT coefficients are also represented with finite word lengths. These factors will affect the final results in processing discrete-time signals using DFT.

There are two ways to deal with the effects of finite precision arithmetic. One, we can use detailed analysis to predict the effect on the final output. The other is to simulate the performance of the system with finite precision arithmetic. We will use the latter approach to determine the effect of finite precision arithmetic on the output DFT of discrete-time sequences. More specifically, we will use MATLAB tools to solve this problem. The MATLAB M-file *fi_FFT.m* implements the decimation-in-time FFT algorithm using finite precision fixed-point arithmetic. The length of the sequence must be a positive integer power of two. The word length for the arithmetic operations is defined by the variable *WordLength_seq* and that for the DFT coefficients by the variable *WordLength_coef*. The program also computes the DFT of the input sequence with full precision and plots the error between the full precision and finite precision DFTs. The input sequence represented with full precision and fixed-point 8-bit precision is shown in the top and bottom plots in Fig. 9.10, respectively. The plots show the first 64 points in the sequence whose length is 256 points.

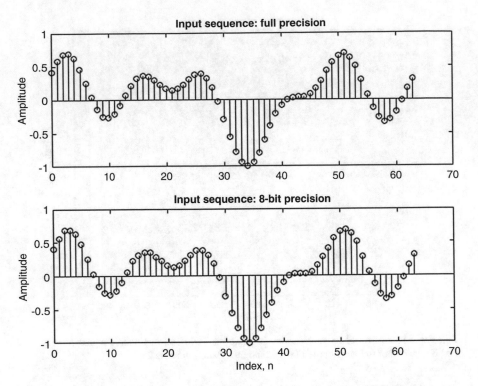

Fig. 9.10 Sequence used to compute the DFT using decimation-in-time FFT algorithm: top, sequence with full precision; bottom, the same sequence represented in fixed-point with 8 bits

The magnitudes of the DFTs of the two sequences are shown in the top plot of Fig. 9.11. As can be seen from the figure, there are some differences between the two DFTs. The difference between the magnitudes of the DFTs is shown in the bottom plot of Fig. 9.11. Again, the error in the DFT due to finite precision arithmetic is seen to be random and has a maximum absolute value of 0.029. The word length used to represent the multipliers in the FFT algorithm is 8 bits. When the coefficient word length is reduced to 4 bits, the maximum absolute error increases to 0.183, which is more than six times the maximum absolute error when the coefficient word length is 8 bits. The magnitude of the DFTs and the corresponding error are shown in Fig. 9.12 for the case where the coefficient word length is 4 bits and the arithmetic word length is 8 bits.

9.5 Sliding Discrete Fourier Transform

In computing the short-time Fourier transform, the long sequence is segmented into either overlapping or non-overlapping blocks, and then the DFT is performed on each finite-length block. In sliding DFT, the DFT is performed on a finite-length

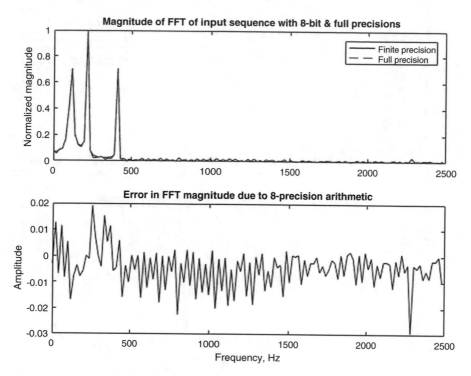

Fig. 9.11 Magnitude of the DFTs of the sequences in Fig. 9.10: top, normalized magnitudes of the DFTs; bottom, error in the magnitude of the DFT between full- and 8-bit precision arithmetic. The word length of the multipliers used in the FFT algorithm is 8 bits

block of a long sequence with a window sliding one sample at a time. For the sake of argument, let us consider the block length to be eight samples. At the time index n, the samples in the block are labeled $\{x[n-7], x[n-6], \ldots, x[n]\}$. At the previous time index n-1, the block sequence is denoted by $\{x[n-8], x[n-7], \ldots, x[n-1]\}$. This ordering of the sequence for n and n-1 is depicted in Fig. 9.13. Using the definition of an eight-point DFT, we have

$$X_k[n] = \sum_{m=n-7}^{n} x[m] W_8^{(m-n+7)k}, 0 \le k \le 7 \qquad (9.27)$$

Note that the DFT is now denoted by two indices – the frequency index k and the time index n. The eight-point DFT at time index n-1 is described by

$$X_k[n-1] = \sum_{m=n-8}^{n-1} x[m] W_8^{(m-n+8)k}, 0 \le k \le 7 \qquad (9.28)$$

By expanding the summation in (9.28), we have

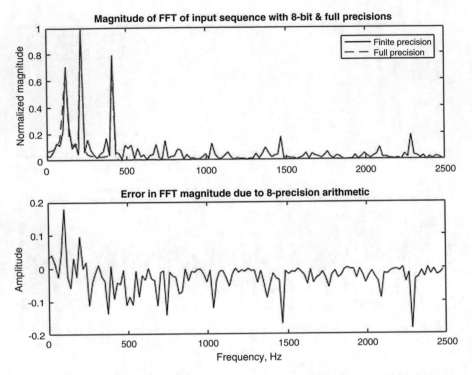

Fig. 9.12 Magnitude of the DFTs of the sequences in Fig. 9.10: top, normalized magnitudes of the DFTs; bottom, error in the magnitude of the DFT between full- and 8-bit precision arithmetic. The word length of the multipliers used in the FFT algorithm is 4 bits

Fig. 9.13 Block of eight samples used to calculate the SDFT at time indices n and n-1

$$X_k[n-1] = x[n-8]W_8^0 + x[n-7]W_8^k + x[n-6]W_8^{2k} + \cdots$$
$$+ x[n-1]W_8^{7k} \tag{9.29}$$

Therefore, from (9.27) and (9.29), we find

$$X_k[n] - W_8^{-k}X_k[n-1] = -x[n-8]W_8^{-k} + x[n]W_8^{7k} \tag{9.30}$$

However

$$W_8^{7k} = W_8^{(8-1)k} = W_8^{8k}W_8^{-k} = e^{-j\frac{2\pi 8k}{8}}W_8^{-k} = W_8^{-k} \tag{9.31}$$

Therefore, using (9.31) in (9.30), we have

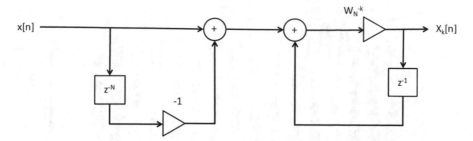

Fig. 9.14 Block diagram of an IIR digital filter structure to implement the SDFT of a discrete-time sequence with a block length N

$$X_k[n] = W_8^{-k}(X_k[n-1] + x[n] - x[n-8]), 0 \le k \le 7 \tag{9.32}$$

What we see from (9.32) is that to compute the eight-point DFT of the sequence at time index n, we need only one complex multiplication for each frequency index k because we already have the eight-point DFT at time index n−1. Equation (9.32) is a recursive equation to compute the eight-point DFT at time index n using the eight-point DFT at the previous time index n-1 and the input samples at the current time index n and the past sample at time index n-8. By generalizing the result in (9.32), the N-point DFT of a long sequence at time index n is expressed as

$$X_k[n] = W_N^{-k}(X_k[n-1] + x[n] - x[n-N]), 0 \le k \le N-1 \tag{9.33}$$

At each time index n, the window slides by one sample, hence the name *sliding discrete Fourier transform*. SDFT is used in characterizing the time-varying spectrums of a signal and is very useful in speech processing. An implementation of the recursive Eq. (9.33) is shown in Fig. 9.14. From Fig. 9.14 we can relate the input and output by

$$X_k[n] = \{x[n] - x[n-N] + X_k[n-1]\}W_N^{-k} \tag{9.34}$$

By denoting the Z-transform of $X_k[n]$ by $X^k[z]$ and applying the Z-transform to (9.34), we find

$$X^k[z] = \{X(z) - z^{-N}X(z) + z^{-1}X^k(z)\}W_N^{-k} \tag{9.35}$$

The digital filter transfer function is then obtained from (9.35) and is given by

$$H^k(z) \equiv \frac{X^k(z)}{X(z)} = \frac{(1 - z^{-N})W_N^{-k}}{1 - W_N^{-k}z^{-1}} \tag{9.36}$$

The frequency response of the transfer function of the IIR digital filter in (9.36) for N = 32 is shown in Fig. 9.15. As can be seen from the figure, the frequency response of the IIR digital filter for each value of k is centered at $\frac{\pi k}{N}$. The MATLAB

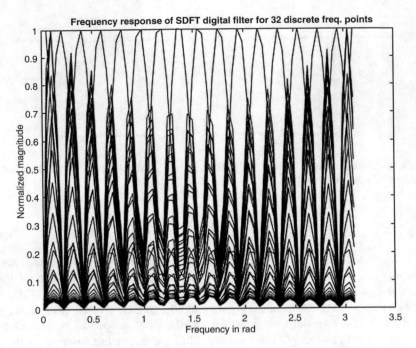

Fig. 9.15 Frequency response of the IIR digital filter that implements the SDFT for a value of N = 32

program to compute the frequency response of the SDFT digital filter is listed in the M-file *SDFT.m*. The same program also computes the SDFT of a discrete-time sequence and compares the result with that obtained by using the FFT routine on each length-N segment of the sequence. As an example, the total length of the input sequence used is 64 samples, and the DFT length used is 32 points. The input signal is shown as stem plot in Fig. 9.16. In Fig. 9.17, the SDFT at time index 32 in dotted red stems in the top plot is shown. The same plot also shows the DFT of that sequence obtained using MATLAB's *fft* routine. They seem to be identical. Similar results are shown in the bottom plot of Fig. 9.17 at time index 47. Again the two are in perfect agreement. As can be observed from the two plots in Fig. 9.17, there is some difference in the two spectra because of the small block length. Remember that SDFT is useful in characterizing a discrete-time sequence with time-varying spectral characteristics. The same M-file also computes the SDFT of a chirp signal. Figure 9.18 displays the input chirp signal as a stem plot over the interval 0–63. The corresponding SDFT at time indices of 32 and 128 are shown in the top and bottom plots in Fig. 9.19, respectively. For the sake of comparison with the SDFT, the two plots in Fig. 9.19 also show the 32-point DFTs at those time indices. Both SDFT and DFT agree perfectly.

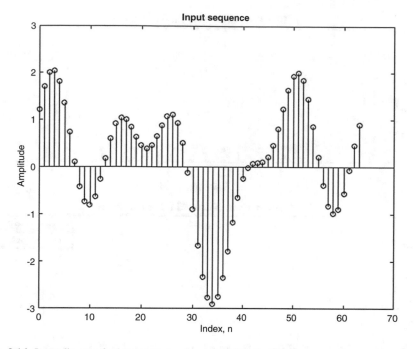

Fig. 9.16 Input discrete-time sequence used in calculating the SDFT

9.6   Energy Compaction Property Revisited

The energy in a finite-length discrete-time sequence is the sum of the square of the absolute value of the sequence. That is,

$$E = \sum_{n=0}^{N-1} |x[n]|^2 \tag{9.37}$$

The energy in the sequence can also be obtained from its DFT sequence according to Parseval. This implies the following:

$$\sum_{n=0}^{N-1} |x[n]|^2 = \frac{1}{N} \sum_{k=0}^{N-1} |X[k]|^2 \tag{9.38}$$

where $\{X[k]\}$ is the DFT of the N-point sequence $\{x[n]\}$. What it really means is that the energy in the finite-length discrete-time sequence is preserved in the DFT domain. Having said that, let us try to understand what we mean by energy compaction.

The right-hand side of (9.38) amounts to the total energy contained in the finite-length sequence in terms of the DFT coefficients. But it does not tell us the

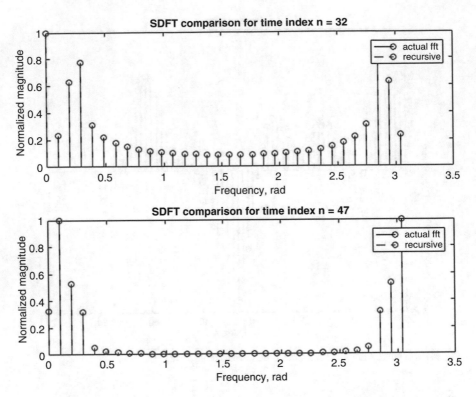

Fig. 9.17 Comparison of SDFT with FFT for an input sequence: top, comparison at time index n = 32; bottom, comparison at time index n = 47

percentage of the total energy carried by the individual DFT coefficient. Some of them may carry a large percentage of the total energy while other coefficients may carry insignificant percent of the total energy. How is that important to us? In fact, it gives us the idea of data compression in the transform domain. Instead of using all the DFT coefficients in the reconstruction of the sequence, we may set a few coefficients to zero and then perform the inverse DFT to reconstruct the sequence. When it comes to the transmission of these coefficients, the zero-valued coefficients need not be sent to the receiver. Of course, we have to know which coefficients have been set to zero. Assuming that the receiver knows which coefficients are zero valued, then one needs to transmit only the non-zero-valued coefficients. For instance, if the total number of DFT coefficients is N and only half of the coefficients is significant, then we achieve 2:1 reduction in the amount of data to be transmitted. There is more than meets the eye in dealing with data compression. Since the topic of compression is not the objective of this book, we will not discuss it any further. However, we will illustrate the energy compaction property of the DFT by an example. Incidentally, energy compaction means what percentage of the total energy in the finite-length sequence is contained or compacted in different coefficients. This property depends on the type of frequency transform being used. Orthogonal

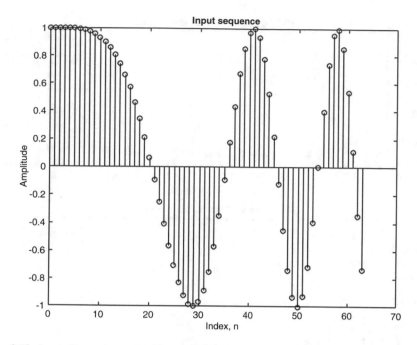

Fig. 9.18 Input chirp sequence used in calculating the SDFT

transforms play an important role in speech and video compression schemes. The discrete Fourier transform is an orthogonal transform. However, not all orthogonal transforms are equally efficient in compacting the energy. There are other transforms such as *discrete cosine transform* (DCT), *discrete sine transform* (DST), *Karhunen-Loeve transform* (KLT), etc. that are more efficient in compacting the signal energy. The interested reader should look at the references to learn more about data compression.

MATLAB Example Let us consider two examples using MATLAB to determine the energy compaction property of the DFT using a known finite-length discrete-time sequence and a gray-scale image. The program details can be found in the M-file *Energy_prop_DFT.m*. The length of the sequence is 128 samples. Figure 9.20 displays the percentage of the total energy in the DFT coefficients against the frequency index. As seen from the figure, except for the first few coefficients, the majority of the DFT coefficients contains insignificant percent of the total energy. By using only the first 12 DFT coefficients in the reconstruction of the sequence, the mean square error (MSE) is found to be about 0.2509, and the SNR is 13.54 dB. The mean square error between the original sequence $\{x[n]\}$ and its reconstructed sequence $\{\tilde{x}[n]\}$ is defined as

$$MSE = \frac{1}{N} \sum_{n=0}^{N-1} (x[n] - \tilde{x}[n])^2 \tag{9.39}$$

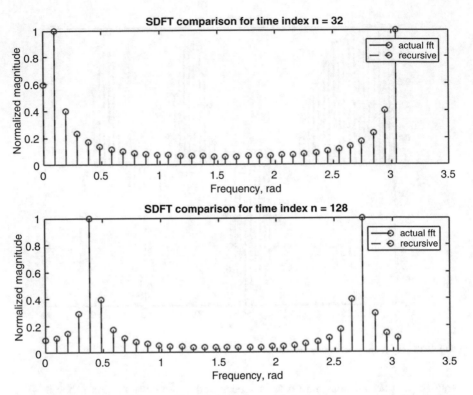

Fig. 9.19 Comparison of SDFT with FFT for an input chirp sequence: top, comparison at time index n = 32; bottom, comparison at time index n = 128. The difference in the two spectra is due to the time-varying frequency of the input chirp signal

Figure 9.21 shows the original and the reconstructed sequence using the first 12 DFT coefficients in the top and bottom plots, respectively. The reconstructed sequence appears to be very close to the input sequence.

The same input signal is used to compute the energy compaction property of DCT. Figure 9.22 shows the percentage of energy contained in the DCT coefficients. As compared to the DFT, the energy compaction property of DCT for 1D signal appears to be different. The MSE between the original and reconstructed sequence with 24 coefficients for the DCT case is found to be 0.1659, and the corresponding SNR is 17.13 dB. The reconstructed signal appears to be quite similar to the original signal as seen from Fig. 9.23.

In the image example, the gray-scale image *cameraman.tif* is used. Its size is 256×256 pixels. Since the image is a two-dimensional signal, a 2D DFT is applied to the image. MATLAB has the function *fft2* to compute the 2D DFT of an image. Its input arguments are the 2D image and the number of pixels in the horizontal and vertical dimensions. It returns the 2D DFT of the input image. In order to plot the percentage of the total energy contained in each DFT coefficient with respect to the frequency points in the horizontal and vertical dimensions, we use the MATLAB

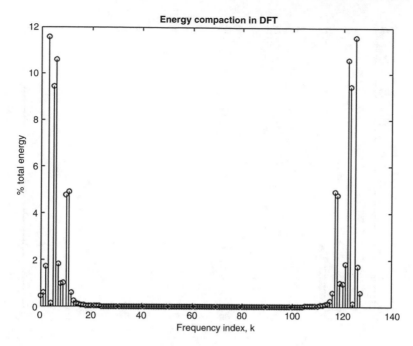

Fig. 9.20 Percentage of total energy contained in the DFT coefficients of an input sequence of length 128 samples. The DFT is of the same length as the sequence

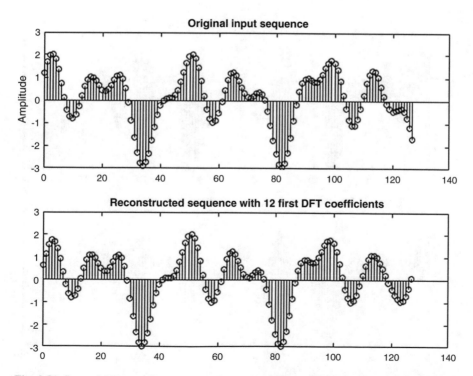

Fig. 9.21 Input sequence and reconstructed sequence using only the first 12 out of 128 DFT coefficients. The resulting SNR is 14.52 dB

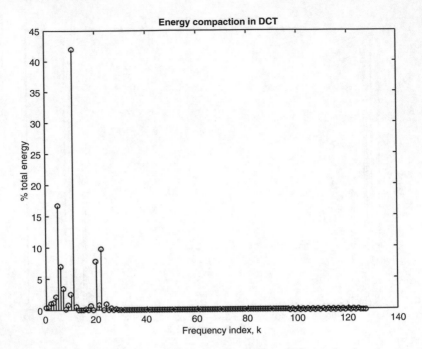

Fig. 9.22 Percentage of total energy contained in the DCT coefficients of an input sequence of length 128 samples. The DCT is of the same length as the sequence

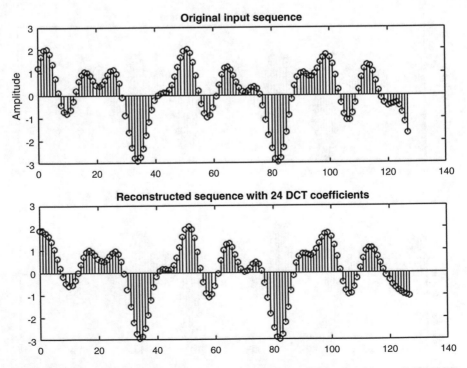

Fig. 9.23 Input sequence and reconstructed sequence using only the first 24 out of 128 DCT coefficients. The resulting SNR is 17.13 dB

Input image

Fig. 9.24 Original cameraman image

function *surf* to obtain a surface plot. This function accepts the points in the x- and y-axes and the function value in the z-axis and plots them as a 3D plot. The input cameraman image is shown in Fig. 9.24. The percentage of total energy in each DFT coefficient is shown in Fig. 9.25 as a surface plot over the first 32×32 points for easy visualization. It is found that the DC coefficient (DFT coefficient X[0,0]} contains about 78% of the total energy. The image reconstructed using only the first 64x64 DFT coefficients is shown in Fig. 9.26. Though one can identify the image, it has a large amount of distortions. The resulting MSE and signal-to-noise ratio (SNR) are found to be about 27.1643 and 7.22 dB, respectively. When using the DCT on the same image, it is found that by using only the first 64×64 DCT coefficients to reconstruct the image, the MSE and SNR are, respectively, 17.5604 and 11.01 dB. The surface plot depicting the percentage of energy contained in the first 32×32 DCT coefficients is shown in Fig. 9.27. The reconstructed image using the first 64×64 DCT coefficients is shown in Fig. 9.28. It is much clearer than that for the DFT case. The resulting compression ratio is 16:1. The DCT appears to be more efficient than the DFT in terms of compacting energy in the coefficients. It is also a real transform, meaning that only real and not complex arithmetic operations are needed. That is a reason why DCT is used extensively in image compression

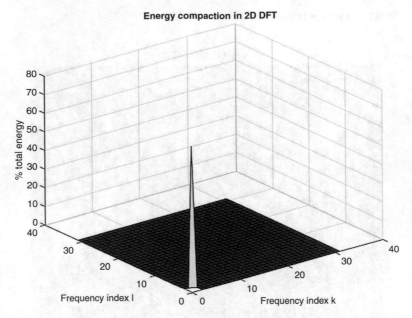

Fig. 9.25 Surface plot depicting the percentage of total energy contained in the first 32×32 DFT coefficients of the image in Fig. 9.22

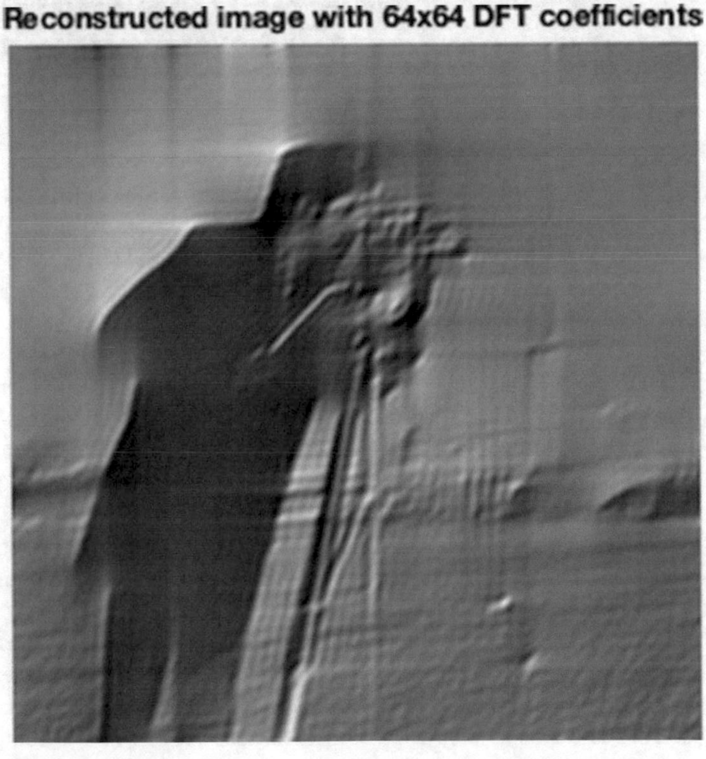

Reconstructed image with 64x64 DFT coefficients

Fig. 9.26 Reconstructed image of cameraman by using only the first 64×64 2D DFT coefficients. The resulting SNR is 5.17 dB

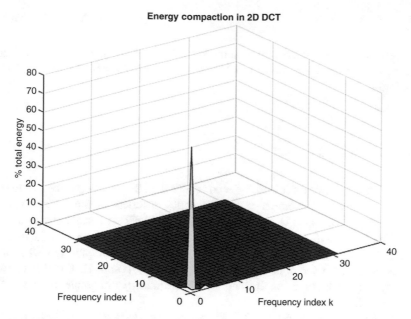

Fig. 9.27 Surface plot depicting the percentage of total energy contained in the first 32×32 DCT coefficients of the image in Fig. 9.22

Fig. 9.28 Reconstructed image of cameraman by using only the first 64x64 2D DCT coefficients. The resulting SNR is 11 dB

standards. It must be pointed out that the same M-file *Energy_prop_DFT.m* is used to compute the energy contained in 1D signal and image for both DFT and DCT transforms.

9.7 Zoom FFT

The FFT that we have discussed so far computes the DFT of a sequence of length-N samples. It gives a panorama of the frequency contents in the input sequence. Sometimes it is desirable to get a closer look at the DFT of a sequence rather than observing it from a distance. In other words, we may want to zoom in to the FFT of a sequence. In this section, we will describe a method to compute the DFT of an N-point sequence over a smaller range of frequency points K. For this to work properly, we will assume that the length of the sequence is an integer multiple of K, i.e., $N = RK$.

The idea behind the zoom FFT is to express the given sequence in terms of its downsampled sequences. We then compute the DFTs of the downsampled sequences and obtain their weighted sum to get the zoomed-in FFT. There are R number of downsampled sequences with each sequence being of length K samples. These downsampled sequences are called the polyphase components. The R downsampled sequences are described by

$$x_i[n] = x[i + nR], 0 \le i \le R - 1 \qquad (9.40)$$

Downsampling by a factor of R is achieved by retaining every R^{th} sample in the input sequence. Since the length of the input sequence is K times R, the length of each downsampled sequence is K. For instance, if $R = 3$ and $K = 8$, the input sequence will be of length 24. Then,

$$x_0[n] = \{x[0], x[3], x[6], \cdots, x[21]\}, \qquad (9.41a)$$

$$x_1[n] = \{x[1], x[4], x[7], \cdots, x[22]\}, \text{and} \qquad (9.41b)$$

$$x_2[n] = \{x[2], x[5], x[8], \cdots, x[23]\} \qquad (9.41c)$$

The Z-transforms of the downsampled sequences can be obtained using the Z-transform definition. For example, if $R = 3$ and $K = 8$, we have

$$Z\{x_0[n]\} \equiv X_0(z) = x[0] + x[3]z^{-1} + x[6]z^{-2} + \cdots + x[21]z^{-7} \qquad (9.42a)$$

$$Z\{x_1[n]\} \equiv X_1(z) = x[1] + x[4]z^{-1} + \cdots + x[22]z^{-7} \qquad (9.42b)$$

Similarly, the Z-transform of the third downsampled sequence can be described by

$$Z\{x_2[n]\} \equiv X_2(z) = x[2] + x[5]z^{-1} + \cdots + x[23]z^{-7}. \tag{9.42c}$$

In other words, the Z-transforms of the downsampled sequences are given by

$$X_r(z) = \sum_{n=0}^{K-1} x_r[r + nR]z^{-n}. \tag{9.42d}$$

The Z-transform of the input sequence can then be expressed in terms of those of the downsampled sequences as

$$X(z) = \sum_{m=0}^{2} z^{-m} X_m\left(z^3\right) \tag{9.43}$$

In general, the Z-transform of the input length-N sequence is defined in terms of the Z-transforms of the downsampled sequences as

$$X(z) = \sum_{m=0}^{R-1} z^{-m} X_m\left(z^R\right), \tag{9.44}$$

where

$$X_m(z) = \sum_{n=0}^{\left\lfloor \frac{N+1}{R} \right\rfloor} x[m + nR]z^{-n}, 0 \le m \le R - 1. \tag{9.45}$$

What we have done so far is to express the Z-transform of the given sequence in terms of the Z-transforms of its downsampled sequences. But our task is to zoom into the DFT of the input sequence. That is, we want to compute the DFT of the input sequence by evaluating its Z-transform over K equally spaced frequency points along the unit circle in the Z-domain starting from the frequency point i. The K frequency points are given by

$$z_k = e^{j\frac{2\pi}{N}k} \equiv W_N^{-k}, i \le k \le i + K - 1 \tag{9.46}$$

Thus, the zoom DFT of the input sequence is described by

$$X[k] = \sum_{n=0}^{R-1} W_N^{nk} X_n\left(W_N^{-kR}\right), i \le k \le i + K - 1, \tag{9.47}$$

where the K-point DFTs are given by

$$X_i[l] = \sum_{m=0}^{K-1} x[i + mR] W_K^{ml}, 0 \le l \le K - 1. \tag{9.48}$$

From (9.47), it is seen that the DFT of the input sequence over K frequency points is the weighted sum of the DFTs of the downsampled sequences. Let us clarify the above discussion by an example. What else can clear the doubt better than a simple example?

Example 9.4 For the given N-point sequence, obtain the R downsampled sequences, and then calculate the DFT of the input sequence over K frequency points. Plot the DFTs of the input sequence and its zoomed-in DFT.

Solution Let the input sequence be described by

$$x[n] = \cos\left(\frac{2\pi f_c}{f_s} n + \beta \sin\left(\frac{2\pi f_m}{f_s} n\right)\right), 0 \le n \le N - 1.$$

Note that the above equation corresponds to an FM sequence, where f_c is the carrier frequency in Hz, f_m is the modulating signal frequency, f_s is the sampling frequency, and β is the modulation index. For this example, $f_c = 1000\ Hz$, $f_m = 105\ Hz$, $f_s = 10000\ Hz$, and $\beta = 2.5$. We also assume that there are R = 3 downsampled sequences, each of length K = 128 samples. Therefore, the length of the input sequence is N = 3*128 = 384 samples. We will use MATLAB to solve the problem. More specifically, the M-file used for this example is named *ZoomZoom. m*. The input sequence is shown in Fig. 9.29. Though the input is discrete in time, a continuous line plot is used for a better visualization of the sequence. As expected, the frequency of the sinusoidal sequence varies periodically. The magnitudes of the DFTs of the three-downsampled sequences of length 128 samples each are shown in

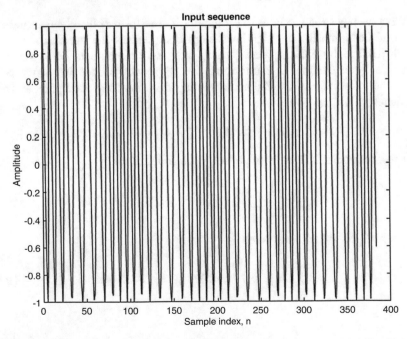

Fig. 9.29 Input FM sequence used in Example 9.4. The length of the sequence is 384 samples

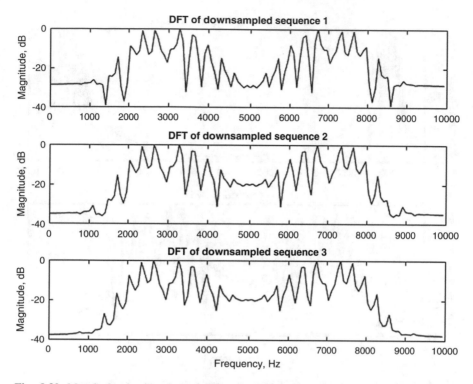

Fig. 9.30 Magnitudes in dB of the DFTs of the three-downsampled sequences of length 128 samples each

Fig. 9.30 in the top, middle, and bottom plots. The magnitude of the DFT of the N-point input sequence is shown in the top plot of Fig. 9.31. It consists of many sidebands. The zoomed-in or close-up of the DFT of the input sequence is shown in the bottom plot of Fig. 9.31. It gives us a close-up look at the spectrum of the input sequence. Incidentally, the zoomed-in DFT is calculated over K points starting from the 10th-frequency point on the unit circle.

9.8   Chirp Fourier Transform

The DTFT of a finite-length sequence can be obtained from its Z-transform by evaluating the Z-transform of the sequence on the unit circle in the z-domain. We have seen this in detail in an earlier chapter. Instead of evaluating the Z-transform of a finite-length sequence for values of z on the unit circle, one can also evaluate the Z-transform on a spiral contour in the z-plane. Because the contour is a spiral, the resulting Z-transform is known as the *chirp* Z-transform (CZT). The purpose behind the CZT is to obtain the DTFT of a finite-length sequence over a limited range of

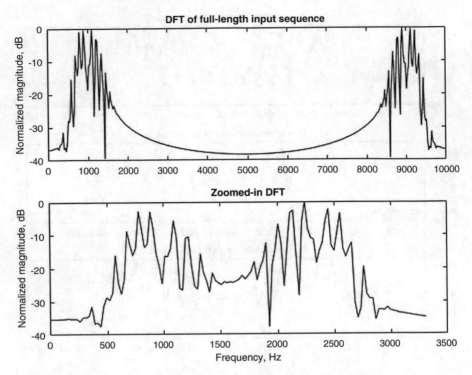

Fig. 9.31 Magnitudes in dB of the full-length input sequence and its zoomed-in DFT: top, magnitude in dB of the input sequence; bottom, magnitude in dB of the 128-point zoomed-in DFT of the input

frequencies rather than the whole range between zero and half the sampling frequency. If the contour is simply the unit circle, then evaluating the CZT over a limited range of frequencies on the unit circle results in chirp Fourier transform (CFT). There are some applications for the chirp Fourier transform, and therefore, we will describe it here. Note that since the CZT is evaluated over a limited frequency range, we actually zoom into the DTFT.

Consider a length-N sequence $\{x[n], 0 \leq n \leq N - 1\}$. Its Z-transform is given by

$$X(z) = \sum_{n=0}^{N-1} x[n]z^{-n}. \tag{9.49}$$

If we evaluate the Z-transform in (9.49) at points z_k, $0 \leq k \leq K - 1$, then

$$X(z_k) = \sum_{n=0}^{N-1} x[n]z_k^{-n}, 0 \leq k \leq K - 1. \tag{9.50}$$

Let $z_k = RU^{-k}$, $0 \leq k \leq K - 1$, where R and U are complex variables. Equation (9.50) can then be written as

$$X(z_k) = \sum_{n=0}^{N-1} x[n] R^{-n} U^{nk}, 0 \le k \le K - 1. \tag{9.51}$$

Let us evaluate (9.51) on the unit circle in the z-domain and also assume that $R = e^{j\Omega_0}$. Then,

$$X\left(e^{j\Omega_k}\right) = \sum_{n=0}^{N-1} x[n] e^{-jn\Omega_0} U^{nk}, 0 \le k \le K - 1. \tag{9.52}$$

Note that we can express $nk = \frac{1}{2}\left[n^2 + k^2 - (k-n)^2\right]$. Then,

$$X\left(e^{j\Omega_k}\right) = \sum_{n=0}^{N-1} x[n] e^{-jn\Omega_0} U^{\frac{n^2}{2}} U^{\frac{k^2}{2}} U^{-\frac{(k-n)^2}{2}}, \tag{9.53}$$

which can be rewritten as

$$X\left(e^{j\Omega_k}\right) = U^{\frac{k^2}{2}} \sum_{n=0}^{N-1} \left\{ x[n] e^{-jn\Omega_0} U^{\frac{n^2}{2}} \right\} U^{-\frac{(k-n)^2}{2}}, 0 \le k \le K - 1. \tag{9.54}$$

Define

$$g[n] \equiv x[n] e^{-jn\Omega_0} U^{\frac{n^2}{2}}. \tag{9.55}$$

In terms of (9.55), we can express the Z-transform of the input sequence as

$$X\left(e^{j\Omega_k}\right) = U^{\frac{k^2}{2}} \sum_{n=0}^{N-1} g[n] U^{-\frac{(k-n)^2}{2}}, 0 \le k \le K - 1. \tag{9.56}$$

By interchanging k and n on both sides of (9.56), we obtain

$$X\left(e^{j\Omega_n}\right) = U^{\frac{n^2}{2}} \sum_{k=0}^{N-1} g[k] U^{-\frac{(n-k)^2}{2}} = U^{\frac{n^2}{2}} \left[g[n] * U^{-\frac{n^2}{2}} \right]. \tag{9.57}$$

That is, the CZT of the N-point sequence is the convolution of the new sequence $g[n]$ and $U^{-\frac{n^2}{2}}$ followed by multiplication by the sequence $U^{\frac{n^2}{2}}$. This depicted as a block diagram in Fig. 9.32.

Example 9.5 Let us elaborate the idea behind CZT by way of an example. This example uses the same input sequence as used in the previous example. The sequence is of length 1024 samples. We will use the MATLAB function *czt*, which calculates the CZT of an input sequence. The actual function call is described by $y = czt(x, N, w, a)$, where x is the sequence whose CZT is y, N is the number of points over which the CZT is calculated, a is the starting point and is defined as

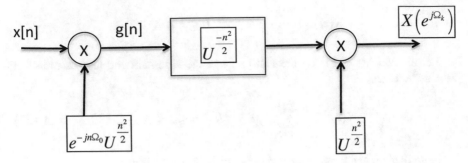

Fig. 9.32 Block diagram showing the calculation of the CZT of an N-point sequence as a filtering operation

$$a = e^{j\frac{2\pi f_1}{f_s}},$$ (9.58)

and w is the set of points starting from f_1, over which the CZT is calculated and is given as

$$w = e^{-j\frac{2\pi(f_2 - f_1)}{N f_s}}.$$ (9.59)

The M-file to run to solve this problem is named *Chirpy.m*. Let us chirp over the frequencies from 950 to 1050 Hz. This range of frequencies is centered at the carrier frequency of 1000 Hz. The normalized magnitude of the DFT of the input sequence is shown in the top plot of Fig. 9.33 over the frequency range from zero to f_s, and the normalized magnitude of the CZT of the same input sequence is shown in the bottom plot of Fig. 9.33 over the frequency range of 100 Hz with the starting frequency of 950 Hz. As can be seen from the figure, the CZT appears to zoom into the DFT of the input sequence. This M-file also calculates the CZT of a Gaussian noise sequence with a standard deviation of 1.25 over the same frequency range as used for the signal. The normalized magnitudes of the DFT and CZT of the noise sequence are shown in the top and bottom plots of Fig. 9.34, respectively.

9.9   Summary

This chapter deals with the signal flow graphs to implement the Cooley-Tukey fast FFT algorithm. Two such algorithms are called decimation-in-time and decimation-in-frequency fast algorithms. Next we discussed spectral analysis of discrete-time signals using DFT. In particular, we learnt how to compute the power spectrum of a finite-length discrete-time signal via its autocorrelation function. We further dealt with short-time Fourier transform, which is useful in characterizing signals with time-varying frequency. The MATLAB function *spectrogram* is a very useful tool in computing the STFT of a discrete-time sequence. Since DFTs are not implemented

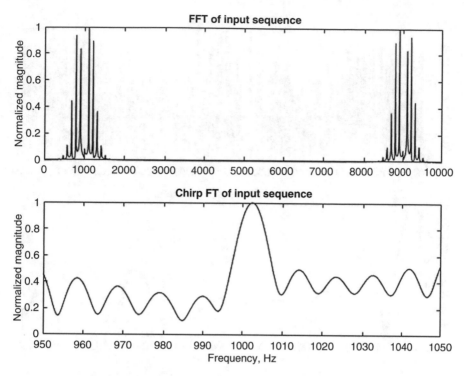

Fig. 9.33 CZT of the sequence used in Example 9.5. Top: normalized magnitude of the DFT of the sequence. Bottom: normalized magnitude of the CZT of the sequence from 950 to 1050 Hz

with infinite precision in practice, we described a fixed-point implementation of the DFT using MATLAB. Sliding DFT is more efficient in computing the DFT of an N-point sequence because it can be achieved in a recursive manner. In fact, the SDFT uses only one complex multiplication per frequency point due to its recursive input-output relation. We then talked about the energy compaction property of DFT. Orthogonal transforms compact the total energy in an input discrete-time signal or images differently in different DFT and DCT coefficients. Typically, a large percentage of the total energy in the signal is compacted into a few first DFT or DCT coefficients, while the rest of the coefficients contain an insignificant amount. This leads to the idea of data compression in the frequency domain. We showed two examples to illustrate the energy compaction property of DFT and DCT using MATLAB. Zoom FFT is used in the fields like Radar and biomedicine. So, we discussed how to calculate the zoom FFT of a given sequence in terms of its downsampled sequences. We illustrated the idea of zoom FFT by an example using MATLAB. Finally, we described the CZT of a sequence and showed how to compute it for an input sequence using MATLAB. In the next chapter, we will describe a few applications of digital signal processing methods in digital communications.

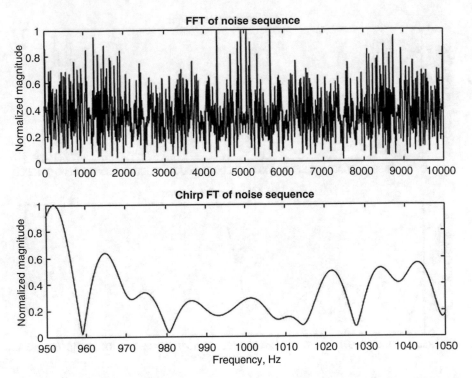

Fig. 9.34 CZT of the Gaussian noise sequence used in Example 9.5. Top: normalized magnitude of the DFT of the noise sequence. Bottom: normalized magnitude of the CZT of the noise sequence from 950 to 1050 Hz. The standard deviation of the noise is 1.25

9.10   Problems

1. Determine the number of real multiplications and real additions required to compute the N-point DFT of a sequence of length N using brute-force method.
2. Consider the linear convolution of the sequences $\{h[n]\}$ of length N and $\{x[n]\}$ of length M via DFT algorithm. Find total number of floating-point complex multiplications and additions required using brute-force method to carry out the DFT in calculating the above linear convolution.
3. The DFT of a 450-point sequence is to be computed over a length of 512 points. How many zero-valued samples should be appended to make the sequence of length, which is a power of two closest to 512? Determine the total number of complex multiplications and additions required to compute the DFT of the zero-padded sequence using (a) brute-force method and (b) FFT algorithm.
4. Consider a discrete-time sequence of length-N samples. If the frequency spacing required in the DFT domain is 2 Hz and the sampling frequency is f_s, determine the value for N.
5. Given a sequence $x[n]$ of length 3333 samples, we want to compute its DFT. The total signal duration is 1.5 seconds. Determine (a) how many zero-valued

samples must be appended to the sequence to compute its radix-2 FFT and (b) what is the frequency resolution after the FFT is performed.

6. In Fig. 9.1c, the radix-2 FFT algorithm to compute the DFT of an eight-point sequence is shown where the input is in bit-reversed order. Show the signal flow graph corresponding to Fig. 9.1c with the input in normal order and the output in bit-reversed order.

7. Repeat problem 6 wherein both the input and output are in normal order.

8. The number of complex multiplications required to compute the DFT of an N-point sequence using the radix-2 FFT algorithm included multiplications by ± 1. Calculate the exact number of complex multiplications required if only non-unity complex multiplications are involved.

9. If we want to compute the linear convolution of a length-32 real sequence and a length-13 real sequence using (a) direct computation of the linear convolution, (b) computation of the linear convolution using a single circular convolution, and (c) computation using radix-2 FFT algorithm, find the least number of real multiplications required for each of the above cases. Exclude the number of multiplications by factors such as ± 1, $\pm j$, and W_N^0 in calculating the number of real multiplications in (c).

10. We want to use the overlap-add algorithm to filter an input sequence of length 1024 samples by a linear-phase FIR filter of length 24, where each linear convolution of the short segments is performed using DFT and the DFTs are calculated via radix-2 FFT algorithm. Determine (a) the necessary power-of-2 transform length which will give the minimum number of multiplications, and (b) find the total number of multiplications required if direct convolution is used.

11. It is required to compute a 512-point DFT of a sequence of length 373 samples. Find (a) the number of zero-valued samples to be appended to the sequence before computing the DFT, (b) total number of complex multiplications and additions required for the direct computation of all DFT samples, and (c) total number of complex multiplications and additions needed to compute the DFT using the FFT algorithm.

12. This problem relates to short-time Fourier transform using MATLAB. Record your speech for a second or two, and then perform the STFT of the recorded speech. Choose the appropriate length of the speech segments, and then perform the FFT. See if you can identify the type of speech such as voiced or fricative from the STFT.

13. Repeat problem 12 using the sliding discrete Fourier transform. Plot the magnitude of the SDFT as a function of time and frequency.

14. Generate a discrete-time sequence of length 1024 samples consisting of say three distinct frequencies and unequal amplitudes. Use a sampling frequency of 5 kHz. Then perform the zoom FFT of the sequence, and plot the magnitude of the DFT. Compare this with the magnitude of the DFT of the original input sequence.

15. Generate an FM sequence with a carrier frequency of 10 kHz, modulating frequency of 1 kHz, and modulation index of 2. Compute the chirp FT of the sequence, and plot the magnitude of the DFT obtained.

References

1. Ahmed N, Natarajan T, Rao KR (1974) Discrete cosine transform. IEEE Trans Comput C-23:90–93
2. Allen JB, Rabiner LR (1977) A unified approach to short-term Fourier analysis and synthesis. Proc IEEE 65:1558–1564
3. Bergland G (1969) Fast Fourier transform hardware implementations – an overview. IEEE Trans Audio Electroacoustics AU-17
4. Blahut RE (1985) Fast algorithms for digital signal processing. Addison-Wesley, Reading
5. Bongiovanni G, Corsini P, Forsini G (1976) One-dimensional and two-dimensional generalized discrete Fourier transform. IEEE Trans Acoust Speech Signal Process ASSP-24:97–99
6. Burrus CS (1977) Index mappings for multidimensional formulation of the DFT and convolution. IEEE Trans Acoust Speech Signal Process ASSP-25:239–242
7. Burrus CS (1988) Unscrambling for fast DFT algorithms. IEEE Trans Acoust Speech Signal process ASSP-36(7)
8. Chen C-T (2001) Digital signal processing: spectral computation and filter design. Oxford University Press, New York
9. Cooley JW, Tukey JW (1965) An algorithm for the machine calculation of complex Fourier series. Math Comput 19:297–301
10. Duhamel P (1986) Implementation of "split-radix" FFT algorithms for complex, real, and real-symmetric data. IEEE Trans Acoust Speech Signal Process ASSP-34:285–295
11. Evans D (1987) An improved digit-reversal permutation algorithm for the fast fourier and Hartley transforms. IEEE Trans Acoust Speech Signal Process ASSP-35(8)
12. Jacobson E, Lyons R (2003) The sliding DFT. IEEE Signal Process Mag 20:74–80
13. Kaiser JF (1980) On a simple algorithm to calculate the "energy" of a signal. Proceedings of the IEEE International Conference on Acoustics, Speech, and Signal Processing, Albuquerque, pp 381–384
14. Kolba DP, Parks TW (1977) A prime factor FFT algorithm using high speed convolution. IEEE Trans Acoust Speech Signal Process 25:281–294
15. Malvar H, Hallapuro A, Karczezwicz M, Kerofsky L (2002) Low-complexity transform and quantization with 16-bit arithmetic for H.26L. Proceedings of the IEEE International Conference on Image Processing, pp. II-489–II-4924
16. Mar A (ed) (1992) Digital signal processing applications using the ADSP-2100 family. Prentice Hall, Englewood Cliffs
17. Marple SL Jr (1987) Digital spectral analysis with applications. Prentice Hall, Englewood Cliffs
18. Nawab SH, Quatieri TF (1988) Chapter 6: Short-time Fourier transform. In: Lim JS, Oppenheim AV (eds) Advanced topics in signal processing. Prentice Hall, Englewood Cliffs
19. Rabiner LR, Schafer RW, Rader CM (1969) The chirp-z transform algorithm. IEEE Trans Audio Electroacoustics AU-17:86–92
20. Regalia PA, Mitra SK, Fadavi-Ardekani J (1987) Implementation of real coefficient digital filters using complex arithmetic. IEEE Trans Circuits Sys CAS-34:345–353
21. Rodriguez JJ (1989) An improved FFT digit-reversal algorithm. IEEE Trans on Acoust Speech Signal Process ASSP-37(8)
22. Sorenson HV, Jones DL, Heideman MT, Burrus CS (June 1987) Real-valued fast Fourier transform algorithms. IEEE Trans Acoust Speech Signal Process ASSP-35(6)
23. Welch PD (1967) The use of fast Fourier transform for the estimation of power spectra: a method based on time averaging over short modified periodograms. IEEE Trans Audio Electroacoustics AU-15:70–73

Chapter 10
DSP in Communications

10.1 Introduction

What we have learnt so far is how to convert an analog signal into a digital signal and to process it using digital filters. The field of digital signal processing has fully matured and has found applications in diverse fields. In this chapter, we will concentrate on the application of DSP in one particular field, namely, the field of digital communications. Radio, telephony, and video are a few of the areas that are completely enveloped by modern digital and wireless communications. Radio broadcast started with analog communications. It used analog modulation techniques such as amplitude modulation (AM) and frequency modulation (FM) to transmit the message signal using radio frequencies (RF). These modulation schemes use the message signal to modulate a carrier signal in its amplitude (AM) or in its instantaneous frequency (FM) before transmission. Later digital modulation methods were introduced to serve the same purpose as the analog counterparts. Digital modulation plays an important role in modern wireless communications. The art of making very large-scale integrated (VLSI) circuits has evolved tremendously. This has enabled the design and fabrication of application-specific integrated circuits (ASIC), which in turn enables the implementation of complex digital techniques in achieving communications successfully as well as lowering the cost. Digital communication systems have many advantages over the analog counterparts. For instance, digital communications has greater immunity to noise. It is also robust to channel impairments. Another advantage is that many different data can be multiplexed and transmitted over a single channel. The various data may include voice, video, and other data, for instance. Since binary digits (bits) are used in digital communications, there is greater security in the transmitted data.

Electronic supplementary material: The online version of this article (https://doi.org/10.1007/978-3-319-76029-2_10) contains supplementary material, which is available to authorized users.

K. S. Thyagarajan, *Introduction to Digital Signal Processing Using MATLAB with Application to Digital Communications*, https://doi.org/10.1007/978-3-319-76029-2_10

This is not feasible in analog communications. Even if there are errors in the received data, they can be detected and corrected by employing what is known as the *channel coding*, in which extra bits are added to the data bits. In analog communications, the noise in the channel will distort the message signal and is, therefore, impossible to recover the original signal. Even though digital modulation as such occupies a higher bandwidth than analog modulation, *source coding* or data compression is used to reduce the message bandwidth to start with. Digital communication link performance can be improved by using encryption and channel equalization techniques. Moreover, *field programmable gate arrays* (FPGA) enable the implementation of digital modulation and demodulation functions purely in software. This has an enormous implication because many handheld devices can perform a variety of functions well in real time using software. These features are certainly a no-no in analog communications.

In this chapter, we will describe a few DSP methods that are used in digital communications. More specifically, we will describe how digital pulses can be shaped at the transmitting side using DSP techniques to cancel what is known as the inter-symbol interference (ISI). At the receiver, another DSP function, namely, *equalization*, is used to mitigate the channel interference. In digital modulation, the receiver has to detect in each bit interval whether a binary "1" or a binary "0" is transmitted. This is achieved in an optimal fashion using the so-called *matched filter* (MF) or equivalently a *correlation filter*. We will learn how to implement such filters as well. We will also learn a few other DSP functions as applied to oversampled ADC and DAC, digital modulation schemes, and *phase-locked loop* (PLL).

10.2   Sampling Rate Conversion

Before the transmission of a message signal such as voice or music using digital modulation, the analog signal must be converted to a digital signal. As we have seen earlier, the analog-to-digital conversion (ADC) is achieved by first sampling the continuous-time signal at a minimum of Nyquist rate and then converting the analog samples to digital numbers. Typical bit widths of an ADC are between 8 and 12 bits. In wireless communications, for instance, the channel bandwidth is an extremely precious thing. Therefore, the service providers do whatever it takes to reduce the data rate of every subscriber. The first thing to do here is to reduce the bit width of the ADC. A one-bit ADC will be super. How is that possible? It is possible by using a very high sampling rate. We will first learn how to change the sampling rate and then describe the various DSP methods that incorporate different sampling rates. This is what is called *multi-rate digital signal processing*. That is, a digital signal processing that involves different sampling rates is termed multi-rate digital signal processing. In multi-rate DSP, sampling rates at certain points are increased from its native rate. This is known as *upsampling*. At other points the sampling rate is decreased from its native rate. This process is termed downsampling. Upsampling or downsampling may use either integer sampling rate or rational sampling rate. Before we discuss ADC with a sampling rate higher than the Nyquist rate, we need to know the effect of upsampling or downsampling on the sampled signal.

10.2.1 Upsampling

Let $\{x[n], -\infty < n < \infty\}$ be a sequence that is sampled at the Nyquist rate. If the original sampling rate is increased by a positive integer factor M, then we can express the upsampled sequence $x_u[n]$ in terms of the original sequence as described by

$$x_u[n] = \begin{cases} x[\frac{n}{M}], n = 0, \pm M, \pm 2M, \cdots \\ 0, otherwise \end{cases} \tag{10.1}$$

From the above equation, we find that upsampling amounts to inserting M-1 zeros between every two consecutive samples. To understand better the process of upsampling, we need to describe the upsampled sequence in the Z-domain. The Z-transform of the upsampled sequence is obtained using the definition of the Z-transform and is given by

$$X_u(z) = \sum_{n=-\infty}^{\infty} x_u[n]z^{-n} \tag{10.2}$$

In terms of the Z-transform of the original sequence, (10.2) reduces to

$$X_u[z] = \sum_{n=-\infty}^{\infty} x\left[\frac{n}{M}\right]z^{-n} \tag{10.3}$$

Define $m = \frac{n}{M}$ and (10.3) becomes

$$X_u[z] = \sum_{m=-\infty}^{\infty} x[m]z^{-mM} = \sum_{m=-\infty}^{\infty} x[m]\left(z^M\right)^{-m} = X\left(z^M\right) \tag{10.4}$$

The DTFT of the upsampled sequence can then be found by using $z = e^{j\Omega}$ in (10.4), which is

$$X_u\left(e^{j\Omega}\right) = X(z)|_{z=e^{j\Omega}} = X\left(e^{jM\Omega}\right) \tag{10.5}$$

The implication of (10.5) is that the spectrum of $x_u[n]$ is the same spectrum of $x[n]$ but repeated M times in the interval $[0, 2\pi]$. In other words, what happens to $x[n]$ in the frequency domain in the interval $[0, 2\pi]$ happens M times to the upsampled sequence in that interval. That is, there are M-1 images of the spectrum $X(e^{j\Omega})$ in the spectrum of the upsampled sequence in the interval $[0, 2\pi]$. In order to preserve the integrity of the original sequence, the upsampled sequence must be lowpass filtered to reject the M-1 images.

Example 10.1 In this example, we will consider a signal consisting of three sinusoids of frequencies 950 Hz, 1800 Hz, and 1917 Hz, which is sampled at a rate of 5000 Hz. This signal is then upsampled by a factor of M = 6. We have to compute the spectra of the original and upsampled sequences as well as the spectrum of the lowpass filtered upsampled signal to verify that the images are removed from the spectrum of the upsampled signal.

Solution The actual input signal before sampling is described by

$$x(t) = \sin(2\pi f_1 t) + 2\sin(2\pi f_2 t) + 1.75\sin(2\pi f_3 t), t \geq 0 \qquad (10.6)$$

where the frequencies are $f_1 = 950Hz, f_2 = 1800Hz$, and $f_3 = 1917Hz$. In Fig. 10.1 is shown the stem plots of the input sequence, upsampled sequence, and upsampled and lowpass filtered sequence in the top, middle, and bottom plots, respectively. As seen in the middle plot, there are M-1 = 5 zeros between every two consecutive samples. The corresponding spectra are shown in Fig. 10.2. There are three frequencies present in the spectrum of the original signal. Since the plot is over the interval

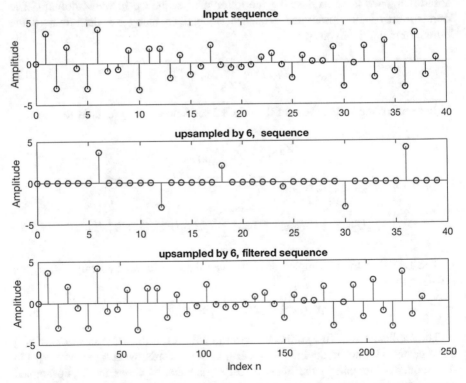

Fig. 10.1 Stem plots of the sequences: top, input sequence sampled as 5000 samples/sec; middle, upsampled sequence using an upsampling factor of 6; bottom, lowpass filtered upsampled sequence; every sixth sample is plotted

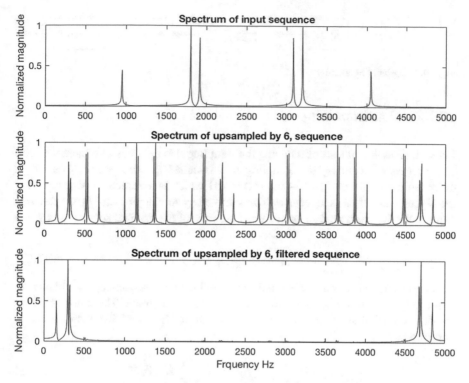

Fig. 10.2 Spectra of the sequences in Example 10.1: top, spectrum of input sequence; middle, spectrum of upsampled sequence; bottom, spectrum of lowpass filtered sequence

$[0, f_s]$, we see the mirror image of the three frequencies in the top plot. The spectrum of the upsampled sequence consists of M-1 = 5 images in the interval between zero and the sampling frequency. The bottom plot depicts the spectrum of the lowpass filtered sequence wherein the images have been removed. Since there are M-1 images in the interval between zero and the sampling frequency, the frequency is compressed. So, if Ω_c is the cutoff frequency of the original sequence, the same frequency will appear at $\frac{\Omega_c}{M}$. Therefore, to remove the images from the upsampled sequence, we should use a lowpass filter having a cutoff frequency $\frac{\Omega_c}{M}$. The MATLAB program used for this example is listed in the M-file named *Example 10_1.m*.

Upsampling Identity The upsampling process consists of first filtering the input sequence by a lowpass filter and then increasing the sampling rate by the positive integer factor M. It is equivalent to first upsampling the input sequence by the factor M followed by lowpass filtering. These two processes are shown in Fig. 10.3. Because of the identity of the two processes, one can use either one to realize upsampling of a sequence. In both cases, the image spectra are removed.

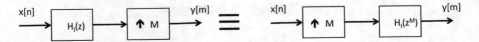

Fig. 10.3 Upsampling identity

10.2.2 Downsampling

Let us look at the process of reducing the sampling rate by a positive integer factor. The process of reducing the sampling rate is called *downsampling*. Consider a discrete-time sequence $\{x[n], -\infty < n < \infty\}$ which is assumed to be sampled at the Nyquist rate. If this sequence is downsampled by an integer factor D, we can then express the downsampled sequence $x_d[n]$ in terms of the original sequence as given by

$$x_d[n] = x[nD], D \in Z \tag{10.7}$$

To understand better the effect of downsampling on the sequence, we will have to describe the downsampled sequence in the frequency domain. The Z-transform of the downsampled sequence is obtained from the definition of Z-transform and is described by

$$X_d(z) = \sum_{n=-\infty}^{\infty} x_d[n]z^{-n} = \sum_{n=-\infty}^{\infty} x[nD]z^{-n} \tag{10.8}$$

Let

$$x'[n] = \begin{cases} x[n], n = 0, \pm D, \pm 2D, \cdots \\ 0, otherwise \end{cases} \tag{10.9}$$

In terms of the new sequence, the Z-transform of the downsampled sequence becomes

$$X_d(z) = \sum_{n=-\infty}^{\infty} x'[nD]z^{-n} \tag{10.10}$$

By using $m = nD$ in the above equation, we have

$$X_d(z) = \sum_{m=-\infty}^{\infty} x'[m]z^{-\frac{m}{D}} = \sum_{m=-\infty}^{\infty} x'[m]\left(z^{\frac{1}{D}}\right)^{-m} = X'\left(z^{\frac{1}{D}}\right) \tag{10.11}$$

We still haven't expressed the Z-transform of the downsampled sequence in terms of the Z-transform of the input sequence. In order to do that, let us define another function given in (10.12):

$$g[n] = \begin{cases} 1, n = 0, \pm D, \pm 2D, \cdots \\ 0, otherwise \end{cases} \tag{10.12}$$

The above sequence can be expressed as the inverse DTFT of the frequency points, $\{W_D^{kn}, 0 \leq k \leq D-1\}$, where, $W_D^k = e^{-j\frac{2\pi k}{D}}$, as given by

$$g[n] = \frac{1}{D} \sum_{k=0}^{D-1} W_D^{kn} \tag{10.13}$$

The sequence $x'[n]$ can be expressed in terms of $x[n]$ and $g[n]$ as

$$x'[n] = g[n]x[n] \tag{10.14}$$

Now using (10.13) and (10.14) in (10.11), we find that

$$X'(z) = \sum_{n=-\infty}^{\infty} g[n]x[n]z^{-n} = \sum_{n=-\infty}^{\infty} \left(\frac{1}{D} \sum_{k=0}^{D-1} W_D^{nk}\right) x[n]z^{-n} \tag{10.15}$$

By interchanging the order of summation in (10.15), we have

$$X'(z) = \frac{1}{D} \sum_{k=0}^{D-1} \sum_{n=-\infty}^{\infty} x[n]\left(zW_D^{-k}\right)^{-n} = \frac{1}{D} \sum_{k=0}^{D-1} X\left(zW_D^{-k}\right) \tag{10.16}$$

Finally, using (10.16) in (10.11), we obtain the Z-transform of the downsampled sequence as

$$X_d(z) = \frac{1}{D} \sum_{k=0}^{D-1} X\left(z^{\frac{1}{D}}W_D^{-k}\right) \tag{10.17}$$

From (10.17), the DTFT of the downsampled sequence is determined to be

$$X_d\left(e^{j\Omega}\right) = X_d(z)\big|_{z=e^{j\Omega}} = \frac{1}{D} \sum_{k=0}^{D-1} X\left(e^{j\left(\frac{\Omega+2\pi k}{D}\right)}\right) = \frac{1}{D} \sum_{k=0}^{D-1} X\left(e^{j\left(\frac{\Omega-2\pi k}{D}\right)}\right) \tag{10.18}$$

From the above equation, we notice that the spectrum of the downsampled sequence is the sum of the shifted and stretched versions of the spectrum of the original sequence. Let us make it clearer by way of an example after we establish the identity of downsampling.

Downsampling Identity The process of downsampling a sequence by a positive integer factor D can be identified as first lowpass filtering the input sequence and then downsampling. This identity is equivalent to first downsampling followed by lowpass filtering. The two processes are shown in Fig. 10.4.

Example 10.2 Use the same sequence as in Example 10.1 and downsample it by a factor of $D = 6$. Compute the spectra of the original, downsampled, and filtered sequences and plot them.

Solution The downsampling and filtering operations are done using the MATLAB M-file named *Example 10_2.m*. The original, downsampled, and filtered sequences

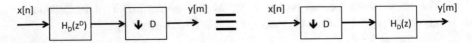

Fig. 10.4 Downsampling identity

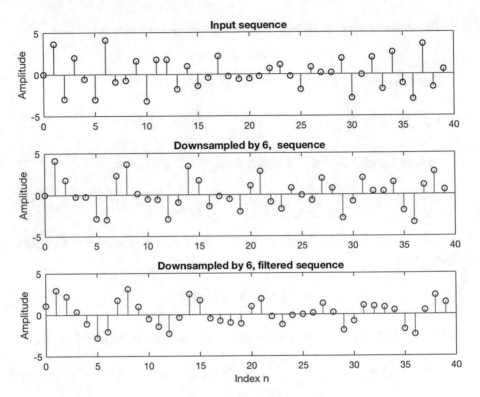

Fig. 10.5 Downsampling the sequence in Example 10.2: top, original sequence; middle, sequence downsampled by a factor of 6; bottom, lowpass filtered downsampled sequence

are shown in Fig. 10.5 in the top, middle, and bottom plots, respectively. We see from the middle plot, the sequence corresponds to every sixth sample of the input sequence. The corresponding spectra are shown in Fig. 10.6.

10.3 Oversampled ADC

As mentioned earlier, an ADC converts a continuous-time signal into a digital signal by first sampling the input signal uniformly at Nyquist rate and then quantizing the analog samples using a B-bit uniform quantizer. The typical bit width of an ADC is

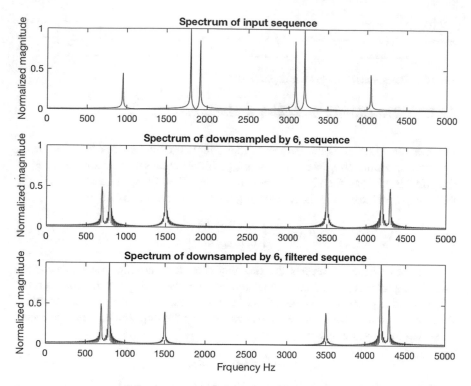

Fig. 10.6 Frequency spectra of the sequences in Fig. 10.5: top, spectrum of the original sequence; middle, spectrum of the downsampled sequence; bottom, spectrum of the downsampled sequence after lowpass filtering

in the range of 8–14 bits. Since the sampling process may introduce aliasing distortion, the input continuous-time signal is filtered by an antialiasing analog filter before sampling. Because of a narrower transition bandwidth, the required analog filter order will be very high. This makes the IC design more complex and it may also lead to instability. In order to ease the requirements on the antialiasing analog filter, one can increase the sampling rate much higher than the Nyquist rate. As we will see below, this will increase the transition bandwidth, thereby lowering the antialiasing filter order. As a result of lowering the filter order, the IC design becomes simpler and more compact. Of course, the output of the ADC must be downsampled to bring the sampling rate back to the Nyquist rate. Figure 10.7 shows the block diagram of an oversampled ADC. The input continuous-time signal is first filtered by an analog lowpass filter of a small order and then input to the ADC. The sampling rate of the ADC is assumed to be M times the Nyquist sampling rate. The output of the ADC is passed through an antialiasing digital filter before it is downsampled by the factor M. In what follows, we will give a brief analysis of the oversampled ADC.

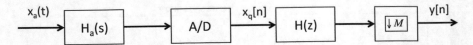

Fig. 10.7 Block diagram of an oversampled ADC

10.3.1　Transition Bandwidth Reduction

Let us assume the maximum frequency in the input continuous-time signal to be f_m. The corresponding Nyquist frequency is $2f_m$. If the actual sampling frequency F_s of the ADC is M times the Nyquist frequency, then $F_s = 2Mf_m$. This amounts to the frequency specification of the antialiasing analog filter as described by

$$H_a(f) = \begin{cases} 1, 0 \le |f| \le f_m \\ 0, Mf_m < |f| < \infty \end{cases} \tag{10.19}$$

The above equation reveals the fact that the passband edge frequency of the antialiasing filter is still the maximum frequency in the analog signal. However, the stopband edge frequency has moved to M times the maximum frequency. Therefore, the transition width, which is the difference between the stopband edge and passband edge, has increased as given by

$$\Delta f = (M - 1)f_m \tag{10.20}$$

If the transition width is large, the filter order will be smaller. That's how the oversampling eases the filter order requirement.

10.3.2　Analysis of Oversampled ADC

Let us go a bit deeper and see the effect of oversampling on the ADC. Let the amplitude range of the input analog signal be

$$|x_a(t)| \le x_{max} \tag{10.21}$$

The corresponding quantization step size of a B-bit ADC is expressed as

$$q = \frac{x_{max} - x_{min}}{2^B} = \frac{2x_{max}}{2^B} = \frac{x_{max}}{2^{B-1}} \tag{10.22}$$

The noise due to quantization is uniformly distributed in the range $[-q/2, q/2]$, and its variance is determined to be

$$\sigma_q^2 = \frac{q^2}{12} = \frac{x_{max}^2}{12 \times 2^{2B-2}} = \frac{x_{max}^2}{3 \times 2^{2B}} \tag{10.23}$$

The output of the ADC is filtered by an antialiasing digital filter, followed by downsampling by the factor M. Let $y[n]$ be the downsampled signal. Since the downsampled signal has the same power as that of the signal before downsampling, we have

$$\sigma_y^2 = G\sigma_q^2 \tag{10.24}$$

where G is the power gain of the antialiasing digital filter, and it is related to the impulse response sequence of the digital filter. If the digital filter is an Nth-order FIR filter, then

$$G = \sum_{n=0}^{N} |h[n]|^2 \tag{10.25}$$

The power gain can also be obtained from the frequency domain since the power is conserved in both domains. Thus,

$$G = \frac{1}{F_s} \int_{-\frac{F_s}{2}}^{\frac{F_s}{2}} |H(f)|^2 df = \frac{1}{F_s} \int_{-\frac{F_s}{2M}}^{\frac{F_s}{2M}} df = \frac{1}{M} \tag{10.26}$$

Therefore, the variance of the noise due to quantization is reduced by the oversampling factor, which is

$$\sigma_y^2 = \frac{\sigma_q^2}{M} \tag{10.27}$$

This is really great! Not only does oversampling reduce the quantization noise, it also spreads it over the frequency range $[0, \frac{F_s}{2}]$. But the cutoff frequency of the antialiasing digital filter is $\frac{F_s}{2M}$. Therefore, it rejects the noise in the range $[\frac{F_s}{2M}, \frac{F_s}{2}]$. This reduction in the quantization noise is related to the reduction in the number of bits of the ADC as compared to the ADC that does not use oversampling. Let B_M and B denote the bit widths of the ADCs with and without oversampling. Since the quantization noise power must be the same in both ADCs for the sake of comparison, the output noise powers can be expressed as

$$\frac{\sigma_q^2}{M} = \frac{x_{max}^2}{3 \times M \times 2^{2B_M}} = \frac{x_{max}^2}{3 \times 2^{2B}} \tag{10.28}$$

From (10.28) we have

$$M \times 2^{2B_M} = 2^{2B} \tag{10.29a}$$

$$B_M = B - \frac{1}{2}log_2M \tag{10.29b}$$

Thus, for the same given quantization noise power, an oversampled ADC needs a bit width that is less than that required by the Nyquist sampled ADC by one half times the logarithm to the base 2 of the oversampling factor.

Oversampling Factor Versus the Filter Order We can relate the oversampling factor M to the filter order for a specific filter as follows. Consider a kth-order Butterworth antialiasing analog filter with a cutoff frequency f_m. The corresponding magnitude of the frequency response is given by

$$|H_a(f)| = \frac{1}{\sqrt{1 + \left(\frac{f}{f_m}\right)^{2k}}} \tag{10.30}$$

The maximum error δ due to aliasing distortion will occur at the *folding* frequency $\frac{F_s}{2}$ and is equal to

$$\delta = \left|H_a\left(\frac{F_s}{2}\right)\right| = \frac{1}{\sqrt{1 + \left(\frac{Mf_m}{f_m}\right)^{2k}}} = \frac{1}{\sqrt{1 + M^{2k}}} \tag{10.31}$$

Then,

$$\frac{1}{\sqrt{1 + M^{2k}}} \leq \delta \tag{10.32}$$

Or

$$1 + M^{2k} \geq \delta^{-2} \Rightarrow M \geq \left(\delta^{-2} - 1\right)^{\frac{1}{2k}} \tag{10.33}$$

For instance, if $\delta = 0.005$ and $k = 3$, then from (10.33), we find that the oversampling factor $M = \lceil 5.848 \rceil = 6$. If the filter order is reduced to 2, then for the same maximum aliasing distortion, the oversampling ratio hikes up to 15. A decrease by one in the filter order increases the oversampling ratio almost by a factor of 3! The maximum aliasing error in dB can be expressed in terms of the oversampling factor and the Butterworth filter order using (10.33) and is given by

$$\delta = -10log_{10}\left(1 + M^{2k}\right) \ dB \tag{10.34}$$

Figure 10.8 shows the maximum aliasing error in dB against the oversampling factor for four values of the Butterworth filter order. As seen from the figure, the oversampling factor increases as the Butterworth filter order decreases for a fixed value of the maximum aliasing error.

Example 10.3 Let us work out an example using MATLAB to digest what we learnt about oversampled ADC. We will use the same input signal that was used in the previous example. The Nyquist sampling rate is 4000 Hz and the oversampling factor used is 15. We will use a 5-bit ADC to quantize the analog samples. The antialiasing digital filter is a fourth-order FIR filter with a normalized cutoff

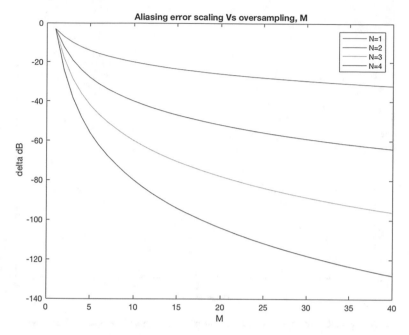

Fig. 10.8 Oversampling factor versus the maximum aliasing error in dB for four different values of the Butterworth analog filter order

frequency of 1/M. The M-file to solve this problem is named *Example 10_3.m*. The program designs a 5-bit uniform quantizer before quantizing the input signal. Figure 10.9 depicts the input-output relationship of the quantizer as a step function. The horizontal axis refers to decision regions and the vertical axis to the reconstruction levels. The actual and 5-bit quantized input sequence is shown in Fig. 10.10 as a stem plot. It appears that the 5-bit quantized version very nearly approximates the full-precision signal. The downsampled versions of the full-precision and 5-bit quantized version of the input signal are shown in Fig. 10.11 in the top and bottom plots, respectively. In Fig. 10.12 is shown the spectra of the input sequence, filtered output, and filtered downsampled output sequences in the top, middle, and bottom plots, respectively. As seen from the figure, the folding frequency of the downsampled signal reverts to 2000 Hz. As pointed out, the quantization noise of the ADC is uniformly distributed in the range $\pm\frac{q}{2}$. Figure 10.13 shows the 5-bit quantizer noise sequence along with its histogram and spectrum in the top, middle, and bottom plots, respectively. As expected, the histogram appears uniform implying that the quantizer noise is uniformly distributed. The program also computes the noise variances of the quantizer before and after filtering by the antialiasing digital filter. The noise variance or power of the 5-bit quantizer is found to be 0.0081. The corresponding SNR is 26.96 dB. The noise variance and the resulting SNR after filtering are, respectively, 0.0037 and 30.33 dB. Because of oversampling, the quantization noise power at the output of the antialiasing digital filter has decreased by more than 3 dB!

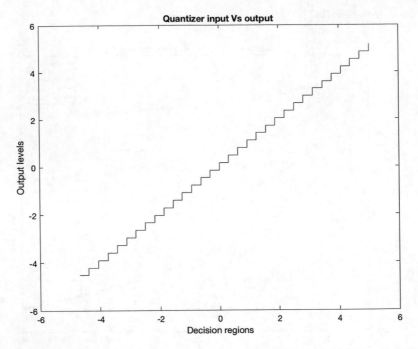

Fig. 10.9 Input/output relationship of a 5-bit uniform quantizer of Example 10.3 as a step function

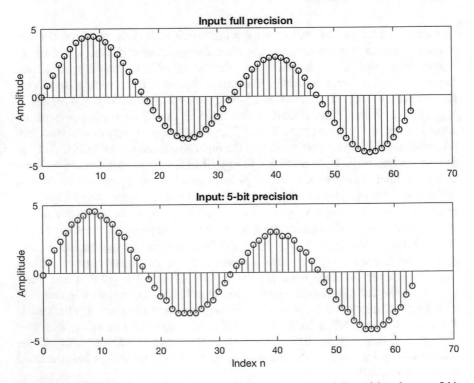

Fig. 10.10 Full-precision and 5-bit quantized input sequence: top, full-precision; bottom, 5-bit quantized sequence

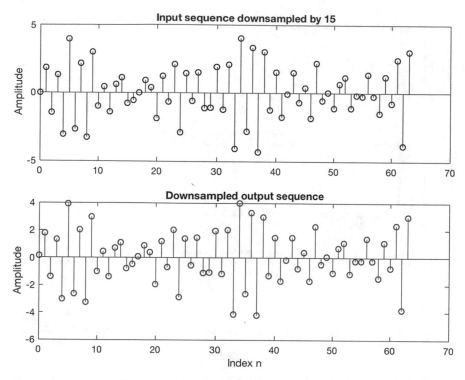

Fig. 10.11 Downsampled sequences: top, full-precision sequence; bottom: 5-bit quantized sequence

10.4 Oversampled DAC

In the previous section, we described an oversampled ADC. The purpose of using a sampling rate much higher than the Nyquist rate is to ease the requirements on the antialiasing analog prefilter by increasing the transition width. Oversampling also reduces the ADC bit width and rejects the out-of-band noise. After transmission and reception of the digital signal, it must be converted back to the analog domain. Since the sampling rate at the ADC is high, it has to be brought back to the Nyquist rate. A block diagram of an oversampled DAC is shown in Fig. 10.14. Since the process of downsampling introduces spectral images, the images must be removed using an anti-imaging lowpass analog filter. The output of the anti-imaging lowpass analog filter is the recovered analog signal. Let us exemplify the oversampled DAC operation by an example using MATLAB.

Example 10.4 This example uses the same signal used in the previous example. To be self-contained, the oversampled ADC is incorporated. Its output is then converted to an analog signal. The anti-imaging lowpass analog filter is incorporated into the DAC. The M-file to execute Example 10.4 is named *Example 10_4.m*. The input to

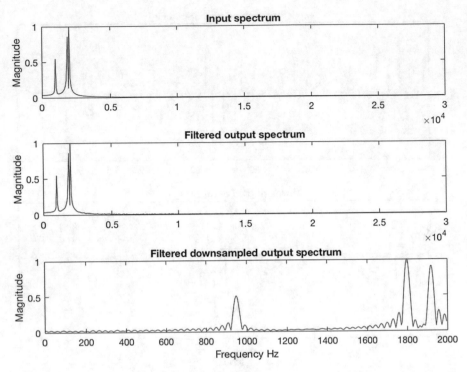

Fig. 10.12 Spectra of the sequences in Example 10.3: top, input sequence; middle, filtered output sequences; bottom, filtered and downsampled sequence

the DAC is a 5-bit quantized sequence that is oversampled by a factor of 6. The anti-imaging filter is a second-order Butterworth filter. The input analog signal and the DAC output signal are shown in Fig. 10.15 in the top and bottom plots, respectively. They seem to be pretty close. The corresponding spectra are shown in Fig. 10.16. Because of quantization, there appears some noise in the output spectrum. The SNR between the input analog signal and the DAC output signal is found to be 25.36 dB. In both oversampled ADC and DAC, DSP plays an important role in filtering the sequences using digital filters.

10.5   Cancelation of Inter-Symbol Interference

In a digital communications system, messages are represented in binary format. Each binary symbol is transmitted as a pulse. These pulses are first lowpass filtered at the transmitter before transmission to confine them to a certain specified bandwidth. As these pulses travel through the channel, they are distorted in their amplitude and phase due to the channel reactance. As a result, the pulses expand in time and so

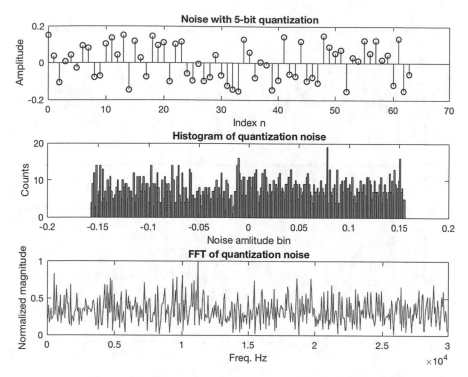

Fig. 10.13 5-bit quantizer noise: top, noise sequence; middle, histogram of the quantizer noise; bottom, spectrum of the quantizer noise

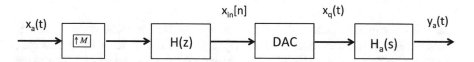

Fig. 10.14 Block diagram of an oversampled DAC

overlap between neighboring pulses. The heart of digital communications is in its precise timing. If pulses from neighboring bit intervals overlap, then error occurs in the detection of the actual transmitted pulse in each bit interval. That is to say that interference between pulses occurs due to distortion in the transmitted pulses. This is known as the *inter-symbol interference* (ISI). The transmitter/channel/receiver chain can be modeled as a cascade of three LTI systems as described by

$$H(f) = H_t(f)H_c(f)H_r(f) \qquad (10.35)$$

where $H_t(f)$ represents the transmitter, $H_c(f)$ the channel, and $H_r(f)$ the receiver. The ISI can be canceled or minimized by adjusting the transmitter and receiver filters. Since the channel is not under our control, we cannot do anything with it. By a proper choice of the transmitter, we can shape the transmitted pulses in such a way

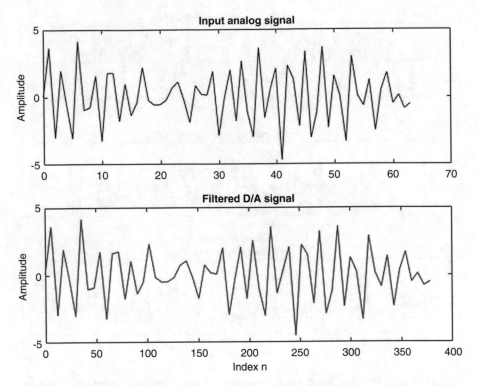

Fig. 10.15 Input analog and DAC output analog signals: top, input analog signal; bottom, DAC output

that the ISI can be minimized. This process is known as *pulse shaping* and is accomplished at the transmitter. Having taken care of the share of the transmitter, the channel effect can be canceled or minimized by a proper choice of a filter at the receiver. This is known as channel *equalization*. We have learnt phase and group delay equalization in the chapter on IIR digital filters. However, when the channel characteristics change slowly, then the equalizing digital filter coefficients should also change. This results in *adaptive equalizers* at the receiver. In any case, digital filtering is involved.

10.5.1 Pulse Shaping

According to Nyquist, if the overall system $H(f)$, from transmitter to receiver in (10.35), amounts to an ideal filter with a cutoff frequency equal to half the transmission symbol rate R_s, then there will be no ISI at the receiver. The ideal filter has the characteristics

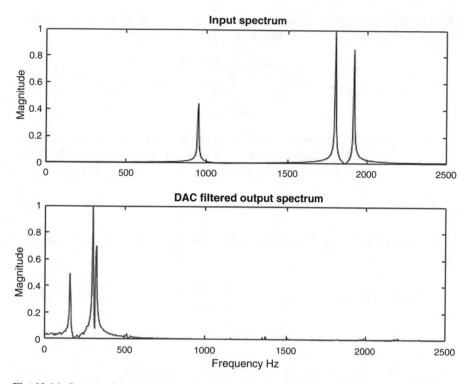

Fig. 10.16 Spectra of the signals in Fig. 10.15: top, spectrum of the input analog signal; bottom, spectrum of the DAC output analog signal

$$H(f) = \begin{cases} 1, |f| \leq W_0 \\ 0, otherwise \end{cases} \qquad (10.36)$$

where

$$W_0 = \frac{1}{2T_s} = \frac{R_s}{2} \qquad (10.37)$$

The corresponding impulse response of the ideal filter is given by

$$h(t) = \frac{\sin(2\pi W_0 t)}{\pi t} = 2W_0 sinc(2W_0 t), \; -\infty < t < \infty \qquad (10.38)$$

The problem with the above ideal filter is that its impulse response is not zero for $t < 0$ and is, therefore, non-causal, meaning that it is not physically realizable as is. It is also susceptible to small timing errors. In effect, what we mean by the above statements is that the Nyquist bandwidth of the filter in (10.37) is not realizable. So, what are we going to do? Fortunately, one can allow some excess bandwidth, thereby making the filter realizable. A small excess bandwidth is tolerable indeed. This is accomplished by using a filter with a frequency response described by

$$H_{RC}(f) = \begin{cases} 1, |f| \leq 2W_0 - W \\ cos^2\left(\dfrac{\pi}{4}\dfrac{|f| + W - 2W_0}{W - W_0}\right), 2W_0 - W < |f| \leq W \\ 0, |f| > W \end{cases} \quad (10.39)$$

Since the frequency response of the filter in (10.39) follows the square of a cosine function, it is called the *raised cosine* (RC) filter. In the raised cosine filter, the quantity $W - W_0$ is the *excess bandwidth* and the *roll-off* factor is defined as

$$r = \frac{W - W_0}{W_0} \quad (10.40)$$

Corresponding to (10.39), the impulse response of the RC filter can be shown to be

$$h_{RC}(t) = 2W_0 sinc(2W_0 t)\left[\frac{\cos(2\pi(W - W_0)t)}{1 - (4(W - W_0)t)^2}\right], \quad -\infty < t < \infty \quad (10.41)$$

Have we solved anything in using RC filter? Even though the impulse response of the RC filter is non-causal, it decays very rapidly, and so it can be truncated without incurring any penalty. Thus, the RC filter becomes causal and realizable. What we are implying is that if the overall frequency response of the communications system from transmitter to receiver corresponds to the raised cosine filter response, i.e.,

$$H(f) = H_{RC}(f), \quad (10.42)$$

then there will be zero ISI because it satisfies Nyquist condition. One way to design the transmitter and receiver filters is to use the following:

$$|H_t(f)| = |H_r(f)| = \sqrt{H_{RC}(f)} \quad (10.43)$$

If (10.43) is satisfied, the overall frequency response of the communications system will not only satisfy Nyquist condition for zero ISI but will also be physically realizable.

Figure 10.17 shows the impulse response of an analog RC filter over the time interval $-0.02 \leq t \leq 0.02$ sec for three roll-off factors of 0, 0.5, and 1. As can be seen from the plots, the impulse response decays very rapidly for the roll-off factor 1. The frequency response of the RC filter corresponding to the three roll-off factors is shown in Fig. 10.18. The excess bandwidth is the largest for the case where the roll-off factor is 1. By discretizing the impulse response of the analog RC filter, we can obtain the impulse response of the corresponding digital filter. Figure 10.19 shows the discrete version of the impulse response of the RC filter, and its frequency response is shown in Fig. 10.20 for the same three values of the roll-off factor. For the discrete-time version of the RC filter, the sampling frequency is assumed to be 4.5 times the Nyquist bandwidth. We further show in Fig. 10.21 the process of

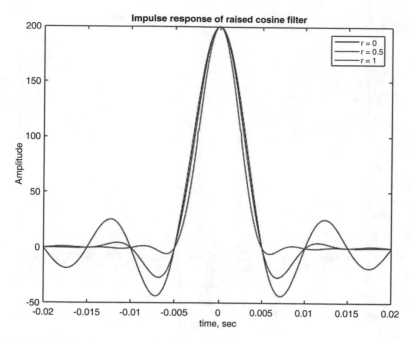

Fig. 10.17 Impulse response of the raised cosine filter in (10.41) with a Nyquist frequency of 100 Hz for three values of the roll-off factor

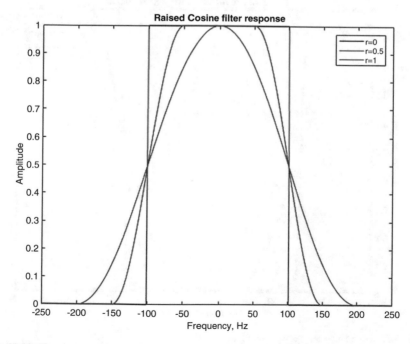

Fig. 10.18 Frequency response of the RC filter whose impulse response is shown in Fig. 10.17

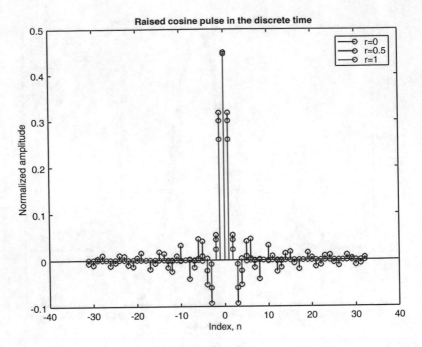

Fig. 10.19 Impulse response of the digital RC filter

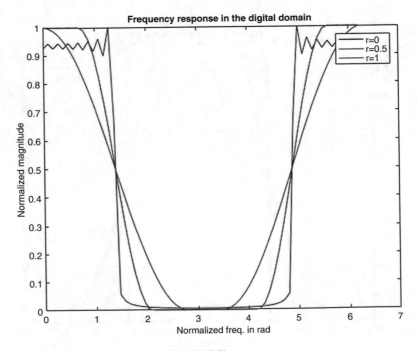

Fig. 10.20 Frequency response of the digital RC filter

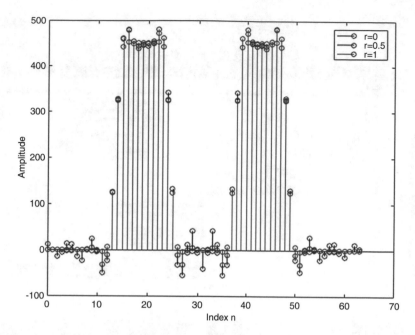

Fig. 10.21   Processing a sequence of pulses through the RC filter

filtering a sequence of rectangular pulses with the RC filter. We see no interference in the main lobe because of the RC shape of the lowpass filter. Even though there is some overlapping of the pulses in adjacent symbol intervals, there is no significant ISI due to the facts that the RC filter response decays very rapidly in the time domain and due to sinc(x) nature of the impulse response. The MATLAB M-file to obtain the impulse and frequency responses of the RC filter in both the continuous-time and discrete-time domains is named *Raised_cosine.m*.

Simulink Example for Pulse Shaping   In this example we simulate the process of shaping a pulse sequence by a raised cosine filter using MATLAB's Simulink. Figure 10.22 shows the block diagram consisting of a pulse generator, whose output is filtered by an RC filter and two scopes to display the respective signals. The parameters of the pulse generator and RC filter are listed by the side of the respective blocks, as shown in the figure. The simulation time is chosen to be 1 sec. After starting the simulation, the respective outputs in the time domain are displayed on the scopes and are shown in Figs. 10.23 and 10.24. The Simulink program is named *RC_filter.slx*.

10.5.2   Equalization

Equalization is the process of correcting the ISI induced by the channel. There are linear and nonlinear equalization techniques available in the literature. We will only

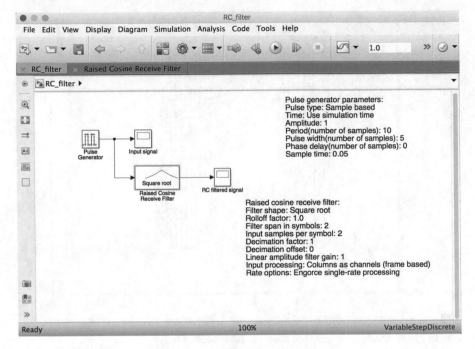

Fig. 10.22 Block diagram to simulate RC filtering of a pulse sequence

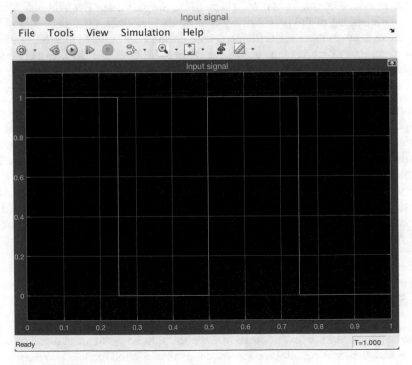

Fig. 10.23 Input pulse sequence to the RC filter

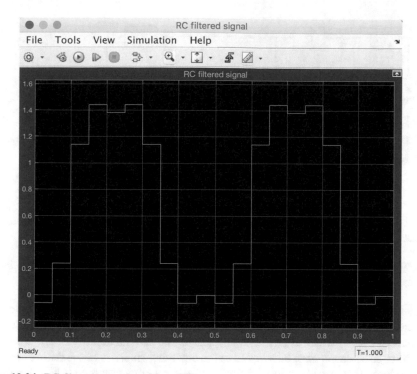

Fig. 10.24 RC-filtered sequence with no ISI

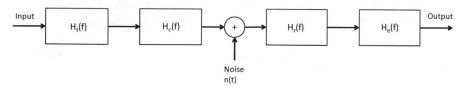

Fig. 10.25 Transmitter-receiver chain with an equalizing filter at the receiver to correct ISI due to channel impairments

deal with linear equalization procedures in this book. Mobile radio channel is nonstationary, meaning that the channel characteristics keep changing from time to time. This happens because the transmitted signal takes different paths as a result of reflection from the nearby tall buildings, hills, towers, etc. Also, since the receiver is not fixed, these reflected signals cause *fading*. This effect is known as *multipath fading*. Due to the relative motion between the transmitter and the receiver, there is the effect of Doppler spread in the received frequency. These impairments cause severe inter-symbol interference, which results in a high rate of bit errors at the receiver. As we saw earlier, if the overall system corresponds to a raised cosine filter function, then there will be no ISI. Therefore, the product of the transmitter and receiver filters equals the raised cosine filter response. In Fig. 10.25 is shown the

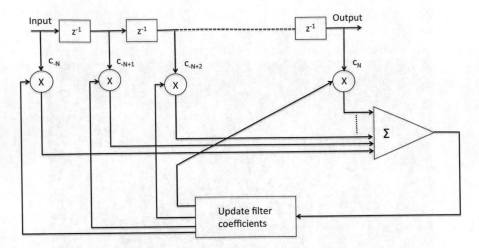

Fig. 10.26 A linear adaptive transversal filter as a channel equalizer

transmitter-channel-receiver tandem. At the receiver, an equalizing filter is employed
to correct the channel impairments. The most common equalization filter is the
transversal filter, which is an FIR filter. Due to the non-stationarity of the channel,
the filter coefficients must be adapted to the changing statistics of the radio channel.

A block diagram of a linear adaptive transversal equalizer with 2 N + 1 taps is
shown in Fig. 10.26. The delay in each delay element corresponds to a symbol
duration. This type of equalizer is termed a *symbol-spaced equalizer*. The response
of the transversal filter $y[k]$ can be expressed in terms of the input $x[n]$ as

$$y[k] = \sum_{n=-N}^{N} c_n x[k-n], \; -N \le k \le N \tag{10.44}$$

In compact matrix form, (10.44) can be written as

$$\boldsymbol{y} = \mathbf{Xc} \tag{10.45}$$

where

$$\boldsymbol{y} = [y_{-N} \cdots \cdots y_N]^T \tag{10.46a}$$

$$\boldsymbol{c} = [c_{-N} \cdots c_0 \cdots c_N]^T \; \text{and} \tag{10.46b}$$

$$\boldsymbol{X} = \begin{bmatrix} x_{-N} \cdots \cdots \cdots 0 \\ x_{-N+1} \; x_{-N} \cdots \\ \vdots \\ \vdots \\ 0 \; 0 \cdots \cdots \cdots x_N \end{bmatrix} \tag{10.46c}$$

The criterion for selecting the filter coefficients $\{c_n\}$ is based on minimizing an objective function such as the mean square error (MSE) or absolute peak error. Since the channel statistics are time-variant, these coefficients are frequently changed using a suitable adaptation scheme. The number of taps in the transversal filter is usually chosen to be larger than the number of symbols involved in the ISI.

Simulink Example for Equalization In order to get a better picture of equalization to cancel ISI in digital communications, let us look at an example using MATLAB's Simulink. In this example, we consider an 8-ary QAM (quadrature amplitude modulation) scheme. In QAM, the amplitude of a carrier is modulated using discrete values. More specifically, the transmitted waveform is described by

$$x_{QAM}(t) = Ap(t)\cos\left(2\pi f_c t\right) + Bp(t)\sin\left(2\pi f_c t\right) \tag{10.47}$$

where $p(t)$ is a rectangular pulse of duration equal to the symbol duration, $\{A\}$ and $\{B\}$ are the sets of amplitudes with M values each, and f_c is the carrier frequency. These amplitudes have $M = 2^k$ discrete values corresponding to k-bit symbols. Let us choose $M = 8$. The simulation will be carried in the baseband, that is, no carrier modulation will be used. The channel introduces additive white Gaussian noise (AWGN) with a signal-to-noise ratio (SNR) of 20 dB. The channel is modeled as a four-tap FIR filter whose impulse response is described by

$$h_e[n] = 1 - 0.3z^{-1} + 0.1z^{-2} + 0.2z^{-3} \tag{10.48}$$

The adaptive transversal equalizer has 8 taps. The simulation uses least mean square (LMS) algorithm to adaptively estimate the filter taps. The signal sets are typically viewed as a constellation, where the two-dimensional vectors are described by

$$d_m = \left(\sqrt{E_s}A \quad \sqrt{E_s}B\right) \tag{10.49}$$

where E_s is the signal energy and $\{A\}$ and $\{B\}$ have $M = 8$ discrete values. In Fig. 10.27 is shown the block diagram of the 8-ary QAM system. The input to the LMS adapter is the signal from the AWGN block, and the desired signal is the output of the QAM modulator. The constellation diagrams of the signals before and after equalization are shown in Fig. 10.28 for an SNR of 20 dB. When the SNR is increased to 40 dB, the clusters appear more focused as seen from Fig. 10.29. The Simulink file is named *Adaptive_Equalizer.slx*. For more details on the parameters used in various blocks in Fig. 10.27, the reader may double-click each block to learn and modify the parameters.

10.5.3 Matched Filter

In digital communications, PCM binary digits are represented by pulses for transmission. In baseband communications, these pulses are transmitted as such, whereas

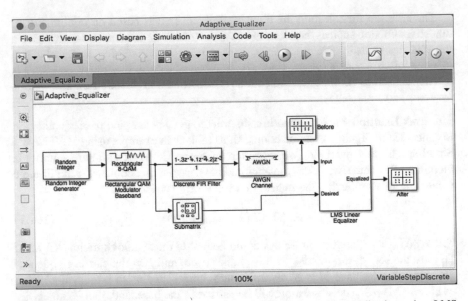

Fig. 10.27 Block diagram to simulate an 8-ary QAM system with equalization using LMS algorithm

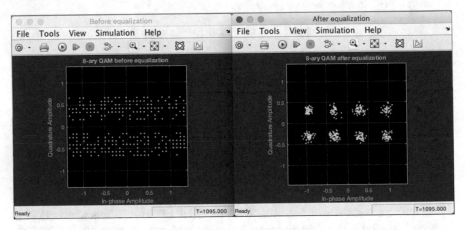

Fig. 10.28 Constellation diagram of the 8-ary QAM system in Fig. 10.27 showing the signal sets before and after equalization for an SNR of 20 dB: left, before equalization; right, after equalization

in carrier communications, these pulses modulate a carrier using different modulation schemes. We will consider baseband transmission here. The binary pulses may take one of non-return to zero (NRZ), return to zero (RZ), and Manchester code. Each one of these pulse types will affect the communications in terms of the DC component, self-clocking, error detection, bandwidth compression, noise immunity, etc. Our task here is to find out what is matched filter, why is it used in digital communications, and can it be realized as a digital filter. As pointed out earlier, the

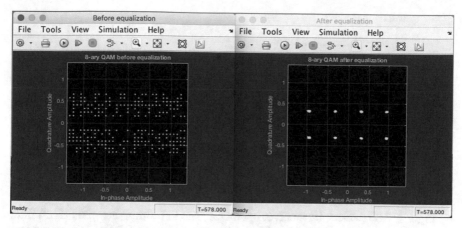

Fig. 10.29 Constellation diagram of the 8-ary QAM system in Fig. 10.27 showing the signal sets before and after equalization for an SNR of 40 dB: left, before equalization; right, after equalization

key factor in digital communications is the timing. The main task of the detector or receiver is to determine in each bit interval which binary digit – a binary "0" or a binary "1" – was transmitted. If there is no channel disturbance such as noise, then there is no problem in deciding which bit is transmitted in a given bit interval. However, the channel adds noise, namely, white Gaussian noise. Since the noise is added to the transmitted signal, this channel-induced noise is called the additive white Gaussian noise (AWGN). The received signal $r(t)$ is the sum of the transmitted signal and noise, as defined by

$$r(t) = s_i(t) + n(t), i = 1, 2; 0 \le t \le T \qquad (10.50)$$

where T is the bit period and the transmitted signal takes the form

$$s_i(t) = \begin{cases} s_1(t), 0 \le t \le T, \textit{for a binary'} 1' \\ s_2(t), 0 \le t \le T, \textit{for a binary'} 0' \end{cases} \qquad (10.51)$$

The processing consists of first filtering the received signal $r(t)$ followed by sampling the filtered signal $z(t)$ at the end of the bit period. Since the filter is LTI, the filtered signal is expressed as

$$z(t) = a_i(t) + n_0(t) \qquad (10.52)$$

The sampled signal value $z(T)$ is then compared against a predetermined threshold value to decide which binary bit was transmitted in that bit interval. The sampled signal is described by

$$z(T) = a_i(T) + n_0(t) \qquad (10.53)$$

If the threshold value is denoted by γ, then the decision amounts to

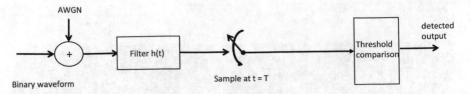

Fig. 10.30 Linear processing at the receiver to detect transmitted binary symbols

$$\widehat{s}(t) = \begin{cases} s_1(t) \ \ if \ z(T) > \gamma \\ s_2(t) \ \ if \ z(T) < \gamma \end{cases} \tag{10.54}$$

Figure 10.30 depicts the receiver operation. The linear time invariant (LTI) filter is implemented as a digital filter. The noise component in (10.53) is a zero-mean Gaussian random variable with a standard deviation σ_0. The probability density function (pdf) of the noise component $n_0(T)$ takes the form

$$p(n_0) = \frac{1}{\sigma_0\sqrt{2\pi}} exp\left(-\frac{n_0^2}{2\sigma_0^2}\right) \tag{10.55}$$

Since $z(T)$ is the sum of Gaussian noise and a signal component, it is also a Gaussian random variable with the same standard deviation as that of $n_0(T)$ but with a mean of $a_i(T)$. Thus, depending on which binary digit is transmitted, the pdf of $z(T)$ is given by

$$p(z|s_1) = \frac{1}{\sigma_0\sqrt{2\pi}} exp\left(-\frac{(z - a_1)^2}{2\sigma_0^2}\right) \tag{10.56a}$$

$$p(z|s_2) = \frac{1}{\sigma_0\sqrt{2\pi}} exp\left(-\frac{(z - a_2)^2}{2\sigma_0^2}\right) \tag{10.56b}$$

The two conditional pdfs in (10.56a) and (10.56b) are illustrated in Fig. 10.31 for the case $a_1 = -a_2 = 2$.

The objective of the detector is to detect which binary digit is transmitted in a given bit interval with the least amount of average bit error. As seen from Fig. 10.30, there are two variables to adjust so as to minimize the probability of a bit error. The first variable is the linear filter. By choosing the right filter, the probability of a bit error is minimized. This results in what is known as the *matched filter* (MF). The second variable is the threshold. Choosing the optimal threshold further minimizes the bit error probability. This results in *maximum likelihood receiver*.

Maximum Likelihood Receiver The decision threshold γ is chosen so as to minimize the probability of a bit error. This is achieved by maximizing the likelihood ratio

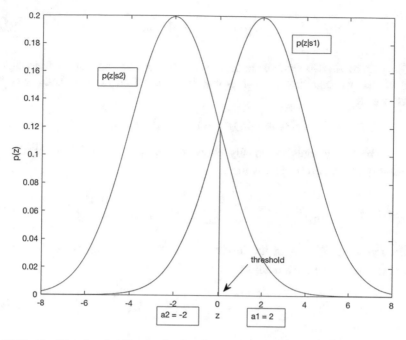

Fig. 10.31 Conditional probability density functions of the linearly processed and sampled signal

$$if \ \frac{p(z|s_1)}{p(z|s_2)} > \frac{P(s_2)}{P(s_1)}, choose \ s_1(t) \tag{10.57a}$$

$$if \ \frac{p(z|s_1)}{p(z|s_2)} < \frac{P(s_2)}{P(s_1)}, choose \ s_2(t) \tag{10.57b}$$

In the above equations, $P(s_1)$ and $P(s_2)$ are the a priori probabilities of the binary waveforms $s_1(t)$ and $s_2(t)$, respectively. Based on (10.57), the optimal threshold corresponds to the intersection of the two conditional pdfs, which is given by

$$\gamma_0 = \frac{a_1 + a_2}{2} \tag{10.58}$$

With the threshold value being determined, the probability of a bit error is determined as follows. The probability of making an error given s_1 was transmitted equals the area under the curve $p(z|s_1)$ from $-\infty$ to γ_0, which is

$$p(e|s_1) = \int_{-\infty}^{\gamma_0} p(z|s_1) dz \tag{10.59}$$

Similarly, the probability of making an error given s_2 was transmitted equals the area under the curve $p(z|s_2)$ from γ_0 to ∞, which is

$$p(e|s_2) = \int_{\gamma_0}^{\infty} p(z|s_2)dz \tag{10.60}$$

Since there are two symbols in the system, the average bit error is the weighted sum of the two conditional error probabilities in (10.59) and (10.60), which is expressed as

$$P_B = P(s_1)p(e|s_1) + P(s_2)p(e|s_2) \tag{10.61}$$

If the binary symbols are equally likely, that is, if $P(s_1) = P(s_2) = \frac{1}{2}$, then the probability of a bit error reduces to

$$P_B = \int_{\gamma_0 = \frac{a_1 + a_2}{2}}^{\infty} p(z|s_2)dz = \frac{1}{\sigma_0\sqrt{2\pi}} \int_{\gamma_0}^{\infty} exp\left(-\frac{(z - a_2)^2}{2\sigma_0^2}\right)dz \tag{10.62}$$

By replacing $\frac{z-a_2}{\sigma_0}$ by x, the lower limit in (10.62) becomes $\frac{a_1-a_2}{2\sigma_0}$. Then the probability of a bit error amounts to

$$P_B = \frac{1}{\sqrt{2\pi}} \int_{\frac{a_1-a_2}{2\sigma_0}}^{\infty} exp\left(-\frac{x^2}{2}\right)dx = Q\left(\frac{a_1 - a_2}{2\sigma_0}\right) \tag{10.63}$$

where $Q(x)$ is called the *complementary error function* and is defined as

$$Q(x) = \frac{1}{\sqrt{2\pi}} \int_{x}^{\infty} exp\left(-\frac{x^2}{2}\right)dx \tag{10.64}$$

In other words, the complementary error function equals the area under the normal curve with zero mean and unit variance from x to ∞. It does not have a closed form solution but is found in most standard textbooks on digital communications. It is also available in MATLAB as a built-in function *erfc*.

Matched Filter   From (10.63), we observe that larger the threshold value, smaller the area under the normal curve or smaller the probability of a bit error. The matched filter maximizes the argument of the complementary error function. Let the input to the LTI filter in Fig. 10.30 be the sum of a known signal $s(t)$ and an AWGN $n(t)$. The output of the filter at the end of the bit interval T equals

$$z(T) = a_i(T) + n_0(T) \tag{10.65}$$

Therefore, the signal-to-noise ratio at $t = T$ is

$$\left(\frac{S}{N}\right)_T = \frac{a_i^2}{\sigma_0^2} \tag{10.66}$$

What we need is a filter that maximizes the SNR in (10.66). The signal component at the filter output can be related to the filter transfer function via Fourier transform by

$$a(t) = \int_{-\infty}^{\infty} H(f)S(f)e^{j2\pi ft} df \qquad (10.67)$$

In (10.67), S(f) is the Fourier transform of the signal s(t). If we assume the two-sided power spectral density of the output Gaussian noise to be $\frac{N_0}{2}$ watts/Hz, then the noise power at the output of the filter is given by

$$\sigma_0^2 = \frac{N_0}{2} \int_{-\infty}^{\infty} |H(f)|^2 df \qquad (10.68)$$

Using (10.66, 10.67, and 10.68), the SNR at the end of the bit interval becomes

$$\left(\frac{S}{N}\right)_T = \frac{\left|\int_{-\infty}^{\infty} H(f)S(f)e^{j2\pi fT} df\right|^2}{\frac{N_0}{2} \int_{-\infty}^{\infty} |H(f)|^2 df} \qquad (10.69)$$

Using Schwartz's inequality, the numerator of (10.69) can be written as

$$\left|\int_{-\infty}^{\infty} H(f)S(f)e^{j2\pi fT} df\right|^2 \leq \int_{-\infty}^{\infty} |H(f)|^2 df \int_{-\infty}^{\infty} |S(f)|^2 df \qquad (10.70)$$

Using (10.70) in (10.69), the expression for the SNR at t = T takes the form

$$\left(\frac{S}{N}\right)_T \leq \frac{2}{N_0} \int_{-\infty}^{\infty} |S(f)|^2 df \qquad (10.72)$$

The maximum SNR is then equal to

$$max\left(\frac{S}{N}\right)_T = \frac{2}{N_0} \int_{-\infty}^{\infty} |S(f)|^2 df = \frac{2E}{N_0} \qquad (10.73)$$

where the signal energy E is given by

$$E = \int_{-\infty}^{\infty} |S(f)|^2 df \qquad (10.74)$$

The equality holds in Schwartz's inequality if the following condition is met:

$$H(f) = H_0(f) = KS^*(f)e^{-j2\pi fT} \qquad (10.75)$$

The interpretation of (10.75) in the time domain from the time reversal property of the Fourier transform is that the impulse response of the matched filter is the time-reversed and right-shifted version of the signal. Mathematically speaking, it amounts to

$$h(t) = \begin{cases} Ks(T-t), 0 \leq t \leq T \\ 0, otherwise \end{cases} \qquad (10.76)$$

From the above equation, we see that the filter impulse response is a replica of the transmitted pulse waveform but for the amplitude and time reversing and shifting. That is the reason the filter is known as the matched filter. Since the MF maximizes the SNR at the end of the bit interval, the maximum SNR is written as

$$\left(\frac{S}{N}\right)_T = \frac{(a_1 - a_2)^2}{\sigma_0^2} = \frac{E_d}{\frac{N_0}{2}} = \frac{2E_d}{N_0} \tag{10.77}$$

where energy difference in (10.77) stands for

$$E_d = \int_0^T (s_1 - s_2)^2 dt \tag{10.78}$$

Using (10.77) and (10.78), the probability of a bit error reduces to

$$P_B = Q\left(\frac{a_1 - a_2}{2\sigma_0}\right) = Q\left(\sqrt{\frac{2E_d}{N_0}}\right) \tag{10.79}$$

Correlation Filter Another interpretation of the MF is as follows. We can express the response of the MF to the received signal at time t as

$$z(t) = \int_0^t r(\tau)s(T - t + \tau)d\tau \tag{10.80}$$

At the end of the bit interval t = T, the MF response becomes

$$z(T) = \int_0^T r(\tau)s(\tau)d\tau, \tag{10.81}$$

which is what is known as the correlation of r(t) with s(t). Therefore, the MF is also a correlation filter. To implement the MF as a correlation filter, we multiply the received signal by a replica of the transmitted signal and integrate the product over the bit interval. The output of the correlation filter at the end of the bit interval is the same as that of the MF.

MF Example Let us consider a simple example based on MATLAB to determine the impulse response and the filter response of a matched filter corresponding to a binary signal set. Let the two signals be defined by

$$s_1[n] = \begin{cases} 1, 0 \leq n \leq 9 \\ 0, otherwise \end{cases} \tag{10.82a}$$

$$s_2[n] = \begin{cases} -1, 0 \leq n \leq 9 \\ 0, otherwise \end{cases} \tag{10.82b}$$

Figure 10.32 shows the signal $s_1[n]$ in the top plot and the impulse response of the corresponding matched filter in the bottom plot. Note that the two sequences look identical because the MF impulse response is flipped and shifted to the right by the

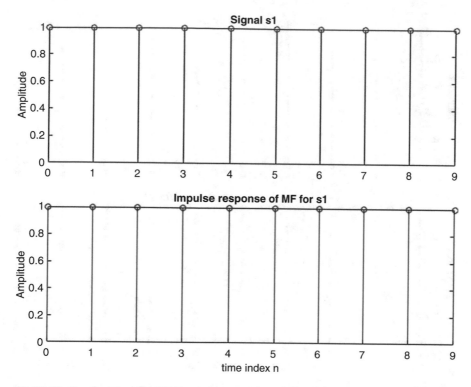

Fig. 10.32 Signal $s_1[n]$ and its MF impulse response: top, signal $s_1[n]$; bottom, corresponding MF impulse response

bit duration, which is of length ten samples. Similarly, the signal $s_2[n]$ and the corresponding MF impulse response are shown in Fig. 10.33. The responses of the two matched filters to input signal plus noise are shown in Fig. 10.34, where the top plot shows the MF response to the signal $s_1[n]$ plus noise and the bottom plot shows the MF response to the signal $s_2[n]$ plus noise. The noise is a Gaussian noise with a standard deviation of 0.25. As expected, the response is a maximum at the end of the bit interval. The responses of the corresponding correlation filters to the two signals are depicted in Fig. 10.35. The responses reach the maximum value at the end of the bit interval. Therefore, the correlation filters are equivalent to the matched filters. The M-file is named *Matched_filter.m*.

10.5.4   *Phase-Locked Loop*

Phase-locked loop, PLL for short, is a closed-loop control system. It acquires and tracks the phase of an incoming carrier signal and follows it so as to enable coherent demodulation. Analog modulation schemes may be amplitude modulation (AM) or

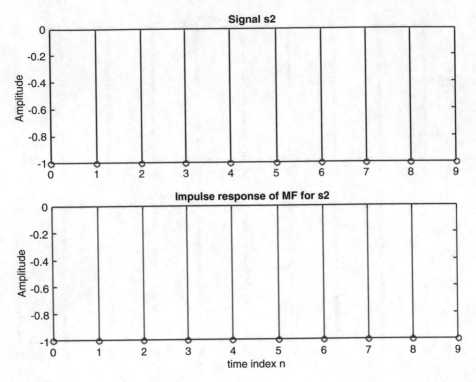

Fig. 10.33 Signal $s_2[n]$ and its MF impulse response: top, signal $s_2[n]$; bottom, corresponding MF impulse response

frequency modulation (FM). In either case, the message waveform can be recovered using coherent demodulation. Coherent demodulation requires a replica of the transmitted carrier to be available at the receiver. That is, the locally available carrier must have the same frequency and phase as that of the received carrier. The aim of the PLL is to enable coherent demodulation.

A PLL can be implemented either in analog form or digital form. We will first describe the analog PLL and then discuss the digital version later. A PLL consists of a phase detector, a loop filter, and a voltage-controlled oscillator (VCO). This is shown in Fig. 10.36. The phase detector produces a signal, which is the difference in phase between the incoming and locally generated signals. The loop filter filters the output of the phase detector to pass the slowly varying phase component and reject the high-frequency component. The VCO generates a replica of the incoming carrier signal based on the output signal of the loop filter. The input to the VCO is a measure of the difference in phase between that of the incoming signal and VCO output. This input then drives the phase of the VCO output in the direction of the phase of the incoming carrier. Since a PLL is a closed-loop system, the phase error tends to zero a little after the start of the PLL. Once the phase error is zero, the PLL is said to be in lock. Then it tracks the phase of the incoming carrier.

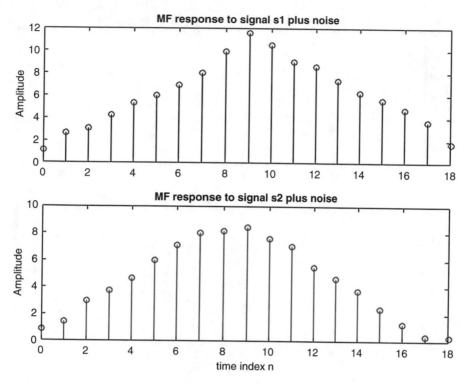

Fig. 10.34 Response of the MF to input signal plus noise: top, response of MF 1 to signal $s_1[n]$ plus Gaussian noise; bottom, response of MF 2 to signal $s_2[n]$ plus Gaussian noise. Noise standard deviation in both cases is 0.25

Analysis of a PLL As seen from Fig. 10.36, $r(t)$ is the received signal; $x(t)$ is the output of the VCO; the output of the phase detector is $e(t)$, which is the product of $r(t)$ and $x(t)$; and $v(t)$ is the output of the loop filter, which is the input to the VCO. Let the incoming carrier signal be described by

$$r(t) = \sin(2\pi f_c t + \theta(t)) \tag{10.83}$$

where the nominal frequency of the carrier is f_c and $\theta(t)$ is its phase, which is a slowly varying signal. Let the VCO output be defined as

$$x(t) = 2\cos(2\pi f_c t + \varphi(t)) \tag{10.84}$$

The phase detector accepts both $r(t)$ and $x(t)$ as inputs and outputs the product of the two input signals as described below.

$$e(t) = r(t)x(t) = 2\sin(\omega_c t + \theta(t))\cos(\omega_c t + \varphi(t)) \tag{10.85}$$

Using the trigonometric identity, the phase detector output can be written as

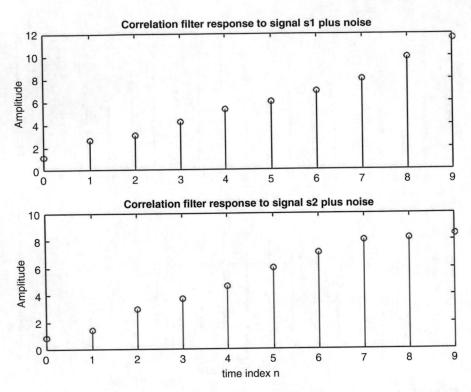

Correlation filter response to signal s1 plus noise

Correlation filter response to signal s2 plus noise

time index n

Fig. 10.35 Response of the correlation filter: top, response of correlation filter 1 to input signal $s_1[n]$ plus noise; bottom, response of correlation filter 2 to input signal $s_2[n]$ plus noise. The standard deviation of the noise is 0.25

Fig. 10.36 Block diagram of a PLL

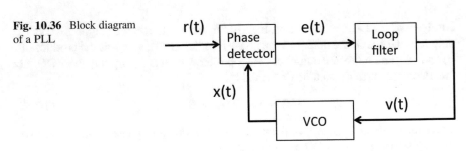

$$e(t) = \sin\left(\theta(t) - \varphi(t)\right) + \sin\left(2\omega_c t + \theta(t) + \varphi(t)\right) \qquad (10.86)$$

Since the aim of the PLL is to track the phase of the incoming signal, the loop filter is designed to be a lowpass filter, which rejects the signal at twice the carrier frequency and passes the lowpass signal $\sin(\theta(t) - \varphi(t))$. The VCO generates a frequency deviation that is proportional to its input voltage $v(t)$. When $v(t) = 0$, the

VCO nominal frequency is ω_c. Denoting the instantaneous frequency deviation from ω_c by $\Delta\omega(t)$, we have

$$\Delta\omega(t) = \frac{d\phi(t)}{dt} = Kv(t) \tag{10.87}$$

where K is a VCO gain in rad/volt. Note that frequency is the derivative of the phase. If the impulse response of the loop filter is denoted by $g(t)$, then, since the loop filter is LTI, its response is the convolution of its input and impulse response. That is,

$$v(t) = e(t)*g(t) \tag{10.88}$$

Under locked condition, the phase error is very small and so

$$e(t) \simeq \theta(t) - \varphi(t) \tag{10.89}$$

Therefore, (10.87) becomes

$$\Delta\omega(t) = K(\theta(t) - \varphi(t))*g(t) \tag{10.90}$$

This gives rise to the linearized PLL, which is shown in Fig. 10.37. Using (10.87) in (10.90), we have

$$\frac{d\varphi(t)}{dt} + K\varphi(t)*g(t) = K\theta(t)*g(t) \tag{10.91}$$

By applying the Laplace transform on both sides of (10.91) and using the differentiation and convolution properties of the Laplace transform, (10.91) can be written as

$$s\Phi(s) + K\Phi(s)G(s) = K\Theta(s)G(s) \tag{10.92}$$

From (10.92) the closed-loop transfer function of the linearized PLL is expressed as

$$H(s) \equiv \frac{\Phi(s)}{\Theta(s)} = \frac{KG(s)}{s + KG(s)} \tag{10.93}$$

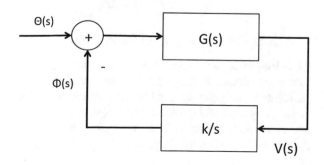

Fig. 10.37 Linearized PLL in the Laplace domain

In (10.92) and (10.93), $\Theta(s)$ and $\Phi(s)$ are the Laplace transforms of $\theta(t)$ and $\varphi(t)$, respectively. The linearized PLL in the Laplace domain is shown in Fig. 10.37. For $s = j\omega$, the frequency response of the linearized PLL takes the form

$$H(\omega) = \frac{KG(\omega)}{j\omega + KG(\omega)} \tag{10.94}$$

Steady-State Phase Error   The Laplace transform of the phase error is given by

$$E(s) = \mathcal{L}\{e(t)\} = \Theta(s) - \Phi(s) \tag{10.95}$$

From (10.93),

$$\Phi(s) = \frac{KG(s)\Theta(s)}{s + KG(s)} \tag{10.96}$$

Using (10.96) in (10.95), the phase error in the Laplace domain is found to be

$$E(s) = \Theta(s)\left\{1 - \frac{KG(s)}{s + KG(s)}\right\} = \frac{s\Theta(s)}{s + KG(s)} \tag{10.97}$$

The steady-state phase error is the phase error as t tends to infinity. Using one of the properties of the Laplace transform, the steady-state phase error is obtained from

$$\lim_{t \to \infty} e(t) = \lim_{s \to 0} sE(s) = \lim_{s \to 0} \frac{s^2\Theta(s)}{s + KG(s)} \tag{10.98}$$

Step Phase Response   Let us assume that the PLL is in phase lock. When a unit step phase is then applied to the PLL at t = 0, the input phase in the Laplace domain is given by

$$\Theta(s) = \frac{1}{s} \tag{10.99}$$

If $G(0) \neq 0$, then the steady-state phase error becomes

$$\lim_{t \to \infty} e(t) = \lim_{s \to 0} \frac{1}{s} \frac{s^2}{s + KG(s)} = 0 \tag{10.100}$$

From (10.100), it is clear that the PLL tracks the input step phase.

PLL Response to a Frequency Step   What happens if an abrupt step in the input frequency occurs? Will the PLL be able to track a frequency step? Let us investigate. Incidentally, a frequency step change could indicate a Doppler shift in the incoming signal frequency. This shift may be due to a relative motion between the transmitter and receiver. Since phase is the integral of the frequency, it changes linearly with respect to time when the frequency change is a step function. Using the integral in time property of the Laplace transform, the input phase due to a frequency step change $\Delta\omega$ is given by

$$\Theta(s) = \frac{\Delta\omega}{s^2} \tag{10.101}$$

The steady-state phase error is found to be

$$\lim_{t\to\infty} e(t) = \lim_{s\to 0} sE(s) = \lim_{s\to 0} \frac{s^2 \frac{\Delta\omega}{s^2}}{s + KG(s)} = \frac{\Delta\omega}{KG(0)} \tag{10.102}$$

The steady-state error depends on G(0). There are three possible types of loop filter, which are allpass, lowpass, and lead-lag filters and are defined below in that order.

$$G(s) = 1 \Rightarrow G(0) = 1 \tag{10.103a}$$

$$G(s) = \frac{\beta}{s + \beta} \Rightarrow G(0) = 1 \tag{10.103b}$$

$$G(s) = \left(\frac{\beta}{\alpha}\right) \frac{s + \alpha}{s + \beta} \Rightarrow G(0) = 1 \tag{10.103c}$$

In any case, G(0) = 1. Therefore, the steady-state phase error becomes

$$\lim_{t\to\infty} e(t) = \frac{\Delta\omega}{K} \tag{10.104}$$

and so the PLL tracks a step change in the input frequency. Even though the steady-state phase error is a constant and not zero, the PLL tracks a step change in input frequency with a constant phase error. Let us clarify the above discussion with a couple of examples.

Example 10.5a Step Phase: In this example, let us calculate the response of the PLL to a step phase input with the following specs. $G(s) = \frac{1}{s+1}$, and the VCO output is $X(s) = \frac{K}{s} V(s)$. Since we are going to implement this PLL in S/W, we will use a digital lowpass loop filter. The integrator, which models the VCO, in the discrete-time domain is simply a delayed accumulator. Let the VCO gain K = 0.1 and let the phase step of 0.95 rad be applied at the time index n = 510. The input phase, the VCO output phase, and the phase error are shown in Fig. 10.38 in top, middle, and bottom plots, respectively. As seen from the figure, the VCO phase undergoes a transient state and reaches a steady state after about ten sampling intervals. Similarly, the phase error reaches a steady state after about ten sampling intervals. The VCO input and output are shown in Fig. 10.39 in the top and bottom plots, respectively.

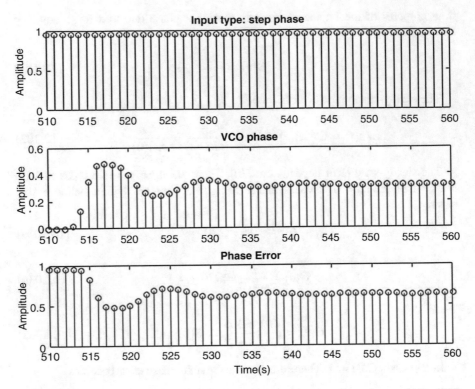

Fig. 10.38 Input phase and VCO phase in Example 10.5a: top, input step phase; middle, VCO output phase; bottom, phase error

Example 10.5b Pulse Phase: In this example we consider an input pulse phase to the PLL with the same loop filter and VCO as in the previous example. In Fig. 10.40 are shown the input phase, the VCO phase, and the phase error in the top, middle, and bottom plots, respectively. The VCO phase is a distorted pulse and so is the phase error. Similar to the previous case, the VCO input and output are shown in the top and bottom plots of Fig. 10.41.

Example 10.5c Ramp Phase: As a third example, let the input to the PLL be a ramp phase. The phase changes from 0 to 1 rad over 20 samples. Note that the instantaneous frequency is the time derivative of the phase. Since the phase varies linearly with time, the frequency will be a constant equal to the slope of the phase. Figure 10.42 shows the input phase, the VCO phase, and the phase error in the top, middle, and bottom plots, respectively. The VCO input and output are shown in the top and bottom plots in Fig. 10.43. The M-file used for the three cases of Example 10.5 is named *Example 10_5.m*.

Simulink Example to Simulate a Linearized Analog PLL To get a hands-on experience in working with PLL, let us look at simulating a linearized analog PLL using MATLAB's Simulink. In this example, the PLL functions in the baseband.

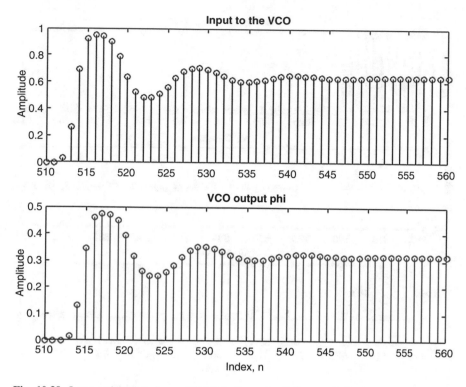

Fig. 10.39 Input and output phase of VCO in Example 10.5a: top, VCO input; bottom, VCO output phase

Under the Communications System Toolbox, we will find the subsystem named Synchronization, and under Synchronization, we have Components in which we will find the block named Linearized Baseband PLL. This block has one input and three outputs. The input is the signal whose phase is to be tracked. The three outputs are the phase detector (PD), the loop filter (Filt), and the voltage-controlled oscillator (VCO). The loop filter parameters to be entered in the PLL Block Parameters are the coefficients of the numerator and denominator polynomials of the lowpass filter transfer function. The coefficients correspond to the descending powers of the Laplace variable s. The filter chosen is a third-order Butterworth analog filter with a passband edge of 100 rad/s. The VCO parameter is its gain or input sensitivity in Hz/V. The input to the PLL block is a baseband sinusoidal source, which is found under Simulink – Source category. We have the option to use either sample based or time based. We will use sample based as the parameter under "sine type." We will also choose 100 samples/period under "samples per period" and an offset of 10 samples. The sample time is 0.01 s. The reader can view all the parameter options available as well as what are selected by double clicking each block in the simulation diagram. We can also add white Gaussian noise to the signal and the sum is fed to the PLL. Figure 10.44 shows the simulation block diagram used in this example.

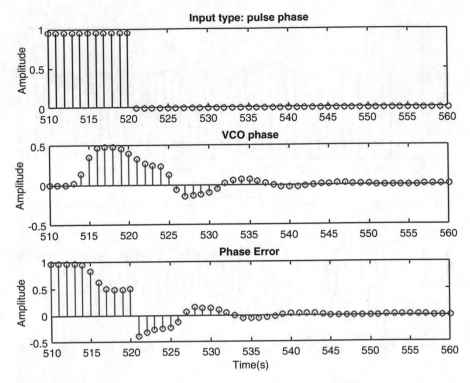

Fig. 10.40 Input phase and VCO phase in Example 10.5b: top, input pulse phase; middle, VCO output phase; bottom, phase error

After connecting all the blocks, we need to save the diagram in a file. The simulation time is chosen to be 10 s. To start the simulation, we have to click the green right arrow. If there are no errors, the simulation will start, and the results will be displayed on the respective scopes. First, let us simulate the PLL without noise. The input signal is shown in Fig. 10.45. As seen from the figure, there are 10 cycles over the 10 sec duration. The VCO output is shown in Fig. 10.46. As expected, it takes a few samples before the locking condition occurs. Next, we add white Gaussian noise with a power of 0.01 W and start the simulation. The VCO output is shown in Fig. 10.47 when noise is added. Due to the presence of noise, the VCO takes more time to track the incoming phase. The MATLAB file to simulate the linearized analog PLL is named *Linear_PLL.slx*.

Digital Phase Lock Loop A digital phase lock loop (DPLL) achieves the same purpose as the analog counterpart but with many advantages. A DPLL has a superior performance over an analog PLL. It is able to acquire and track much faster than an analog PLL. It is much more reliable and has lower size and cost. The VCO of an analog PLL is highly sensitive to temperature and power supply variations. Therefore, it needs not only an initial calibration but also frequent adjustments. This will

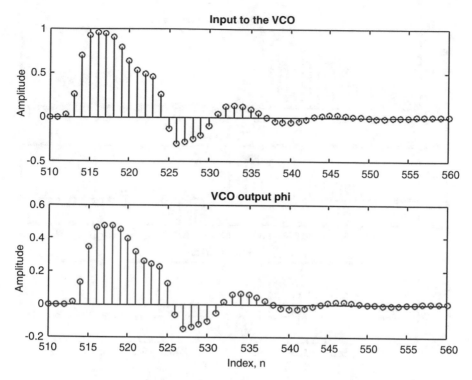

Fig. 10.41 Input and output phase of VCO in Example 10.5b: top, VCO input; bottom, VCO output phase

be a problem in consumer products such as a cellular phone. DPLL has no such problem. The phase detector in an analog PLL uses analog multipliers, which are sensitive to drift in DC voltage, whereas DPLL does not suffer from this problem. An analog PLL does not function well at low frequencies because the lowpass loop filter is analog. A larger time period is necessary for a better frequency resolution, which in turn reduces the locking speed. DPLL does not have this problem either. Moreover, since an analog PLL uses analog multipliers and analog filter, self-acquisition is slow and unreliable. DPLL has a faster locking speed. With so many advantages over an analog PLL, it is certainly desirable to use a DPLL instead. This also gives us the motivation to look into DPLL.

A simple block diagram of a DPLL is shown below in Fig. 10.48. There are different DPLLs available in the literature. One of them is called the Nyquist DPLL. In this type of PLL, the input analog sinusoidal signal is uniformly sampled at least at the Nyquist rate, and the analog samples are quantized to B-bits to form the input digital signal. It is then digitally multiplied by the output of the digital-controlled oscillator (DCO) to form the error sequence. This error sequence is then filtered by a lowpass digital filter. The output of the lowpass digital filter then controls the DCO frequency. Figure 10.49 shows the block diagram of a Nyquist DPLL.

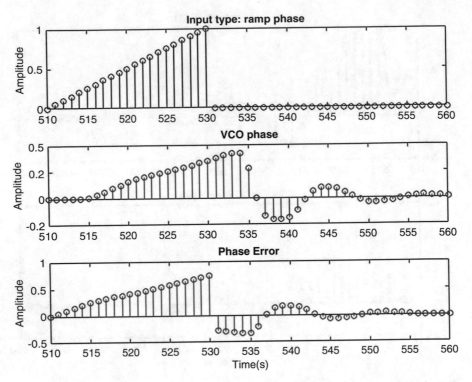

Fig. 10.42 Input phase and VCO phase in Example 10.5c: top, input ramp phase; middle, VCO output phase; bottom, phase error

Software-Based DCO Unlike the analog VCO, the DCO in the DPLL in Fig. 10.49 uses software or algorithm to generate the sinusoid. It uses the basic idea behind the analog VCO in the following manner. In the continuous-time or analog domain, the VCO output is described by

$$y(t) = B \cos \left(\omega_c t + K \int_0^t v(\tau) d\tau \right) \tag{10.105}$$

In the discrete-time domain, we can write the above VCO output as

$$y[n] = B \cos \left(\frac{2\pi f_c}{f_s} n + K \sum_{i=0}^{n-1} v[i] \right) \tag{10.106}$$

where f_s is the sampling frequency and the summation inside the argument of the cosine function corresponds to the integral of the VCO input. The sequence in (10.106) is then converted to a square wave, which is obtained by

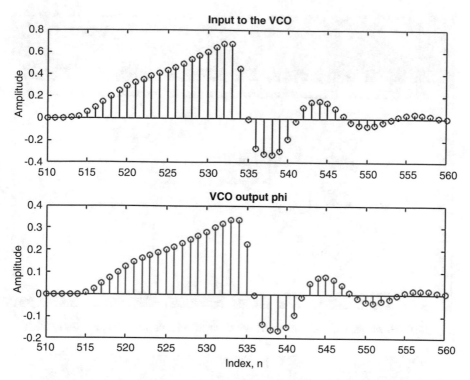

Fig. 10.43 Input and output phase of VCO in Example 10.5c: top, VCO input; bottom, VCO output phase

$$y[n] = sq\left(\frac{2\pi f_c}{f_s}n + K\sum_{i=0}^{n-1} v[i]\right) \tag{10.107}$$

where the function sq(x) is defined as

$$sq(x) = \begin{cases} 1, 0 \le x < \pi \\ -1, \pi \le x < 2\pi \end{cases} \tag{10.108}$$

and it is periodic as indicated below.

$$sq(x + 2m\pi) = sq(x), m \in Z \tag{10.109}$$

An Example to Illustrate the Function in Equation 10.108 Let us illustrate the conversion of the sequence in (10.107) into a square sequence using MATLAB. For the sake of illustration, we choose the sinusoidal frequency to be 100 Hz with a sampling frequency of 1000 Hz. Let the DCO gain be equal to 1 rad/volt. The program to run is named *SQ.m*. The discrete square sequence is shown in Fig. 10.50

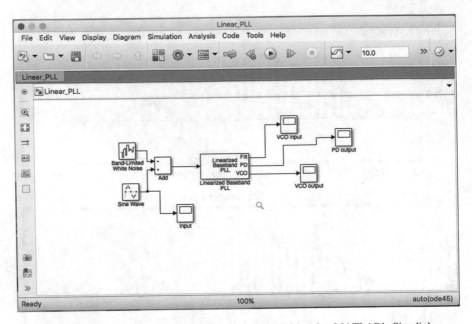

Fig. 10.44 Block diagram to simulate a linearized analog PLL using MATLAB's Simulink

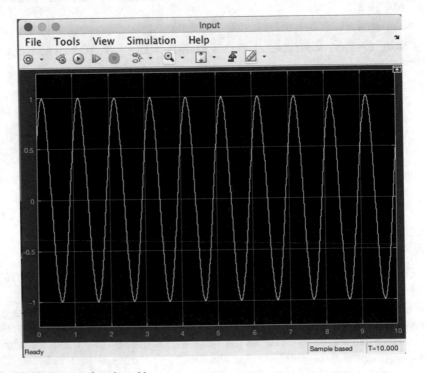

Fig. 10.45 Input analog sinusoid

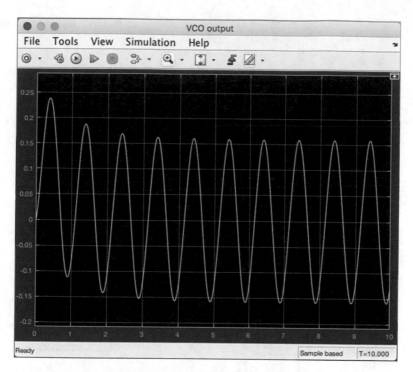

Fig. 10.46 VCO output with no input noise

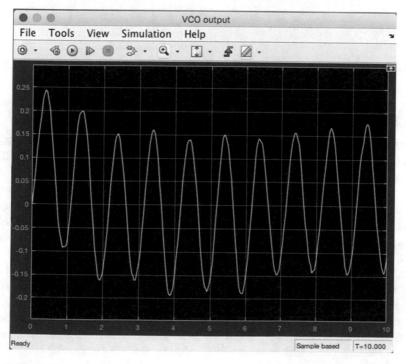

Fig. 10.47 VCO output with a white Gaussian noise of 0.01 W of power

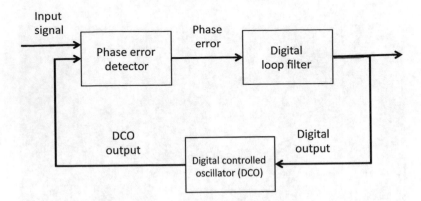

Fig. 10.48 Block diagram of a typical DPLL

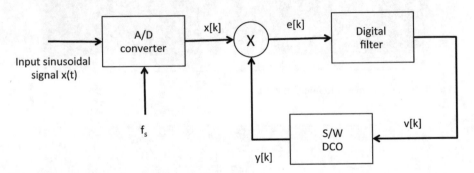

Fig. 10.49 Block diagram of a Nyquist DPLL

as a stem plot. The same sequence when the DCO gain is K = 2 is shown in Fig. 10.51 for comparison.

Simulink Example of a DPLL In this example, we will simulate a DPLL using MATLAB's Simulink. The block diagram for the simulation is shown in Fig. 10.52. It is similar to the one used in the simulation of an analog PLL. The main difference is in the PLL block. It is called charge-pump PLL. The phase detector outputs a square waveform as opposed to a continuous-time signal. The loop filter is a lowpass analog filter. The input signal is the same sinusoid used in the previous Simulink example. The loop filter is a third-order Butterworth lowpass filter with a passband edge of 10 rad/s. The VCO gain is 1.25 rad/volt. The block diagram for the simulation is shown in Fig. 10.52. The details of the parameters can be found by double clicking the respective block. The outputs of the phase detector, loop filter, and the VCO are shown in Figs. 10.53, 10.54, and 10.55, respectively. The VCO seems to track the incoming phase. The same three outputs are displayed in Figs. 10.56, 10.57, and 10.58 when the input signal is corrupted by an additive white Gaussian noise with a power of 0.01. Again, the VCO tracks the phase of the incoming signal.

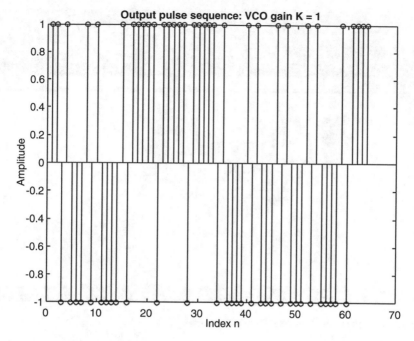

Fig. 10.50 The output sequence of the sq(x) function with the DCO gain equal to 1

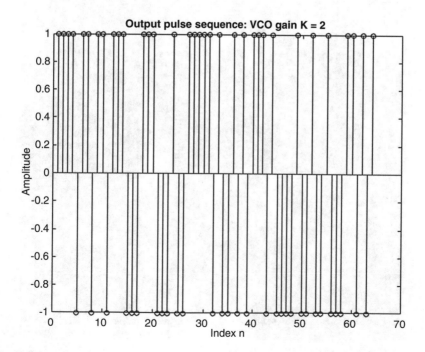

Fig. 10.51 The same output sequence of the sq(x) function with the DCO gain equal to 2

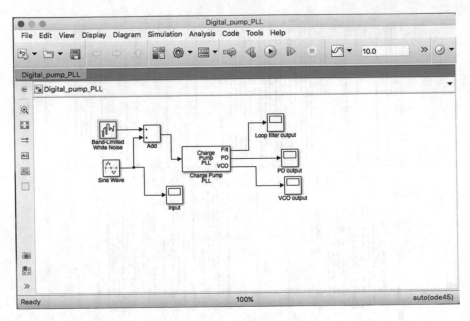

Fig. 10.52 Simulation block diagram of a charge-pump PLL

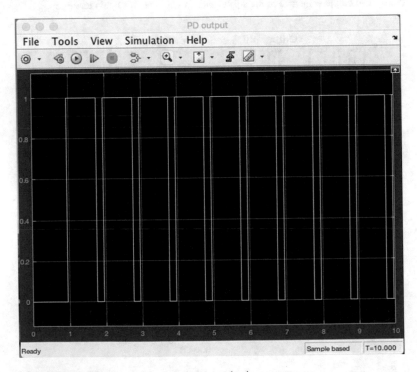

Fig. 10.53 Output of the phase detector when no noise is present

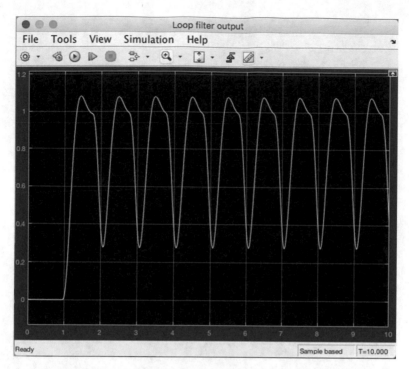

Fig. 10.54 Output of the loop filter when no noise is present

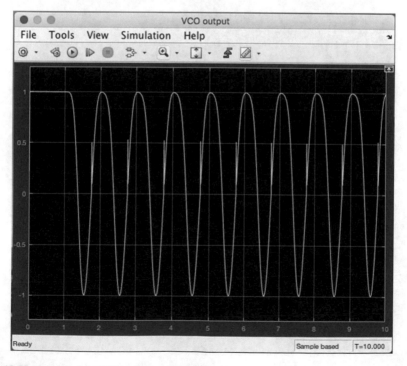

Fig. 10.55 Output of the VCO when no noise is present

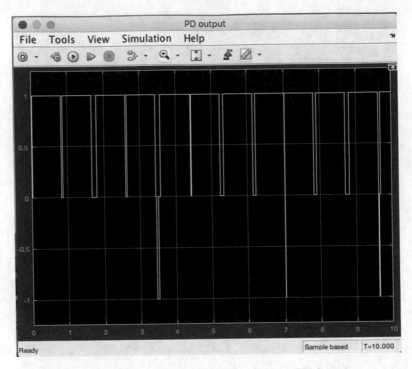

Fig. 10.56 Output of the phase detector when AWGN with power 0.01 is present

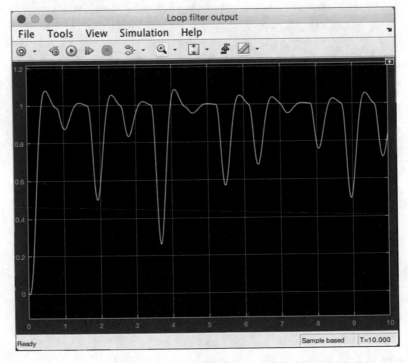

Fig. 10.57 Output of the loop filter when AWGN with power 0.01 is present

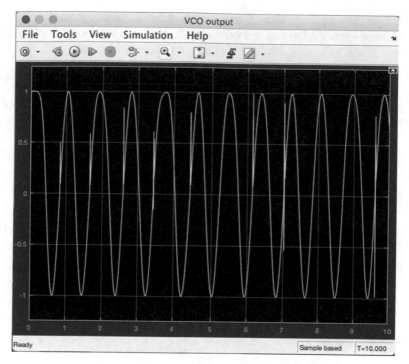

Fig. 10.58 Output of the VCO when AWGN with power 0.01 is present

10.5.5   OFDM

OFDM stands for orthogonal frequency-division multiplexing. It is also a multicarrier modulation scheme. In a conventional frequency-division multiplexing, data from different subscribers each modulate a subcarrier to form a non-overlapping spectrum. These subcarriers then modulate a final carrier for transmission. The problem with the conventional frequency-division multiplexing is the ISI at the receiver due to the dispersive fading channel effect and requires complex equalization. On the other hand, OFDM uses subcarriers that are orthogonal, which does not cause ISI. It also supports high data rates. Therefore, OFDM has gained popularity and is used in standards such as digital audio broadcasting and digital TV. It is also used in high data rate transmission in mobile wireless channels. Remember that our objective here is to show that DSP is used in OFDM.

OFDM Basics OFDM is a digital multicarrier modulation. The symbols from different users to be transmitted simultaneously are denoted by $\{X(k), 0 \leq k \leq N - 1\}$. The OFDM signal in the time domain can be described by

$$x(t) = \sum_{k=0}^{N-1} X(k)e^{j2\pi f_k t}, 0 \leq t \leq T_s \tag{10.110}$$

where T_s is the symbol duration, $f_k = f_0 + k\Delta f$, and Δf is the sub-channel spacing. If we choose the symbol duration and sub-channel spacing to satisfy $T_s\Delta f = 1$, then the subcarriers will be orthogonal. To prove this statement, let us use

$$\psi_k(t) = e^{j2\pi f_k t}, 0 \leq t \leq T_s, 0 \leq k \leq N - 1 \tag{10.111}$$

If the set $\{\psi_k(t), 0 \leq k \leq N - 1\}$ is orthogonal, then the following condition must be satisfied.

$$\frac{1}{T_s} \int_0^{T_s} \psi_k(t)\psi_l^*(t)dt = \delta(k - l) \tag{10.112}$$

Using (10.111) in (10.112), we obtain

$$\frac{1}{T_s} \int_0^{T_s} e^{j2\pi f_k t} e^{-j2\pi f_l t} dt = \frac{1}{T_s} \int_0^{T_s} e^{j2\pi(f_k - f_l)t} dt \tag{10.113}$$

But,

$$f_k - f_l = (k - l)\Delta f = \frac{(k - l)}{T_s} \tag{10.114}$$

Therefore, we have

$$\frac{1}{T_s} \int_0^{T_s} \psi_k(t)\psi_l^*(t)dt = \delta(k - l) \tag{10.115}$$

Thus, the subcarriers in the OFDM signal are orthogonal. This property also enables us to demodulate the OFDM signal to obtain the transmitted symbols $\{X(k), 0 \leq k \leq N - 1\}$. To demodulate, we multiply the received signal $x(t)$ by $e^{-j2\pi f_k t}$ and integrate over the symbol interval. Thus,

$$\frac{1}{T_s} \int_0^{T_s} x(t)e^{-j2\pi f_k t} dt \tag{10.116}$$

Substituting for x(t) from (10.110), we have

$$\frac{1}{T_s} \int_0^{T_s} \left\{ \sum_{m=0}^{N-1} X(m)e^{j2\pi f_m t} \right\} e^{-j2\pi f_k t} dt \tag{10.117}$$

By interchanging the order of integration and summation in the above equation, we get

$$\sum_{m=0}^{N-1} X(m)\left\{ \frac{1}{T_s} \int_0^{T_s} e^{j2\pi(f_m - f_k)t} dt \right\} = \sum_{m=0}^{N-1} X(m)\delta(m - k) = X(k) \tag{10.118}$$

Thus, the symbol corresponding to the kth user has been recovered. So far, our discussion on OFDM has been on continuous-time or analog domain. But we are interested in the discrete-time domain. So, let us see how we can modulate and demodulate the OFDM signal using DSP. To this end, let us sample uniformly the OFDM analog signal described in (10.110) at intervals $T = \frac{T_s}{N}$. Then the sampled OFDM signal is described by

$$x[n] = x(t = nT) = \sum_{k=0}^{N-1} X(k) e^{j2\pi f_k n \frac{T_s}{N}} \tag{10.119}$$

Without loss of generality, let $f_0 = 0$. Then,

$$f_k = k\Delta f = \frac{k}{T_s} \tag{10.120}$$

Using (10.120) in (10.119), we get the expression for the discrete version of the OFDM signal as given by

$$x[n] = \sum_{k=0}^{N-1} X(k) e^{j2\pi n \frac{k T_s}{T_s N}} = \sum_{k=0}^{N-1} X(k) e^{j\frac{2\pi}{N} nk} = \sum_{k=0}^{N-1} X(k) W_N^{-nk} \tag{10.121}$$

Thus, the OFDM sequence is the inverse DFT (IDFT) of the N-point DFT X (k) except for the 1/N factor. Here, N corresponds to the number of users or subscribers. As we know, the N-point DFT can be implemented efficiently using the FFT algorithm. Note that X(k) may be complex.

MATLAB Example Let us exemplify the idea of OFDM using MATLAB. The frequency response of the analog OFDM signal in (10.110) is obtained by taking the Fourier transform of x(t) in (10.110) and is given below.

$$X(f) = \int_{-\infty}^{\infty} x(t) e^{-j2\pi f t} dt = \int_0^{T_s} \left\{ \sum_{k=0}^{N-1} s_k e^{j2\pi f_k t} \right\} e^{-j2\pi f t} dt \tag{10.122}$$

In the above equation, we have used s_k to represent the symbols used in OFDM to avoid confusion with the Fourier transform of x(t), which is denoted by X(f). By interchanging the integration and summation in (10.122) and after simplification, we have

$$X(f) = \sum_{k=0}^{N-1} s_k T_s e^{-j\pi((f-f_0)T_s - k)} \mathrm{sinc}[(f - f_0)T_s - k] \tag{10.123}$$

where we have used the facts $f_k = f_0 + k\Delta f$ and $T_s \Delta f = 1$. In this example the following values are used: $N = 32$, $f_0 = 5000\ Hz$, and $\Delta f = 50$. The M-file to run is named *OFDM_example.m*. The 32 symbols are generated as random integers using the MATLAB function *randi(N,1,N)*. The function generates a 1xN integer vector whose values range from 1 to N. The frequency response of the analog OFDM signal is shown in Fig. 10.59, and the frequency response of the corresponding discrete-time sequence is shown in Fig. 10.60. The M-file also recovers the symbols by

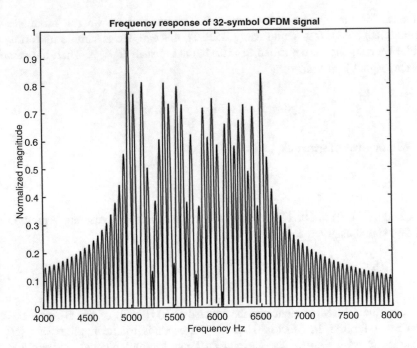

Fig. 10.59 Frequency response of an analog OFDM with N = 32

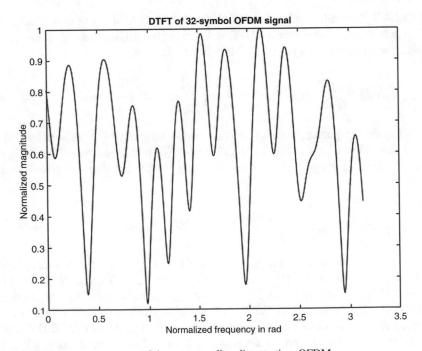

Fig. 10.60 Frequency response of the corresponding discrete-time OFDM

performing the FFT on the discrete-time OFDM sequence. It is found that the recovered symbols are identical to the transmitted symbols.

10.5.6 Software-Defined Radio

In this section, we will describe the process of sampling an RF bandpass signal at a much lower frequency without incurring aliasing distortion so that the hardware in a software-defined radio (SDR) can process the received signal without undue computational burden. But before we take up this task, it is necessary to know what an SDR is. There are a few definitions for the SDR in the literature as follows. The SDR Forum, for instance, defines an SDR as one that receives fully programmable traffic and control information and supports a broad range of frequencies, air interfaces, and applications software. Not a useful definition, is it? Another definition of an SDR is that it is a radio that performs in software its modulation, error correction, and encryption functions, exhibits some control over the RF hardware, and can be reprogrammed. A third definition is that an SDR can be defined mostly in software and whose physical layer behavior can be significantly altered through software changes. This definition makes more sense. Yet another definition of an SDR is that it defines a fully configurable radio that can be programmed in software to reconfigure the physical hardware. In other words, an SDR is a system that emulates the functions of a hardware-defined radio to receive not just one type but several types of modulated signals, such as AM, FM, etc.

Need for an SDR Wireless communications is evolving at a rapid rate from 2G to 3G to 4G and now to 5G, all within two decades. There is a high and ever-increasing demand for wireless Internet connectivity for audio, video, etc. Integrated seamless global coverage implies (a) the ability to roam globally and (b) be able to interface with different systems and mobile standards to provide seamless service at a fixed location. Because of these ever-increasing demands, fixed hardware-defined radio is nearly impossible to accommodate all the abovementioned requirements, considering the fact that most consumer wireless devices are handheld. If a radio is defined mostly by software, then it is flexible and can perform all the abovementioned tasks seamlessly. It can easily adapt to changes by upgrading its software as often as is necessary. A block diagram of an SDR is depicted in Fig. 10.61. As mentioned earlier, we will only describe the ADC at the receiver front end in what follows.

Factors Influencing an SDR Following are the factors that influence the widespread acceptance of SDR. It is multifunctional. For example, a Bluetooth-enabled fax machine may be able to send a fax to a nearby laptop computer equipped with SDR that supports the Bluetooth interface. Another factor is global mobility, which is to be able to support standards, such as GSM, UMTS, CDMA, etc., as well as military standards. Compactness and power efficiency are very important for an SDR to be accepted. Obviously, SDR is compact because it has a small and fixed hardware. It is power efficient especially when the number of functions is large. It

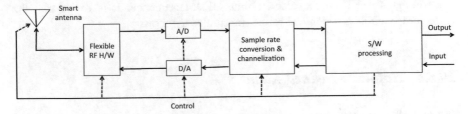

Fig. 10.61 Block diagram of a software-defined radio

must be easy to manufacture. Due to digitization of the signal taking place early in the receiver chain, the hardware is made small and compact, and so it is power efficient. Last but not the least important factor is the ease of upgrades. Since the SDR uses software to perform most of the receiver functions, it requires only software upgrade, which can be done without disrupting the current function. For all these reasons, SDR is gaining popularity in wireless communications.

Sampling Bandpass Signals A bandpass signal is one, which is centered at a high frequency with a narrow bandwidth. Bandpass signals are used typically in communications to carry message or information signals. These signals may be amplitude modulated (AM), frequency modulated (FM), and digitally modulated as in PSK (phase shift keying), FSK (frequency shift keying), etc. An AM signal in the continuous-time domain is generated by multiplying a carrier signal with a modulating or message waveform. Thus,

$$x_{AM}(t) = A_c m(t) \cos(\omega_c t) \tag{10.124}$$

where A_c is the amplitude of the carrier signal, whose frequency is ω_c rad/s, and m (t) is the message waveform such as speech, music, etc. If the message waveform has zero DC value, then the resulting AM signal has frequencies above and below the carrier frequency, but no carrier frequency per se. Hence this AM signal is called a double-sideband suppressed carrier (DSB-SC). Note that the message waveform is a baseband signal, meaning that its frequency spectrum is centered at zero frequency. When this message waveform modulates a carrier in its amplitude, the message spectrum shifts to above and below the carrier frequency. The frequencies lying above the carrier frequency are called the *upper sideband*, and those below the carrier frequency are likewise called the *lower sideband*. A frequency-modulated signal, on the other hand, is expressed as

$$x_{FM}(t) = A_c \cos(\omega_c t + \varphi(t)) \tag{10.125}$$

with

$$\frac{d\varphi(t)}{dt} = \beta m(t) \tag{10.126}$$

Unlike in the AM, the amplitude of the carrier of the FM signal is constant, but its frequency deviates from the carrier frequency, which is proportional to the message waveform as described in (10.126). The parameter β controls the frequency deviation. Again, the frequency spectrum of the FM signal is centered at the carrier frequency with upper and lower sidebands. Therefore, it is also a bandpass signal.

An AM radio operates over the frequency range of 550 kHz to 1600 kHz. Each carrier is separated by 10 kHz in frequency. For instance, if we were to sample an AM signal with carrier frequency of 550 kHz at the Nyquist rate, then the minimum required sampling frequency would be equal to twice the highest frequency, which is $2 \times (550 + 5)$ kHz $= 1.1$ MHz. Similarly, the minimum required sampling frequency of the AM signal at the upper end of the AM spectrum would be 3.21 MHz. On the other hand, FM radio has a carrier frequency range from 88 MHz to 108 MHz with a channel separation of 300 kHz. The minimum required sampling frequencies for the FM signal would be 176.3 MHz at the lower end of the FM spectrum and 216.3 MHz at the upper end of the spectrum. These sampling frequencies are very high. These high sampling rates put a heavy computational load on the whole SDR receiver. Is there a way to sample these BP signals at a much lower frequency without incurring aliasing distortions? Yes, there is a way out. Let us look into it.

Let the center frequency of a BP signal be f_c Hz with a bandwidth B Hz. Then the sampling frequency f_s to be used must satisfy the following condition.

$$\frac{2f_c - B}{N + 1} \leq f_s \leq \frac{2f_c - B}{N} \tag{10.127}$$

In (10.127), N is the number of replications of the spectrum of the BP signal in the frequency range $2f_c - B$, which is due to the process of sampling. From (10.127), we notice that the actual sampling frequency will be much less than the Nyquist frequency. Further, one of the replications near the zero frequency can be filtered and processed further with a much lower computational burden. Let us demonstrate the statements by way of an example using MATLAB.

Example of BP Down Conversion of an AM Signal For the sake of argument, consider an AM DSB-SC signal at a carrier frequency of 10 kHz. Let the message waveform be a single sinusoid at a frequency of 1 kHz. Let us also choose the factor in (10.127) to be N = 3. Then, the sampling frequency must be, according to (10.127), between 4.5 kHz and 6 kHz. Let us choose the BP sampling frequency to be 5.250 kHz, which is midway between the two frequencies. Let us then compare the signal sampled at 5.25 kHz with that sampled at 30 kHz, which is slightly higher than the Nyquist frequency. The 10 kHz signal and the signal sampled at 5.25 kHz are shown in Fig. 10.62. The corresponding spectra are shown in Fig. 10.63. As seen from Fig. 10.63, the BP spectrum has shifted to a much lower frequency, thereby reducing the computational load.

Example of BP Down Conversion of an FM Signal For the sake of argument, let the carrier frequency of the FM signal to be used be 10 kHz and let the frequency of the modulating signal be 1.0 kHz. Let the modulation index $\beta = 2$. Then the bandwidth of

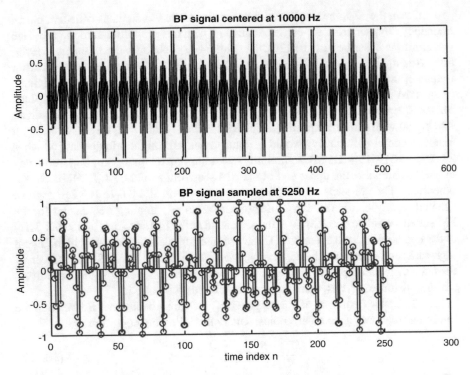

Fig. 10.62 An AM signal sampled at 30 kHz and 5.25 kHz. Top, sampling frequency of 30 kHz; bottom, sampling frequency of 5.25 kHz

the FM signal according to Carson's empirical formula is given by $2(\beta + 1)f_m = 6\,kHz$. If we choose N = 2, then according to (10.127), the BP sampling frequency must be between 4.666 kHz and 7 kHz. Let the sampling frequency be 5.833 kHz. Then the FM signals with sampling frequencies of 30 kHz and 5.833 kHz are shown in the upper and lower plots in Fig. 10.64. The corresponding spectra are shown in Fig. 10.65. Due to sampling the BP FM signal at a lower rate, the frequencies are shifted toward the zero frequency. The M-file for both examples is named *BP_downconversion.m*.

10.6   Summary

The overall objective of this chapter has been to describe the application of digital signal processing methods in digital communications. To this end, we started with a brief introduction to multi-rate signal processing, which are upsampling and downsampling of a discrete-time signal. With this description, we learnt the design of oversampled ADC and DAC. Since an ADC is in the front and DAC at the end of a digital communications system, we described them to start with. We used

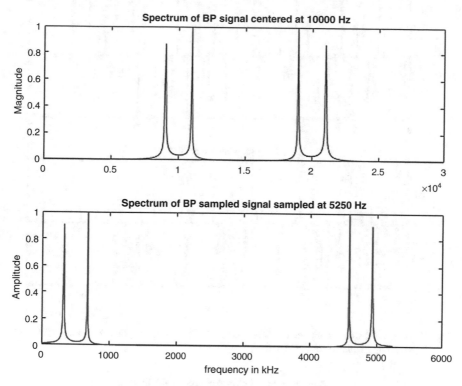

Fig. 10.63 Spectra of the signals shown in Fig. 10.62: top, sampling frequency of 30 kHz; bottom, sampling frequency of 5.25 kHz

MATLAB-based examples to illustrate these conversion techniques. Next we introduced the method of optimal detection of a binary signal. That gave rise to the discussion on inter-symbol interference. The processes to cancel ISI are pulse shaping at the transmitting end and equalization at the receiver end. MATLAB-based and Simulink examples were shown to help the reader understand these techniques better. We then described matched filter and correlation filter used in the optimal detection of binary symbols in digital communications. Phase-locked loop is an important ingredient in any digital communications system. So, we indulged a bit in describing linear PLL and exemplified the concept with MATLAB-based examples. Orthogonal frequency modulation or multiplexing is a bandwidth-efficient digital modulation scheme. Moreover, OFDM can be performed very efficiently using FFT technique. We worked out a MATLAB-based example to illustrate the orthogonality of the subcarriers used in OFDM. The chapter ended with a discussion of software-defined radio. Since sampling takes place at the RF end of an SDR, we showed an example, where AM and FM signals are sampled at a much lower rate than the Nyquist rate and yet preserve the integrity of the signals.

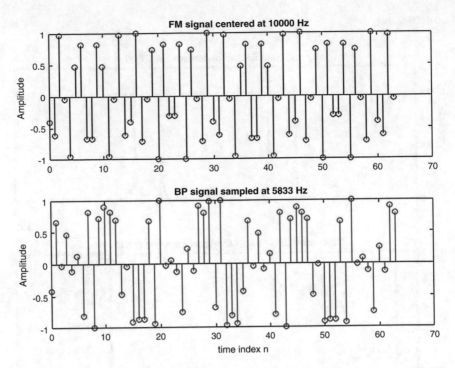

Fig. 10.64 An FM signal sampled at 30 kHz and 5.833 kHz. Top, sampling frequency of 30 kHz; bottom, sampling frequency of 5.833 kHz

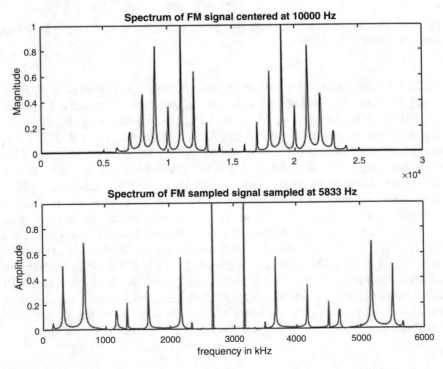

Fig. 10.65 Spectra of the signals shown in Fig. 10.64: top, sampling frequency of 30 kHz; bottom, sampling frequency of 5.833 kHz

10.7 Problems

1. We want to transmit the word "Why me" using an 8-ary system. (a) Encode the word "Why me" into a sequence of bits using 7-bit ASCII coding, followed by an eighth bit for error detection per character. The eighth bit is chosen so that the number of ones in the 8 bits is an even number. How many total bits are there in the message? (b) Partition the bit stream into $k = 3$ bit segments. Represent each of the 3-bit segments as an octal number (symbol). How many octal symbols are there in the message? (c) If the system were designed with 16-ary modulation, how many symbols would be used to represent the word "Why me"? (d) If the system were designed with 256-ary modulation, how many symbols would be used to represent the word "Why me"? Note: There is a space "Why" and "me."

2. If you want to transmit 600 characters/s, where each character is represented by its 7-bit ASCII codeword followed by an eighth bit for error detection, per character as in Problem 1, using a multilevel PCM format with $M = 16$ levels, (a) what is the effective transmitted bit rate? (b) What is the PCM symbol rate?

3. A bipolar binary signal, $s_i(t)$, is a $+ 1$- or -1- V pulse during the interval $(0, T)$. AWGN having two-sided power spectral density of 0.5×10^{-3} W/Hz is added to the signal. If the received signal is detected with a matched filter, determine the maximum bit rate that can be sent with a bit error probability of $P_B \leq 10^{-4}$.

4. Bipolar pulse signals, $s_i(t)$, $i = 1,2$, of amplitude ± 1 V are received in the presence of Gaussian noise with variance $\sigma^2 = 0.2$ V^2. Determine the optimum detection threshold, γ_0, for matched filter detection if the a priori probabilities are (a) $P(s_1) = 0.5$, (b) $P(s_1) = 0.6$, and (c) $P(s_1) = 0.3$. (d) Explain the effect of the a priori probabilities on the value of γ_0.

5. Determine the theoretical minimum system bandwidth needed for a 12 Mbits/s signal using 16-level PCM without ISI.

6. A binary waveform of 9200 bits/s is converted to an octal waveform that is transmitted over a system having a raised cosine roll-off filter characteristic. The system has a conditioned or equalized response out to 2.3 kHz. (a) Determine the octal symbol rate. (b) Calculate the roll-off factor of the filter characteristic.

7. A signal in the frequency range 500 to 3500 Hz has a peak-to-peak swing of 10 V. It is sampled at 8000 samples/s, and the samples are quantized to 128 evenly spaced levels. Calculate and compare the bandwidths and ratio of peak signal power to rms quantization noise if the quantized samples are transmitted either as binary pulses or as four-level pulses. Assume that the system bandwidth is defined by the main spectral lobe of the signal.

8. A coherent BPSK system, which operates continuously, makes errors at the average rate of 120 per day. The data rate is 1000 bits/s. The single-sided noise power spectral density is $N_0 = 10^{-9}$ W/Hz. (a) If the system is ergodic, what is the average bit error probability? (b) If the value of the received average signal power per bit is adjusted to be 10^{-5} W, will this received power be adequate to maintain the error probability found in part (a)?

9. If the main performance criterion of a digital communications system is the bit error probability, find out which of the following two modulation schemes would be selected for an AWGN channel: (a) binary coherent orthogonal FSK with $E_b/N_0 = 10$ dB and (b) binary noncoherent orthogonal FSK with $E_b/N_0 = 12$ dB. Show all the steps involved in the computations.

10. Calculate the probability of bit error for the coherent matched filter detection of the equally likely binary FSK signals $s_1(t) = 0.5 \cos(2010\pi t)$ and $s_2(t) = 0.5 \cos(2050\pi t)$, where the two-sided AWGN power spectral density is $N_0/2 = 10^{-3}$. Assume that the symbol duration is $T = 0.02$ s.

11. A digital communications system uses MF detection of equally likely BPSK signals, $s_1(t) = \sqrt{(2E/T)} \cos(\omega_0 t)$ and $s_2(t) = \sqrt{(2E/T)} \cos(\omega_0 t + \pi)$, and operates in AWGN with a received E_b/N_0 of 6.7 dB. Assume that $E\{z(T)\} = \pm\sqrt{E}$. (a) Find the minimum probability of bit error for this signal set and Eb/N0, and (b) if the decision threshold is $\gamma = 0.15\sqrt{E}$, find the probability of bit error.

12. A coherent orthogonal MFSK system with $M = 8$ has the equally likely waveforms $s_i(t) = A \cos(2\pi f_i t)$, $i = 1,2,\ldots,M$, $0 \le t \le T$, where $T = 0.25$ ms. The received carrier amplitude, A, is 1 mV, and the two-sided AWGN spectral density, $N_0/2 = 10^{-10}$ W/Hz. Calculate the probability of bit error.

13. Design a first-order delta modulator and calculate the SNR with and without the lowpass filter at the decoder using (a) a sinusoid with amplitude and frequency of your choice as the input and (b) the sinusoid in (a) plus a uniformly distributed noise in the range $[-1,1]$ with a standard deviation of -12 dB with respect to the amplitude of the input sinusoid.

14. Consider a sine wave of frequency 1150 Hz with unit amplitude. Sample this signal at a sampling frequency equal to 2.2 times the Nyquist frequency. Quantize the samples to 3 bits using a uniform quantizer. Upsample this sequence by a factor of 3 and then filter it with a suitable lowpass FIR filter. This forms the input to the DAC. Plot the various sequences and the magnitude of their FFTs. You may use the MATLAB code in Example 10.1.

15. Consider an ideal lowpass filter with a passband gain of $A \ge 1$ and a cutoff frequency of $F_c < f_s/2$. For what value of F_c is the power gain equal to one?

16. Consider the 10-bit oversampled ADC shown in Fig. 10.7 with analog inputs in the range ± 4. (a) Find the average power of the quantization noise of the quantized input $x_q[n]$. (b) Suppose a second-order Butterworth filter is used for the analog antialiasing prefilter. The objective is to reduce the aliasing error by a factor of $\varepsilon = 0.001$. Find the minimum required oversampling factor M. (c) Find the average power of the quantization noise at the output, y[n], of the oversampling ADC. (d) Suppose $f_s = 1500$ Hz. Sketch the ideal magnitude response of the digital antialiasing filter H(z). (e) Design a linear-phase FIR filter of order 60 whose frequency response approximates $H(e^{j\omega})$ using Hanning window.

17. A 12-bit oversampled ADC oversamples by a factor of $M = 32$. To achieve the same average power of the quantization noise at the output, but without oversampling, how many bits are required?

18. Suppose an analog signal in the range ± 4 is sampled with a 10-bit oversampled ADC with an oversampling factor of $M = 16$. The output of the ADC is passed

through an FIR filter H(z), where $H(z) = 1 - 2z^{-1} + 3z^{-2} - 2z^{-3} + z^{-4}$. Determine (a) the quantization step size q, (b) the power gain of the filter H(z), and (c) the average power of the quantization noise at the system output, y [n]. (d) To get the same quantization noise power, but without using oversampling, how many bits are required?

19. A transmitter transmits an unmodulated tone of constant energy (a beacon) to a distant receiver. The receiver and transmitter are in motion with respect to each other such that $d(t) = K[1-\sin(nt)] + K_0$, where d(t) is the distance between the transmitter and receiver and K, n, and K_0 are constants. This relative motion will cause a Doppler shift in the received transmitter frequency of $\Delta\omega_D(t) = \frac{\omega_0 \nu(t)}{c}$, where $\Delta\omega_D$ is the Doppler shift, ω_0 the nominal carrier frequency, v(t) the relative velocity between the transmitter and receiver, and c is the speed of light. Assuming that the linearized loop equations hold and that the receiver's PLL is in lock (zero phase error) at t = 0 show that an appropriately designed first-order loop can maintain frequency lock.

20. Let us suppose that a transmitter and receiver are in relative motion as in Problem 19. Assume that the linearized loop equations hold. Under this assumption, determine the PLL phase error as a function of time for the allpass, and lowpass loop filters of Equations (10.103a) and (10.103b). Demonstrate that the validity of the assumption of the linearized loop equations depends on the value of the gain K_0.

21. A high-performance aircraft is transmitting an unmodulated carrier signal to a ground terminal. The ground terminal is initially in phase lock with the signal. The aircraft performs a maneuver whose dynamics are described by the acceleration, $a(t) = Kt^2$, where K is a constant. Assuming that the linearized equations apply determine the minimum order of the PLL required to track the signal from this aircraft.

22. Show that the loop bandwidth of a first-order PLL is given by $B_{loop} = H_0/4$, where H_0 is the loop gain.

23. Assume that the loop filter's transfer function is given by
$G(s) = \frac{1+0.01s}{1+s}$
with K = 1. Determine (a) the closed-loop transfer function H(s) and (b) the phase response to an input unit step phase. You may use MATLAB to solve this problem.

References

1. Ansari R, Liu B (1993) Multirate signal processing. In: Mitra SK, Kaiser JF (eds) Handbook for digital signal processing, Chapter 14. Wiley-Interscience, New York, pp 981–1084
2. Burns P (2003) Software defined radio. Artech House, Boston
3. Couch LW II (1983) Digital and analog communication systems. Macmillan, New York
4. Crochiere RE, Rabiner LR (1983) Multirate digital signal processing. Prentice Hall, Englewood Cliffs
5. Fliege NJ (1994) Multirate digital signal processing. Wiley, New York

6. Freking ME (1994/2005) Digital signal processing in communication systems. Van Nostrand Reinhold/Cambridge University Press, New York
7. Gardner FM (1979) Phaselock techniques, 2nd edn. Wiley, New York
8. Jayant NS, Knoll P (1984) Digital coding of waveforms. Prentice Hall, Englewood Cliffs
9. Kenington PB (2005) Software defined radio. Artech House, Boston
10. Lindsey WC, Simon MK (1977) Detection of digital FSK and PSK using a first-order phase-locked loop. IEEE Trans Commun COM-25(2):200–2014
11. Lindsey WC, Simon MK (eds) (1977) Phase locked loops and their applications. IEEE Press, New York
12. Luthra A, Rajan G (1991) Sampling rate conversion of video signals. SMPTE J 100:869–879
13. Mitra SK (2011) Digital signal processing: a computer-based approach, 4th edn. McGraw-Hill, New York, NY
14. Papoulis A (1965) Probability, random variables, and stochastic processes. McGraw-Hill, New York
15. Pickholtz RL, Schilling DL, Milstein LB (1982) Theory of spread-spectrum communications – a tutorial. IEEE Trans Commun COM30(5):855–884
16. Proakis JG (1983) Digital communications. McGraw-Hill, New York
17. Rappaport TS (2002) Wireless communications: principles and practice, 2nd edn. Prentice Hall, Upper Saddle River
18. Reed JH (2002) Software radio – a modern approach to radio engineering. Prentice Hall, Upper Saddle River
19. Schilling RJ, Harris SL (2012) Digital signal processing using MATLAB, 3rd edn. Cengage Learning, Boston
20. Scholtz RA (1982) The origins of spread spectrum communications. IEEE Trans Commun COM30(5):822–854
21. Sklar B (1988) Digital communications: fundamentals and applications. Prentice Hall, Englewood Cliffs
22. Viterbi AJ (1966) Principles of coherent communications. McGraw-Hill, New York
23. Viterbi AJ (1995) CDMA: principles of spread spectrum communication, Addison-Wesley Wireless Communications Series. Addison-Wesley Publisher, Reading

Index

© Springer International Publishing AG, part of Springer Nature 2019   495
K. S. Thyagarajan, *Introduction to Digital Signal Processing Using MATLAB
with Application to Digital Communications*,
https://doi.org/10.1007/978-3-319-76029-2

Printed in the United States
By Bookmasters